107 Advances in Polymer Science

Biopolymers I

Editors: N. A. Peppas and R. S. Langer

With contributions by
S. Amselem, E. Doelker, A. J. Domb,
J. Heller, R. W. Lenz, M. Maniar,
M. V. Sefton, J. Shah, W. T. K. Stevenson

With 131 Figures and 32 Tables

Springer-Verlag
Berlin Heidelberg GmbH

Guest Editors

Professor Robert S. Langer
Department of Chemical Engineering
Massachusetts Institute of Technology
25 Ames Street, Cambridge, MA 02139, USA

Professor Nicholas A. Peppas
School of Chemical Engineering, Purdue University
West Lafayette, IN 47907-1283, USA

ISBN 978-3-662-15941-5   ISBN 978-3-540-47478-4 (eBook)
DOI 10.1007/978-3-540-47478-4

Springer-Verlag Berlin Heidelberg 1993
Originally published by Springer-Verlag Berlin Heidelberg New York in 1993
Softcover reprint of the hardcover 1st edition 1993

Library of Congress Catalog Card Number 61-642

Typesetting: Macmillan India Ltd., Bangalore-25

02/3020 - 5 4 3 2 1 0 Printed on acid-free paper

Preface

Significant developments have occurred in recent years in the field of biopolymers and biomaterials. New synthetic materials have been synthesized and tested for a variety of biomedical and related applications from linings for artificial hearts to artificial pancreas devices and from intraocular lenses to drug delivery systems. Of particular interest for the future are the development and understanding of degradable polymers. At present, there are few acceptable degradable polymers for either clinical or environmental use. The next several decades should see a significant change in this area and the novel degradable polymers discussed in this volume, and the research metholodologies used for their study, should play a crucial role in this change. In addition, new uses of polymers such as their ability to encapsulate mammalian cells which aid in organ replacement therapy represent cutting edge advances in the development and use of biomaterials. We focus in this volume, in particular, on these important new areas of biomaterials research.

The present volume comprises five review articles written especially for this series by leading authorities in the field. The first three adresses major structural characteristics of biodegradable polymers. Robert Lenz provides a thorough review of biodegradable polymers. Jorge Heller offers a critical analysis of the structure, properties and medical applications of polyorthoesters, whereas Abraham Domb and his associates offer the same critical analysis of polyanhydrides. Eric Doelker discusses the structure and properties of cellulose derivatives. Finally Michael Sefton presents the use of polyacrylates for the microencapsulation of live animal cells.

Robert Langer
Nicholas A. Peppas

Editors

Table of Contents

Biodegradable Polymers

Robert W. Lenz

Polymer Science and Engineering Department, University of Massachusetts, Amherst, MA 01003, USA

The mechanisms of the biodegradation of natural and synthetic polymers is the principal subject of this review, but important concepts related to this subject are included, especially the types of microorganisms involved, the types and reactions of enzymes of importance in the biodegradation of polymers, the structures and properties of important biopolymers, the tests used to evaluate biodegradation, and practical applications of the biodegradation of polymers. In this review, discussions on "biodegradation" are confined to those chemical degradation processes which are caused by biochemical reactions, especially by reactions catalyzed by enzymes produced by microorganisms.

Advances in Polymer Science, Vol. 107
© Springer-Verlag Berlin Heidelberg 1993

1 Introduction

The term "biodegradable polymers" has different meanings for different people
[1]. For some, the term is used loosely to refer to polymers which will degrade in
a biological environment, including any environment where biological or biochemical processes are occurring, regardless of the degradation mechanism. For
others, the term is applied only to polymers which can degrade by biochemical
reactions, especially reactions catalyzed by enzymes. In this review, the latter
meaning is used as the basis for the selection and discussion of the subjects
included, and the following, perhaps over simplified, definition of the process
involved is applied throughout:

Biodegradation – chemical degradation caused by biochemical reactions,
especially those catalyzed by enzymes produced by microorganisms under either
aerobic or anaerobic conditions.

According to this definition, then, the biodegradation of polymers involves,
primarily, enzyme-catalyzed chemical reactions, and such reactions can occur
by either random attack along the polymer chain backbone or by specific attack
at the polymer chain ends. The former results in random chain cleavage with
a concomitant substantial decrease in molecular weight, while the latter results
in removal of only terminal units, which are generally either monomers, dimers
or trimers (M_1, M_2, M_3, respectively in the illustrations below), and in that case
the immediate effect on molecular weight of the residual polymer is much less.
These two degradation processes are illustrated schematically below with an
indication of the general types of enzymes associated with each process; M represents both the internal repeating units and the terminal units and the subscript
represents the number of such units in the product formed:

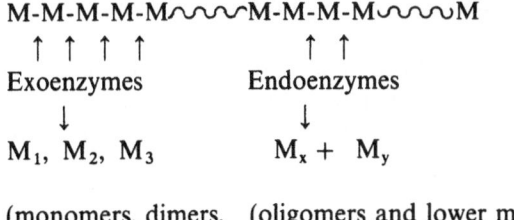

$$M\text{-}M\text{-}M\text{-}M\text{-}M \sim\!\sim\!\sim M\text{-}M\text{-}M\text{-}M \sim\!\sim\!\sim M$$

Exoenzymes	Endoenzymes
↓	↓
M_1, M_2, M_3	$M_x + M_y$
(monomers, dimers, trimers)	(oligomers and lower molecular weight polymers)

The microorganisms which can provide the enzymes for such processes
include bacteria, fungi, yeasts, algae and others. However, it should be noted
that such microorganisms can also secrete reactive reagents into the environment which can cause degradation reactions (especially acids), and similarly,
some enzymes can catalyze the formation of reactive reagents in the environment (especially peroxides) that can degrade polymers. Both types of degradation processes are included in the definition above because, even though en-

zymes are not directly involved in degradation, the reactions of biochemical products are important, integral components in each process.

Up to this point, the discussion has emphasized the biochemical aspects of biodegradation, but for completeness we must also consider the degradation component; that is, what is meant by degradation and how is it observed and measured. For polymers, degradation can be described in terms of either, or both, the changes in molecular structure and/or physical and mechanical properties. The structural changes, which occur in virtually all types of biodegradation processes, must include a decrease in molecular weight. This decrease, depending on the polymer and the process, may be caused by a variety of chemical reactions on the polymer which can often be observed spectroscopically (e.g. oxidation to form carbonyl groups), but the ultimate requirement to satisfy our definition of biodegradation is for main chain cleavage to occur with fragmentation of the polymer molecule. The resulting molecular weight loss may also be associated with a variety of chemical and physical effects, which can be readily measured (e.g. by-product formation, weight loss, surface erosion, etc.), but the single most important parameter involved is the molecular weight.

The most significant property changes associated with the degradation of polymers also relate to the molecular weight decrease. That is, physical properties such as glass temperature, degree of crystallinity and viscosity, and mechanical properties such as tensile strength, modulus and impact strength, which are directly related to molecular weight, will generally be adversely affected by the decrease in molecular weight, and that decrease is the fundamental characteristic of polymer degradation.

Because of the complexity of the relationships between, and the consequences of, the chemical, physical, and property changes for different polymers in different types of biochemical degradation processes and environments, it is exceedingly difficult to specify unique or even standard tests and specific or standard descriptions of "biodegradation". Interestingly, too, biodegradation, in contrast to most other types of polymer degradation processes, is now considered a desirable characteristic in many applications, and the subject of the biodegradability of polymers has taken on complex attitudinal, commercial and political connotations, all of which greatly complicate the use and understanding of the term "biodegradable polymers."

2 Enzymes and Enzymatic Reactions [2]

2.1 Enzymes

Enzymes are, first and foremost, catalysts. More specifically, enzymes are catalysts made by cells for chemical reactions which occur in biochemical processes. Structurally, enzymes are complex and highly specialized proteins, which are produced by the cell in order to catalyze specific types of chemical reactions of biological importance. That is, in general in a biological system, each enzyme performs one chemical function, and while there are a great many different types of enzymes (over 2000), essentially all of them can be classified within one of the following categories according to the general type of reaction they catalyze:

1. hydrolase – catalyzes hydrolysis reactions, especially those of esters, amides and acetals;
2. esterase or amidase – catalyzes esterification or amidation reactions, or ester interchange and amide interchange reactions;
3. isomerase (transferase) – catalyzes molecular reorganization reactions by transferring atoms or groups within molecules;
4. reductase or oxidoreductase – catalyzes electron transfer reactions which result in either oxidation or reduction processes;
5. hydrogenase or dehydrogenase (lyase) – catalyzes hydrogen addition or removal reactions;
6. ligase – catalyzes condensation reactions which form new carbon–carbon, carbon–sulfur, carbon–oxygen and carbon–nitrogen bonds.

In each case, the enzyme may be highly specific for a given type of chemical structure or substrate, and it can increase the rate of the reaction of the substrate by a factor of from 10^6 to 10^{20} fold without creating undesirable reaction products. Even so, enzymes usually function best in dilute aqueous solutions under mild conditions of temperature and pH. Indeed, mild conditions are generally required to prevent adverse chemical and physical changes of the proteins which constitute the enzymes. These proteins can have number average molecular weights ranging from 10 000 to as high as 1 000 000. In each case, they form their active sites by assuming a specific tertiary structure, and the disruption of that physical state will generally result in the loss of enzyme activity. The active site, which can contain a variety of different types of amino acid repeating units, interacts with the substrate molecule (or molecules) to position and activate it (them) for the reaction involved as discussed in the following sections.

2.2 Enzymatic Reactions

Two successive, reversible reactions are involved in most enzymatically catalyzed processes, and the equilibria of both reactions are important as shown in the equation below, in which E represents the enzyme molecule, S the substrate molecule on which the reaction occurs, and P the product of the reaction:

$$E + S \underset{k_{-1}}{\overset{k_1}{\rightleftharpoons}} ES \underset{k_{-2}}{\overset{k_2}{\rightleftharpoons}} P + E$$

The overall rate of product formation is generally controlled by the second step, so the formation of P can be described kinetically as a unimolecular reaction of

the enzyme-substrate complex, ES. Hence, the maximum rate of product formation and substrate utilization will occur when all of the enzyme molecules are
bound to substrate molecules.

Enzymatic reactions are usually characterized by a parameter, the
Michaelis-Menten constant or K_M, which is determined by the efficiency of the
first equilibrium reaction for the formation of ES. That is, K_M is the concentration in mM of S at which the initial rate of the overall process, V_o, is one-half of
the maximum rate, V_{max}, possible. The maximum rate occurs when all of E is
converted to ES. Each particular type and concentration of E and S, and each
set of reactions conditions, has its own K_M, and the Michaelis-Menten equation
describes the relationship between V_o, [S] (the concentration of S), V_{max} and
K_M for a given amount of E under a fixed set of conditions, as follows:

$$V_o = \frac{V_{max}[S]}{K_M + [S]}$$

As a result, the more effective the enzyme, the lower will be the K_M.

Of course, many enzymatic reactions involve two different molecules or
substrates, S_1 and S_2, and in those cases each substrate will have a K_M.
A common example of such a process is the esterification reaction of an acid, S_1,
and an alcohol, S_2, catalyzed by an esterase, E, represented by the equation
below:

$$E \ + \ \underset{S_1}{RCOOH} \ + \ \underset{S_2}{R'OH} \ \rightleftharpoons \ E\overset{S_1}{\underset{S_2}{\diagdown}} \ \rightleftharpoons \ \underset{P_1}{RCOOR'} \ + \ \underset{P_2}{H_2O} \ + \ E$$

2.2.1 Bimolecular Reaction Mechanisms [2]

The type of reaction which is probably of most importance in the enzymatic
degradation of polymers is the bimolecular reaction illustrated above, in which
the enzyme catalyzes the interaction of the polymer and a low molecular reagent
(such as water in a hydrolysis reaction). These reactions can occur by either
a single displacement or a double displacement mechanism. In the former, both
substrates, A and B below, are bound to the enzyme by consecutive, reversible
reactions, after which the final complex, EAB, dissociates into the products,
C and D, and the free enzyme, as follows:

$$E + A \rightleftharpoons EA \overset{+B}{\rightleftharpoons} EAB \rightleftharpoons C + D + E$$

In the double displacement reaction only one substrate at a time is bound to
the enzyme, but the complex of the first bound substrate, EAX in the equation
below, undergoes a unimolecular dissociation at an appropriate functional
group to form a new complex between the enzyme and a fragment, X, of AX, as
shown below. This intermediate complex combines with the second substrate, B,

and transfers the X fragment to B, as follows:

$$E + AX \rightleftharpoons EAX \rightleftharpoons EX + A$$
$$\updownarrow$$
$$EXB \rightleftharpoons XB + E$$

2.2.2 Enzyme Activity [2]

As discussed above, the most common indicator of enzyme activity is K_M, which is determined by measuring V_{max} at enzyme saturation conditions and V_o at a concentration of S, [S], near to the saturation concentration of S for a given concentration of E, [E]. V_o is taken as the initial slope of the plot of reaction conversion vs time for that value of [S], and a series of such plots and V_o values are obtained at different values of [E]. The slopes of a plot of V_o vs [E] then gives $V_o/[E]$, which can be used to calculate K_M from the Michaelis-Menten equation.

Three other parameters are also used to evaluate enzyme activity; these are:

(1) Units of Enzyme Activity – the amount of E that causes the transformation of 1 µmol of S per minute at 25 °C under optimal pH conditions.

(2) Specific Activity – the number of units of enzyme activity/mg of protein (which, of course, is also an indication of the purity of the enzymes).

(3) Turnover Number – the number of molecules of S transformed/minute by each molecule of E (or by each active site on E) when E is the limiting variable (this number can easily be greater than 10^6).

2.2.3 Enzyme Specificity [2]

Enzymes can be either totally specific for a given substrate, to such an extent that they will not tolerate any structural or configurational changes in the substrate, or they can be broadly specific for a given type of functional group, and they will still operate on substrates with structural variations around that functional group. An example of the former is the enzyme aspartase, Asp, which catalyzes both the addition of ammonia to fumaric acid and the elimination of ammonia from L-aspartic acid by the following reactions:

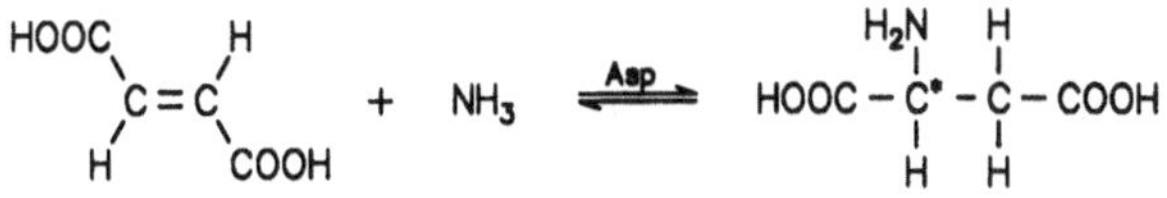

Asp will not catalyze the addition of NH_3 to the *cis* acid, maleic acid, and it will not catalyze the elimination of NH_3 from D-aspartic acid. Chymotrypsin, on the other hand, is an example of an enzyme with broad activity, and it can catalyze the hydrolysis of many different types of both amides (especially peptides) and esters.

3 Microorganisms

Two types of microorganisms are of particular interest in the biodegradation of natural and synthetic polymers; these are bacteria and fungi. Many species or types of microorganisms are found broadly in nature.

Bacteria and fungi can be divided into autotrophs, which can use CO_2 as the sole carbon source, and heterotrophs, which must obtain their carbon from more complex organic compounds. Most microorganisms are of the latter type (all fungi are heterotrophic) and they subsist off either the metabolic byproducts from autotrophic cells or the many and different types of organic compounds found in nature. Both categories of microorganisms include either aerobic, anaerobic or facultative (can grow in the presence or absence of oxygen) species. Of the heterotrophs, the facultative microorganisms may be the most common.

4 Biopolymers

Biopolymers are polymers formed in nature during the growth cycles of all organisms; hence, they are also referred to as natural polymers. The biopolymers of interest in this review are those that serve in nature as either structural or reserve cellular materials. Their syntheses always involve enzyme-catalyzed, chain-growth polymerization reactions of activated monomers, which are generally formed within the cells by complex metabolic processes. The most prevalent structural and reserve biopolymers are the polysaccharides, of which many different types exist, but several other more limited types of polymers exist in nature which serve these roles and are of particular interest for materials applications. The latter include the polyesters and proteins produced by bacteria and the hydrocarbon elastomers produced by plants (e.g. natural rubber). In almost all cases (natural rubber is an exception), all of the repeating units of these biopolymers contain one or more chiral centers and the repeating units are always present in optically pure form; that is, biopolymers with asymmetric centers are always 100% isotactic.

4.1 Polysaccharides

For materials applications, the principal polysaccharides of interest are cellulose and starch, but increasing attention is being given to the more complex carbohydrate polymers produced by bacteria and fungi, especially to polysaccharides such as xanthan, curdlan, pullulan and hyaluronic acid. These latter polymers generally contain more than one type of carbohydrate unit, and in many cases these polymers have regularly arranged branched structures. Starch, for example, is a physical combination of branched and linear polymers (amylopectin and amylose, respectively), but it contains only a single type of carbohydrate, glucose.

Both cellulose and starch are composed of hundreds or thousands of D-glucopyranoside repeating units. As shown below, these units are linked together by acetal bonds formed between the hemiacetal carbon atom, C_1, of the cyclic glucose structure in one unit and a hydroxyl group at either the C_3 (for cellulose and amylose) or the C_6 (for the branch units in amylopectin) carbon atoms in the adjacent unit. This type of structure occurs because in aqueous solution glucose can exist in either the acyclic aldehyde or cyclic hemiacetal form, as shown below, and the latter form is the structure that becomes incorporated into the polysaccharide. Also, the cyclic form can exist in either one of two isomers, the α-isomer with an axial OH group on the ring of the β-isomer with an equatorial OH group, as follows:

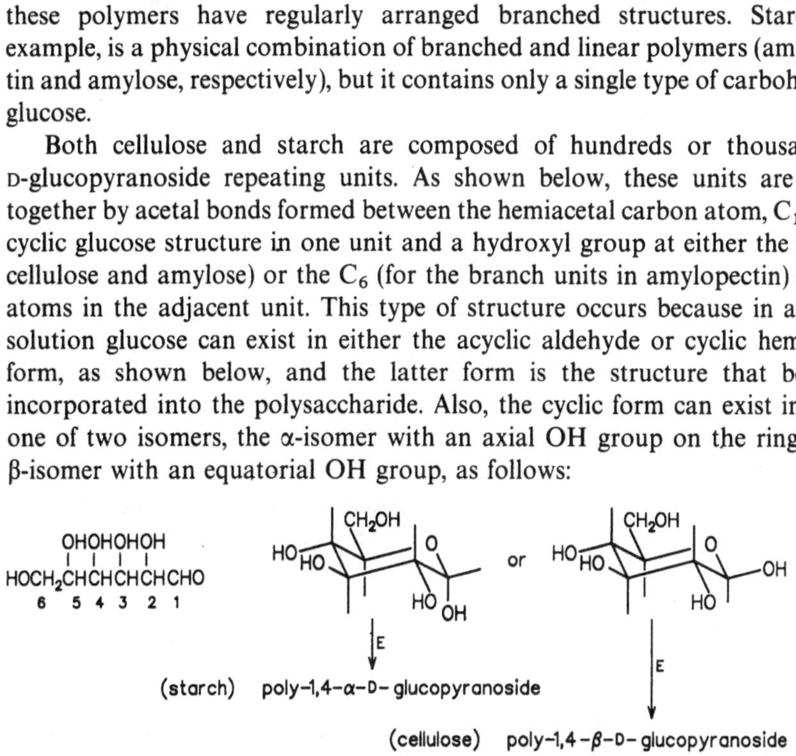

(starch)   poly-1,4-α-D- glucopyranoside

(cellulose)   poly-1,4-β-D- glucopyranoside

As shown above, in starch the glucopyranoside ring is present in the α-form, while in cellulose the repeating units exist in the β-form. Because of this difference in structures, the enzymes that catalyze the acetal hydrolysis reactions in the biodegradation reactions of each of these two polysaccharides are different for each and are not interchangeable.

4.1.1 Starch and Cellulose

Starch is a reserve polymer which occurs widely in plants. The principal crops used for its production are potatoes, corn and rice. In all of these plants, starch is produced in the form of granules, which vary in size and somewhat in composition from plant to plant. In general, the linear polymer, amylose, makes up about 20 weight percent of the granule, and the branched polymer, amylopectin, the remainder. Amylose is crystalline and can have a number average molecular weight as high as 500 000, but it is soluble in boiling water. Amylopectin is insoluble in boiling water, but in their use in food, both fractions are readily hydrolyzed at the acetal link by enzymes, E. The α-1, 4-link in both components of starch is attacked by amylases, as shown below, and the α-1,6-link in amylopectin is attacked by glucosidases.

In its application in biodegradable plastics, starch is either physically mixed, with the native granules kept intact, or melted and blended on a molecular level with the appropriate polymer as described in Sects. 5.7 and 5.8. In either form, the fraction of starch in the mixture which is accessible to the enzymes can be degraded by either, or both, amylases and glucosidases.

Cellulose is primarily a structural polymer in plants (especially in cotton, ramie and hemp) and trees. In the latter, cellulose is the principal structural material and constitutes about 50 weight percent of wood. Cellulose is also produced by bacteria in the form of exocellular microfibrils. In all forms, cellulose is a very highly crystalline, high molecular weight polymer, which is infusible and insoluble in all but the most aggressive, hydrogen-bond breaking solvents such as N-methylmorpholine N-oxide. Because of its infusibility and insolubility, cellulose is usually converted into derivatives to make it processable.

All of the important derivatives of cellulose are reaction products of one or more of the three hydroxyl groups, which are present in each glucopyranoside repeating unit, including: (1) ethers – e.g. methyl cellulose and hydroxyethyl cellulose; (2) esters – e.g. cellulose acetate and cellulose xanthate, which is used as a soluble intermediate for processing cellulose into either fiber or film form, during which the cellulose is regenerated by controlled hydrolysis; and (3) acetals – especially the cyclic acetal formed between the C_2 and C_3 hydroxyl groups and butyraldehyde. Unlike starch, cellulose cannot be made melt processable by plasticization with water, but the useful derivatives of cellulose are either fusible or soluble in common solvents, or both. Like starch, however, cellulose, even though very highly crystalline, is readily hydrolyzed at the acetal position by specific enzymes (cellulases), which are secreted by a wide variety of microorganisms.

Only a relatively small amount of information is available on the effect of substitution or derivatization of cellulose on its enzymatic degradation [3]. However, it has been suggested that if at least one hydroxyl group in each repeating units is substituted, the modified cellulose is not degraded by microorganisms [4].

4.1.2 Bacterial Polysaccharides [5]

Many different types of polysaccharides are produced and excreted by bacteria into their immediate surroundings. These exopolysaccharides serve a variety of functions for the microorganisms ranging from food and energy reserve materials, to protective gels, to adhesives. Several of the exopolysaccharides are produced at a commercial level by industrial fermentation processes, including xanthan (which has a cellulose-type main chain and trisaccharide graft chains containing glucuronic acid), dextran (primarily an α-1,6-glucopyranoside polymer), gellan (a tetrasaccharide polymer with glycerol substituents), curdlan (a β-1,3-glucopyrannoside polymer) and pullulan (which is composed of 1,6-linked maltotriose units and produced by fungi).

4.1.3 Enzymatic Polymerization

Both starch and cellulose are prepared in nature by enzymatic, chain growth polymerization reactions of glucose nucleotide monomers [6]. In both cases, the monomer precursor is glucose-1-phosphate, which is enzymatically converted to the nucleotide derivative. The latter, in turn, complexes with an enzyme to form the activated monomer at the active site on the enzyme, which also contains the growing polymer molecule, as schematically illustrated below for the enzymatic polymerization of cellulose:

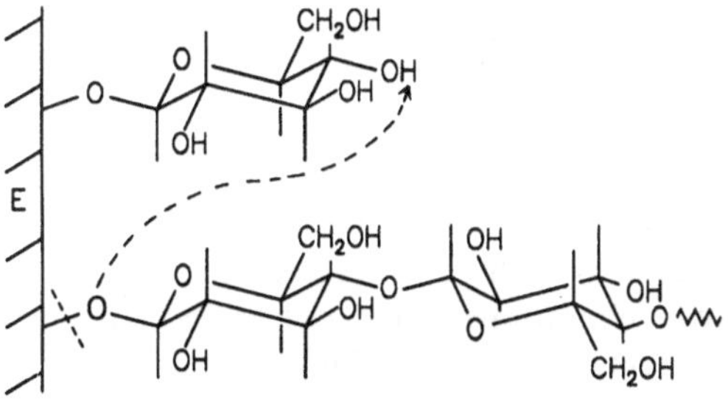

4.2 Proteins

The proteins which have found applications as materials are, for the most part, neither soluble nor fusible without degradation, so they are used in the form in which they are found in nature. This description is especially true for the fibrous proteins: wool, silk and collagen. All proteins are specific copolymers with regular arrangements of many different types of α-aminoacids, so the biosynthesis of proteins is an extremely complex process involving many different types of enzymes. In contrast, the enzymatic degradation of proteins, with general purpose proteases, is a relatively straightforward, amide hydroysis reaction.

4.3 Polyesters

The natural polyesters, which are produced by a wide variety of bacteria as intracellular reserve materials, are receiving increased attention for possible application as biodegradable, melt processable polymers which can be produced from renewable resources. The members of this family of thermoplastic biopolymers, which have the general structure shown below, can vary in materials properties from rigid brittle plastics, to flexible plastics with good impact properties, to strong tough elastomers, depending on the size of the pendant alkyl group, R, and the composition of the polymer [7–9]:

$$\left[OCHCH_2\overset{\overset{O}{\|}}{C} \right]; \quad R = -(CH_2)_x-CH_3; \quad X = 0-8 \text{ or higher}$$

with R below the OCHCH₂C group.

All of these polyesters contain units which are 100% optically pure at the β-position, so all are 100% isotactic. The polymer with R = CH$_3$, poly-β-hydroxybutyrate or PHB, is highly crystalline with a melting temperature of 180 °C and a glass temperature, T$_g$, of approximately 5 °C [10]. This combination of very high crystallinity and relatively high T$_g$ makes films and plastics of PHB very brittle, so copolymers with units containing other alkyl groups, especially R = C$_2$H$_5$, are preferred. To obtain the latter, the bacteria used for polymer production are grown with combinations of substrates that permit the cells to produce the copolyesters containing random combinations of β-hydroxybutyrate, HB (x = 0), and β-hydroxyvalerate, HV (x = 1) units. Such P(HB/HV) copolymers have much better mechanical properties, and they are now produced in relatively large quantities by large scale fermentation processes for the industrial development of useful plastics, films, and fibers. All of these materials are inherently biodegradable, and with proper selection of the bacterium, the substrate and the growth conditions, these polymers can have number average molecular weights varying from several hundred thousand to well over one million.

Polyesters with longer alkyl substituents, with x = 3–6 or so, are also produced by a variety of bacteria, generally in the form of copolymers which have low degrees of crystallinity and low T$_m$ and T$_g$ values. As a result, these longer alkyl chain polyesters are useful as thermoplastic elastomers, which can have excellent strength and toughness, and are also inherently biodegradable [9].

All of these bacterial polyesters are produced as intracellular reserve materials when the cells are grown under stressed conditions, which can result from a limitation of some vital component needed by the microorganisms for their normal metabolic processes. That is, under normal or balanced growth conditions, the appropriate bacterium would completely utilize the organic substrates available for energy and for creating cellular materials, but under biologically stressed conditions, the cells will interrupt their normal metabolic process to convert the intermediates generated from such substrates into the polyester. A photomicrograph of cells of an anaerobic bacterium, Rhodobacter spheroides, containing PHB granues (white areas) is shown in Fig. 1.

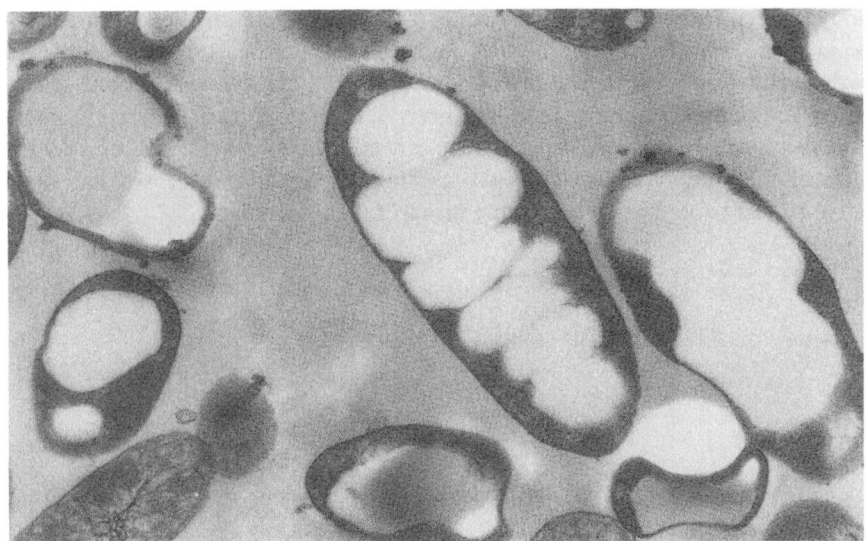

Fig. 1. Electron photomicrograph of cells of the facultative photosynthetic bacterium *Rhodobacter sphaeroides* which contain granules of PHB; overall magnification is approximately 50 000X

The monomer and polymer forming processes are shown in the following series of enzymatic reactions involved in the production of PHB:

$$\text{Organic Substrates} \xrightarrow[\text{Processes}]{\text{Normal Metabolic}} CH_3COOH \xrightarrow[\text{Growth}]{\text{Balanced}} \text{Energy} + \text{Cellular Materials}$$

$$\downarrow \xrightarrow[\text{Growth}]{\text{Unbalanced}} PHB$$

PHB Formation:

$$2CH_3COOH \xrightarrow{E_1} 2CH_3COCoA \xrightarrow{E_2} CH_3\overset{\overset{O}{\|}}{C}CH_2\overset{\overset{O}{\|}}{C}CoA$$

$$\xrightarrow{E_3} CH_3^*\overset{\overset{OH}{|}}{C}HCH_2\overset{\overset{O}{\|}}{C}CoA \xrightarrow{E_4} \left[\overset{*}{O}CHCH_2\overset{\overset{O}{\|}}{C}\right]_{\overset{|}{CH_3}}$$

PHB and the other polyesters are water insoluble and, as such, they form a separate, granular phase inside the cell which is referred to as an inclusion body. Afterwards, when the cells can no longer subsist on the organic compounds in their environment, they utilize these granules by enzymatically hydrolyzing the polymer and reversing the series of enzymatic reactions shown above to obtain energy and food for survival. In nature, many microorganisms also secrete hydrolase enzymes, which can perform the same function as the intracellular depolymerase enzymes used in the reverse cycle so that these polyesters are totally biodegradable in natural environments.

5 Biodegradation of Natural Polymers

5.1 Cellulose

Cellulose is the most abundant biopolymer, and it has been the most studied for its biodegradation, partly because of the practical importance of its degradation, including: (1) the attack of microorganisms on paper and wood (rotting); (2) the industrial production of chemicals such as ethanol, acetic acid and acetone by the enzymatic degradation of cellulose and hemicellulose; (3) the digestion of cellulose by cattle and other ruminant animals; and more recently, (4) the global warming and greenhouse effects due to the generation of CO_2 and CH_4 in the aerobic and anaerobic degradation of cellulose [11].

5.1.1 Cellulollytic Microorganisms

Many fungi and bacteria, both aerobic and anaerobic, secrete cellulolytic enzymes under a wide range of conditions, some at temperatures up to 85 °C and some at a pH as high as 9 [12, 13]. At least fifty different species of fungi, known as "rot fungi", are known to degrade cellulose, and usually several different types of fungi operate together in the total biodegradation of the three components of wood: cellulose, hemicellulose (about 25% by weight) and lignin (about 25% by weight). Different combinations are found for different types of wood and in different locations (above ground or below ground). Similarly, at least ten different types of anaerobic and thirty different types of aerobic bacteria are known to have cellullolytic activity.

5.1.2 Fungi

The wood-degrading fungi are divided into three types: (1) white-rot fungi, which can secrete both exocellular peroxidases to degrade lignin and, to a lesser extent, cellulases to degrade the polysaccharides in order to produce simple sugars which serve as nutrients for these microorganisms; (2) brown-rot fungi, which produce, primarily, enzymes for the degradation of cellulose and the hemicelluloses; and (3) soft-rot fungi, which also act principally on these two types of polysaccharides [12].

The lignin-degrading peroxidases operate by generating hydrogen peroxide, which dissociates into hydroxyl radicals that react with lignin by free radical substitution reactions to break down the network structure of the highly crosslinked, three-dimensional polymer. This reaction sequence is illustrated below for a representative section of the complex lignin macromolecule:

$$E + O_2 \longrightarrow H_2O_2 \longrightarrow 2\,HO\cdot$$

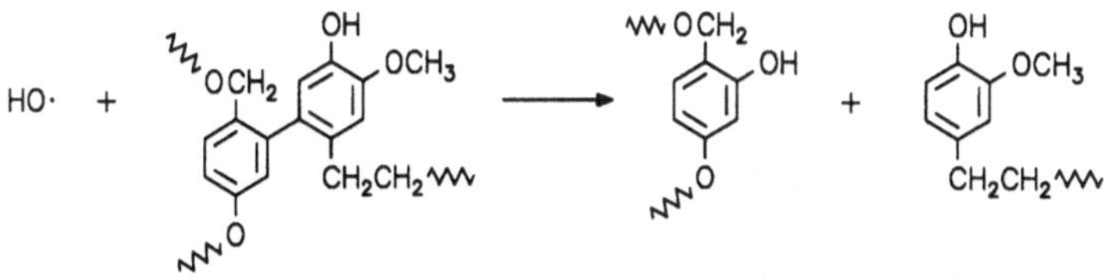

The cellulases produced by fungi for degrading the cellulose component of wood include three principle types of enzymes: (1) endo-1,4-β-glucanases, which randomly attack internal 1,4-β-glucosidic bonds along the main chain to generate lower molecular weight polymers and oligomers (cellodextrins); (2) exo-1,4-β-glucanases, which selectively hydrolyze terminal units on the chain, always at the non-reducing end of the chain, to form monomers (glucose), dimers (cellobiose), trimers (cellotriose) and possibly also the four- and five-unit oligomers; and (3) 1,4-β-glucosidases, which catalyze the hydrolysis of cellobiose and the higher, water-soluble cellodextrins to glucose to provide nutrients for the fungi [12, 13]. For almost all wood-degrading fungi, the first two types of enzymes are always present together, and they work in consort to degrade polymer molecules on the surface of this highly crystalline polymer. The overall reactions of exo- and endo-enzymes with cellulose are shown schematically below:

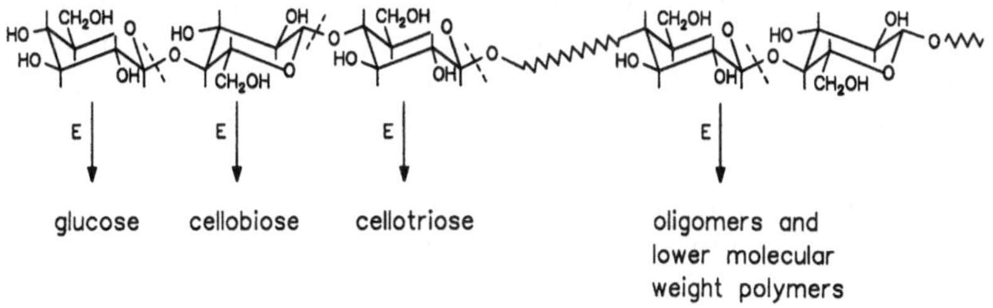

glucose   cellobiose   cellotriose   oligomers and
                                            lower molecular
                                            weight polymers

Some fungi can also secrete enzymes that catalyze oxidation reactions of either cellulose itself or the lower molecular weight oligomers from the enzymatic hydrolysis of cellulose. Of these, the peroxidases can provide hydrogen peroxides for free radical attack on the C_2–C_3 positions of cellulose to form "aldehyde" cellulose, which is very reactive and can hydrolyze to form lower molecular weight fragments, as shown below, while other oxidative enzymes can oxidize glucose and related oligomers to glucuronic acids.

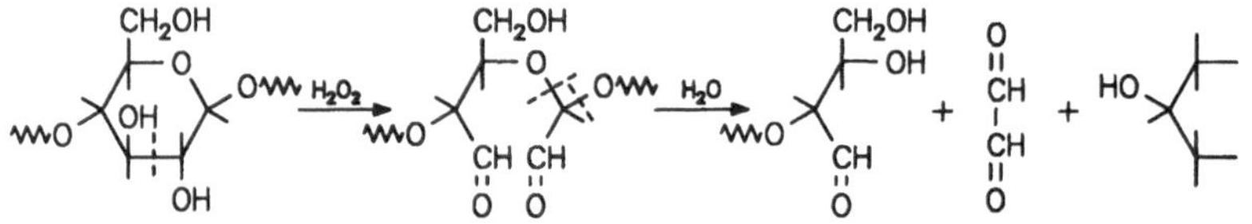

5.1.3 Bacteria

Bacteria also secrete both endo- and exoenzymes, some of which form complexes that act jointly in degrading cellulose to form the carbohydrate nutrients which the microorganisms need for survival [12, 13]. Both the enzyme complexes, which generally contain from two to four different enzymes, and the individual enzymes can attack cellulose either from a dissolved state in aqueous solution or from a state in which the enzyme is bound to the outer cell wall of the bacteria. In most cases, the principal product is cellobiose. In contrast, while multiple enzymes may be involved, enzyme complexes are seldom found with fungi.

Aerobic soil environments generally contain consortia of several different types of degrading bacteria and fungi. In these colonies the different types of microorganisms can operate cooperatively, including combinations of: (1) primary microorganisms, which provide enzymes that degrade cellulose to glucose and cellodextrins, a portion of which they utilize, and (2) secondary microorganisms, which provide enzymes that degrade the cellodextrins to glucose, which they consume. By consuming glucose the latter assist in the growth of the primary microorganisms because they prevent the build-up of the cellodextrins, which can inhibit glucanases if they are present in the environment at too high a concentration. The final products from the aerobic biodegradation are CO_2 and water.

Anaerobic environments, which exist either below the surface of soil or in muds or sediments, also contain combinations of primary and secondary microorganisms to facilitate the biodegradation of cellulose. In this case, however, a variety of final products are formed, including CO_2, hydrogen, methane, hydrogen sulfide, and ammonia. CO_2 can be formed by oxidative reactions which utilize inorganic compounds, such as sulfate and nitrate ions, in the environment as oxidizing agents, as shown below:

$$Glucose + SO_4^{2-} \xrightarrow{E} CO_2 + S^{2-} \qquad H_2S$$

$$Glucose + NO_3^- \xrightarrow{E} CO_2 + NH_3$$

Hydrogen produced by some anaerobic bacteria can be utilized by autotrophic bacteria to reduce oxidized organic compounds and CO_2 to form either acetic acid (acetogenic bacteria) or methane (methanogenic bacteria), as follows:

$$2CO_2 + 4H_2 \xrightarrow{E} CH_3COOH + 2H_2O$$

$$CO_2 + 4H_2 \xrightarrow{E} CH_4 + 2H_2O$$

5.2 Starch

The enzymatic degradation of starch is applied on an industrial scale to the production of "corn syrup" (a solution of oligomers) and "dextrose" (glucose) by utilizing thermally stable endoenzymes that can operate in an aqueous medium at temperatures up to 70–90 °C [14]. These temperatures are desired because the starch granules rupture in water at about 60 °C. The use of enzymatic hydrolysis for the conversion of starch to sugars is a much cleaner (more selective) process than is an acid-catalyzed hydrolysis.

Both endoamylases (α-amylases) and exoamylases (glucoamylases) are produced by microorganisms, and these enzymes are also utilized by plants and mammals for the biodegradation of starch. The endoamylases generally hydrolyze only the main chain acetal bonds in either amylose or amylopectin and are not active on the branch points of the latter, but many of the exoamylases can cleave either main chain or branch bonds. Some exoamylases, however, cannot catalyze the hydrolysis of the 1,4-glycosidic bonds on units containing a 1,6-glycosidic branch point.

The exoamylases, as with the exocellulases, can generate either glucose or the dimer (maltose) or the trimer (maltotriose) by attacking the non-reducing end of the starch molecules. Enzymes which attack only the branch points ("debranching enzymes") in amylopectin are also known.

While little more is known in detail about the mechanism of the enzymatic degradation reactions of cellulose and starch, it is likely that, in many cases, once an enzyme molecule associates with a polymer chain and causes the first chain cleavage, either by an exo- or an endo-cleavage, it remains associated with the fragmented chain, and it can catalyze the hydrolysis of several more units before dissociating. At present, however, little quantitative information is available about this type of process.

5.3 Bacterial Polysaccharides

Each of the many different types of exopolysaccharides produced by bacteria has specific enzymes capable of degrading it; that is, the degrading enzymes for these polymers show a very high degree of substrate specificity [7, 15]. That specificity is not only for the particular polymer, but also for a particular bond or functional group site on that polymer, and many of these polysaccharides have highly complex molecular structures involving different types of carbohydrate repeating units substituted at different points in the ring.

Xanthan, the best studied of the exopolysaccharides, provides a good example of the complexity and specificity of enzymatic attack. Xanthan has a β-1,4-glucosidic (Glu) backbone containing branch points at the 3-positions of every fourth unit. The branches are trisaccharides with α-1,2-mannose acetate, β-1,4-glucuronic acid and β-mannoside (Man) units. Each of the different main chain, branch and side chain bonds can be hydrolyzed by only a specific enzyme, as illustrated below:

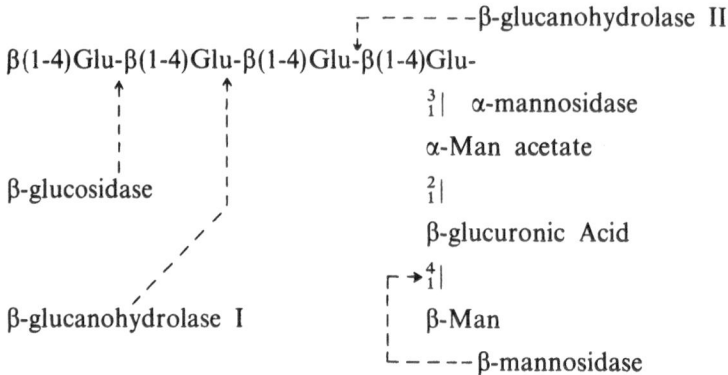

5.4 Bacterial Polyesters

The enzymatic hydrolysis of poly-β-hydroxybutyrate, PHB, by several different bacteria, which are known to secrete active esterases, has been studied in some detail by several research groups [7, 8]. As with the polysaccharides, the final products of these degradation reactions are the monomers, dimers and trimers, which are removed by hydrolysis only from hydroxyl-end of the polymer chain, as follows:

These products are water soluble and can diffuse through the surrounding aqueous environment until they come in contact with, and are taken up by, the cells and used as nutrients.

The exocellular enzymes produced by three different polyester-degrading bacteria have been characterized using native granules of PHB as the substrate. The bacteria studied were the following: (1) two *Pseudomonas* strains which secreted non-specific enzymes; the depolymerase from one of the strains, which was designated *Pseudomonas PI*, could also hydrolyze a variety of simple esters as well as PHB [16]; (2) *Pseudomonas lemoignei*, which secreted four different depolymerase exoenzymes, each with molecular weights of about 50 kDa, that varied in activity and product distribution (i.e. in the types of oligomers formed) [17], and (3) *Alcaligenes faecalis*, which secreted a depolymerase only in the presence of PHB granules [18].

The hydrolysis reaction catalyzed by the depolymerases from *P. lemoignei* generated some monomer and dimer, but the major products were the oligomers higher than the dimer, mostly the trimer. In contrast, the depolymerase from *A. faecalis* catalyzes a hydrolytic reaction to form mostly the dimer along with small amounts of monomer, and it can also hydrolyze higher oligomers to the dimer. The four different esterases (A_1, A_2, B_1, B_2 below) secreted by *P. lemoignei* were characterized for their specific activities and Michaelis-Menten constants, K_M, both with the PHB granules and with the water-soluble, β-hydroxybutyrate dimers, M_2, trimers, M_3, tetramers, M_4, and pentamers, M_5, with the results tabulated below [16b]:

Enzyme	PHB		K_M for Oligomers, mM			
	Specific Activity units, mg	K_M, mM	M_2	M_3	M_4	M_5
A_1	135	93	0	2.4	0.26	0.079
A_2	86	73		(not tested)		
B_1	34	105	0	1.8	0.22	0.091
B_2	70	131		(not tested)		

As can be seen from these data, the order of activity was $A_1 > A_2 > B_1 > B_2$ for PHB as the substrate, and for the oligomer substrates, the order was $M_5 > M_4 > M_3$ (M_2 unreactive) for enzymes A_1 and B_2. However, the hydrolysis products from PHB with both A_1 and B_2 were only the monomer and the M_2 and M_3 oligomers, and no higher oligomers were formed. With A_1 the product distribution was: 23% M_1 + 57% M_2 + 20% M_3, and with B_2: 34% M_1 + 19% M_2 + 47% M_3. For each of the oligomers as substrates, the relative frequencies of attack (relative rates of hydrolysis) at the internal ester groups with each of these two enzymes, A_1 and B_2, is shown schematically below:

A_1

M_3: HO-HB-HB-HB-COOH
            ↑1.0

M_4: HO-HB-HB-HB-HB-COOH
            ↑0.8 ↑0.2

M_5: HO-HB-HB-HB-HB-HB-COOH
            ↑0.65↑ 0.35

B_2

HO-HB-HB-HB-COOH
        ↑1.0

HO-HB-HB-HB-HB-COOH
        ↑0.5 ↑0.5

HO-HB-HB-HB-HB-HB-COOH
        ↑0.25↑ 0.75

The depolymerase secreted by *A. faecalis* in the presence of PHB also had a molecular weight of 50 kDa, but it was much more active than the enzymes from *P. lemoignei* as indicated by a much lower K_M value for the former with the native PHB granules [17a]. Furthermore, the hydrolysis of PHB catalyzed by this enzyme produced almost entirely the dimer, and the dimer and monomer were the only products obtained from the hydrolysis of the M_3, M_4 and M_5 oligomers with this enzyme.

A careful study of the enzymatic degradation of films of PHB and P(HB/HV) copolymers, using the purified exocellular depolymerase secreted by *A. faecalis*, was carried out by Doi and coworkers to compare the mechanisms of enzymatic hydrolysis with base-catalyzed chemical hydrolysis [19]. The enzymatic hydrolysis was carried out in water at 37 °C and pH 7.5 and the chemical hydrolysis in water at 55 °C and pH 7.4. By continuously following both the weight loss of the films and molecular weight decrease of the polymers with time, they were able to show that chemical hydrolysis and enzymatic hydrolysis occurred by entirely different mechanisms.

In chemical hydrolysis, the base and water could penetrate throughout the film and hydrolyze the polymer randomly at all points within the film. As a result, molecular weight of the polymer decreased continuously with time, and very little or no weight loss occurred until the polymer molecules throughout the entire film had been severely degraded. In contrast, in enzymatic hydrolysis, the high molecular weight protein enzyme could not penetrate into the film, so degradation was confined to the surface. As a result, only the surface layer of polymer was degraded, and the low molecular products formed were continuously removed by solution in water, so the film sample lost weight continuously, but the remaining polymer showed no measurable molecular weight decrease. This contrast in degradation mechanisms can be used as a diagnostic test for the occurrence of enzymatic biodegradation with all types of polymers in many different environments in which the polymer is present as a solid [19]. Furthermore, the relationship between the weight loss and the molecular weight decrease can be used to estimate the relative contributions of chemical and enzymatic degradation in environments where both can occur.

6 Biodegradation of Synthetic Polymers [20–22]

Broad interest in the possible biodegradation of synthetic polymers has developed only in recent years and primarily in response to the growing problem of the waste disposal of plastics. Essentially all biopolymers are susceptible to enzymatic degradation because the enzymatic polymerization reactions responsible for their synthesis in nature have closely related counterparts in nature for their enzymatic depolymerization: "what nature creates, nature can destroy". If it were not so, polymers could not be utilized as reserve materials and waste polymers from dead microorganisms, plants and animals would not be removed from the environment (and their chemical components would not be recycled). However, synthetic polymers are new to our environment so natural mechanisms for their destruction have not developed. At best, we can only hope that some enzymes which catalyze the degradation of natural organic matter will be sufficiently non-specific to do the same for synthetic polymers with similar structures or functional groups.

Until recently the ability of synthetic polymers to be degraded and/or utilized by microorganisms was assessed by qualitative or inferential methods. The most widely used of these in the past was an ASTM test in which a sample of molded plastic or film was placed in contact with a standard set of microorganisms, especially fungi, in a favorable medium for growth, and the density of growth of the microorganisms on the surface of the sample was estimated visually and assigned a number from 0 (no visible growth) to 4 (from 60 to 100% of the surface covered) [23]. Unfortunately, this test did not reveal whether the plastic was being degraded by biochemical reactions or whether the proliferation of the cells on the surface was an artifact. Other tests which have often been used in the past to evaluate biodegradation were just as inconclusive (and even inappropriate), particularly sample weight loss and reduction in mechanical properties, because forms of degradation or responses to the environment (e.g. environmental stress cracking) other than enzymatic attack could have been the cause of the effects observed.

Nevertheless, a considerable amount of qualitative and semi-quantitative information has been accumulated in recent years to draw some conclusions on what the important factors are that affect the rate of degradation of synthetic polymers in a biological environment. These factors include [3, 21]:

(1) polymer structure – especially hydrophilicity and the presence of functional groups in, or immediately on, the main chain; also molecular weight and molecular weight distribution;

(2) physical and morphological state of the polymer – particularly whether it is crystalline or amorphous and, for the former, the degree and form of crystallinity, for the latter, the glass transition temperature;

(3) environmental conditions (temperature, pH, humidity, oxygen availability, light, etc.) and the presence of other nutrients for growth of the microorganisms;

(4) microbial population – number, types and interactions of fungi and bacteria, in particular;

(5) surface-to-volume ratio, sample size and purity of the polymer sample;

(6) duration of the test;

(7) method of contact between the sample and the microorganisms (or enzymes) – whether in solution, as a gel or slurry, by soil burial of a solid, etc.

In very few cases, however, has consideration been given to the types of products formed by biodegradation and to their effect on the environment. That is, are the products toxic to either microorganisms, animal life or humans? Is the polymer, in the end, converted completely to CO_2 and water? In many studies the disintegration of the sample into very small solid particles ("plastic sand") has been taken as a sufficient response in order to label the plastic "biodegradable". Certainly, much needs to be done in this area, and "fate analysis" must become an integral part of the evaluation of the biodegradability of synthetic polymers (see Sect. 7.3).

6.1 Polyesters

The most widely evaluated, and the first systematically studied, of the types of synthetic polymers for biodegradability were the polyesters. The ASTM test described above was applied to several different polyesters as early as 1972 [23]. In 1975 that type of evaluation was repeated and expanded to include soil burial on a much larger series of polymers. The ASTM test in both of these studies was based on a standard set of four fungi, and some of the results obtained from the later study are shown in Table 1 [25].

Table 1. Biodegradation of polyesters evaluated by the ASTM test and by soil burial [25]

Polyester Number	Structure	ASTM Growth Rating	Weight Loss in Soil Burial, (%[a])	TOC[b] A	B
I	$\left[O(CH_2)_5\overset{\displaystyle O}{\overset{\displaystyle \|}{C}}\right]$	4	15	310	3610
II	$\left[OCH_2CH_2O\overset{\displaystyle O}{\overset{\displaystyle \|}{C}}(CH_2)_2\overset{\displaystyle O}{\overset{\displaystyle \|}{C}}\right]$	1 +	19	–	
III	$\left[OCH_2CH_2O\overset{\displaystyle O}{\overset{\displaystyle \|}{C}}(CH_2)_7\overset{\displaystyle O}{\overset{\displaystyle \|}{C}}\right]$	3 –	17	3080	3770
IV	$\left[O(CH_2)_{10}O\overset{\displaystyle O}{\overset{\displaystyle \|}{C}}(CH_2)\overset{\displaystyle O}{\overset{\displaystyle \|}{C}}\right]$	3 –	9	–	
V	$\left[O(CH_2)_6O\overset{\displaystyle O}{\overset{\displaystyle \|}{C}}(CH_2)_8\overset{\displaystyle O}{\overset{\displaystyle \|}{C}}\right]$	2 –	40	380	1160

[a] After 4 weeks. [b] Total organic carbon of water soluble products from the degradation processes with lipases secreted by two different fungi: A – *R. delemar* and B – *R. arrbizus* [26].

Of the polyesters in Table 1, polycaprolactone, I, was the most responsive to the fungal growth test in the high surface density of microbial growth, but it did not lose the most weight in the four-week soil burial test. These conflicting results were an early indication of the complexity and unreliability of such semi-quantitative tests, which were the principal tests used during that period to evaluate biodegradability.

Three of the polyesters in Table 1 (Polyesters I, III, and V) were recently evaluated for their degradation by specific lipases secreted by two fungi: *Rhizopus delemar* and *Rhizopus arrbizus* [26]. The primary natural function of lipases is for the hydrolysis of fatty acid esters of glycerol (triacyl glycerides). The rates of solubilization by degradation of powders of each of the polymers, which were evaluated as suspensions in aqueous media containing each of the lipases separately, was determined by measuring the water-soluble total organic carbon (TOC) concentration at 30 °C after 16 hours using a TOC analyzer. The results for the three polymers studied are also included in Table 1.

All three polyesters evaluated were observed to be substantially degraded by enzymatic hydrolysis reactions, but their extents of degradation by the two different enzymes differed, and the results, again, did not correlate well with the ASTM and weight loss tests. Nevertheless, it was clear for these studies that synthetic polyesters could be biodegraded in the true sense of the term used in this review, so the commercial claim that polycaprolactone, I, is a "biodegradable polymer" is appropriate, at least as far as the capability for attack by microorganisms and enzymatic hydrolysis of the polymer is concerned [27]. However, there are no reports of the fate of the degradation products, which are presumably ω-hydroxycaproic acid and its oligomers, and their possible toxicity. Also, it should be noted that the results of soil burial tests of films of polycaprolactone, which are reported in commercial product brochures [27], showed that a continuous decrease in the molecular weight and very rapid decreases in elongation to break and tensile strength of the sample occurred with time to accompany the weight loss observed. Because this polymer does not swell appreciably in water and enzymes could not penetrate into the interior of the films, the decreases in molecular weight and mechanical properties must be an indication that chemical hydrolysis occurred throughout the film at the same time that enzymatic hydrolysis occurred on the film surface.

In a careful study of the enzymatic degradation of polycaprolactone by fungi (including yeast), using a scanning electron microscope, Huang and coworkers observed that attack on the amorphous regions occurred preferentially [28]. After the polymer in the amorphous regions was hydrolyzed, the crystallites were degraded from the edges inward. Included in this study was the effect of enzymatic hydrolysis on the molecular weight of the polymer in the sample. The molecular weight and molecular weight distribution did not change during the enzymatic hydrolysis, as was observed by Doi and coworkers for the degradation of bacterial polyesters. These results confirmed that in this type of process only the polymer on the sample surface is degraded, and the low molecular weight degradation products are removed from the sample by solubilization in the surrounding aqueous medium.

6.2 Polyethers

Many different studies, dating back to 1962, have been carried out on the biodegradability of aliphatic polyethers, especially poly(ethylene glycol), PEG [29]. The ability of different microorganisms to degrade PEG and the rates of degradation have been found to be a function of the molecular weight, and polymers with number average molecular weights, M_n, as high as 20 000 are reportedly degraded, at least to some extent. Kawai observed that PEG samples with M_n values below 1000 daltons were assimilated by several kinds of bacteria individually, but PEG samples with an M_n above 6000 were only degraded and utilized by symbiotic mixed cultures.

At least three types of enzymatic reactions have been identified in the biodegradation of PEG; these are described and their reaction equations are shown below [29]:

(1) The enzymatic dehydrogenation of a primary alcohol to an aldehyde, in this case at the hydroxyl end groups of PEG:

$$\sim\!\!\sim\!\!\sim OCH_2CH_2OCH_2CH_2OH \xrightarrow{E_1} \sim\!\!\sim OCH_2CH_2OCH_2\overset{\overset{\displaystyle O}{\|}}{C}H$$

(2) The enzymatic oxidation of this aldehyde to a terminal carboxylic acid:

$$\sim\!\!\sim\!\!\sim OCH_2CH_2\overset{\overset{\displaystyle O}{\|}}{C}CH_2\overset{\overset{\displaystyle O}{\|}}{C}H \xrightarrow{E_2} \sim\!\!\sim OCH_2CH_2OCH_2\overset{\overset{\displaystyle O}{\|}}{C}OH$$

(3) An enzymatic oxidation reaction, presumably to form the hemiacetal at the α-position, which is followed by hydrolysis to release glyoxalic acid and decrease the size of the PEG molecule by one unit:

$$\sim\!\!\sim\!\!\sim OCH_2CH_2OCH_2\overset{\overset{\displaystyle O}{\|}}{C}OH \xrightarrow{E_3} \sim\!\!\sim OCH_2CH_2O\overset{\overset{\displaystyle OH}{|}}{\underset{\underset{\displaystyle H}{|}}{C}}COOH$$

$$\xrightarrow{H_2O} \sim\!\!\sim OCH_2CH_2OH + H\overset{\overset{\displaystyle O}{\|}}{C}COOH$$

(4) The enzymatic oxidation of internal methylene groups along the PEG main chain to form ester groups, which can be hydrolyzed to substantially

reduce the molecular weight of the polymer and create new, reactive end groups:

$$\text{\textasciitilde OCH}_2\text{CH}_2\text{OCH}_2\text{CH}_2\text{O}\text{\textasciitilde} \xrightarrow{E_4} \text{\textasciitilde OCH}_2\overset{\overset{\textstyle O}{\|}}{\text{C}}\text{OCH}_2\text{CH}_2\text{O}\text{\textasciitilde}$$

$$\text{\textasciitilde} \xrightarrow{H_2O} \text{\textasciitilde OCH}_2\overset{\overset{\textstyle O}{\|}}{\text{C}}\text{OH} + \text{HOCH}_2\text{CH}_2\text{O}\text{\textasciitilde}$$

An enzyme responsible for Reaction (1) above, a PEG dehydrogenase, has been isolated and evaluated for its activity with a wide range of low molecular weight, primary alcohols. The relative activities of these substrates were based on the use of a PEG sample with a molecular weight of 6000 as the standard. The relative activities (in %) of this enzyme in the degradation of the alcohols under a given set of conditions, in decreasing order, were as follows: 1-hexanol (115%), 1-pentanol (111%), PEG 6000 (100%), 1-butanol (87%), PEG 400 (49%), PEG 1000 (26%), 1,4-butanediol (26%), 1-propanol (18%), ethanol (1%). It is difficult to rationalize this order of activity, especially the effect of PEG molecular weight on enzyme activity, and it is apparent that the effect of molecular and property variables on degradation processes involved are quite complex.

6.3 Polyethylene

There is no convincing evidence that unmodified, high molecular weight poly-ethylene can be utilized by microorganisms as a substrate and degraded at any measurable rate by enzymatic reactions [3, 30]. The biological utilization of hydrocarbons by microorganisms is well documented, and the enzymatic reactions involved in their total assimilation are known, but molecular weight plays a decisive role in these processes, and the upper limit for this variable at which a linear alkane can be attacked remains questionable.

Early studies, again using the ASTM growth rating test, recognized the importance of the molecular weight of polyethylene on its biodegradability, and the results from those and subsequent tests indicated that the molecular weight limit for biodegradation of linear alkanes, and for the low molecular weight products obtained from the pre-degradation of polyethylene, was probably in the order of 500 daltons [22, 31]. That is, while it has been observed that microorganisms can grow on, and apparently utilize, samples of pre-degraded polyethylenes with molecular weights as high as about 20 000, most likely only the lower molecular weight fractions of approximately 500 daltons or so were effective substrates for growth [32, 33]. Indeed, in one study in which the ASTM test was applied to the growth of microorganisms on specific linear alkanes, a sharp diversion was observed between very good growth (a rating of 4) for the $C_{32}H_{66}$ and lower alkanes, on the one hand, and no growth (a rating of 0) for the $C_{36}H_{74}$ and higher alkanes, on the other. Linear alkanes of this size have melting points in the range of 72 to 78 °C.

Numerous studies have shown that lower aliphatic hydrocarbons containing up to at least 20 carbon atoms (molecular weights below 300 daltons) can be assimilated by a wide variety of microorganisms, including twenty or so different types of bacteria and 150 or so different types of yeasts, which are found in soil and marine sediments. In aqueous media, some of these microorganisms secrete surfactants to facilitate contact and the utilization of the hydrocarbon substrates.

The reactions involved in the degradation and consumption of hydrocarbons include several different types of enzymatic oxidation, dehydrogenation and carbon–carbon bond breaking processes that can degrade the hydrocarbons to form the true substrates for the growth of microorganisms. The "true substrates" are the two- to four-carbon organic compounds that can be metabolized by the cell for growth. The principal utilizable substrate so formed in most cases is acetic acid, which is generated by the reaction sequence shown below:

$$CH_3(CH_2)_xCH_3 \xrightarrow{E_1} CH_3(CH_2)_xCH_2OH$$

$$\xrightarrow{E_2} \xrightarrow{E_3} CH_3(CH_2)_xCOOH$$

$$\xrightarrow{E_4} CH_3(CH_2)_{x-2}CH{=}CHCOOH$$

$$\xrightarrow{E_5} CH_3(CH_2)_x\overset{\overset{\displaystyle OH}{|}}{C}HCH_2COOH$$

$$\xrightarrow{E_6} CH_3(CH_2)_{x-2}\overset{\overset{\displaystyle O}{\|}}{C}CH_2COOH$$

$$\xrightarrow{E_7} CH_3(CH_2)_{x-2}COOH + CH_3COOH$$

Presumably, this same sequence of enzymatic reactions can occur with either low molecular weight fractions or partially degraded polyethylenes if the microorganisms can associate with the end groups [32, 34, 35]. Depending on the type of the initial degradation process (oxidative, thermal, photolytic), the end groups could be either methyl, hydroxymethyl, aldehyde or carboxyl groups similar to those shown in the sequence above. The surface density or surface availability of such groups should vary inversely with sample molecular weight, unless there is a preference for such groups to either associate in the interior or migrate to the surface of the sample. However, an important unanswered question in this entire process is whether all of the reactions in the sequence above can occur at the cell surface, or must the hydrocarbon substrate be taken into the cell at some stage in order to carry out the final sequence of enzymatic reactions and fragmentation. If transport into the cell is required, then, of course, the molecular weight of the substrate could be the determining property.

6.4 Poly(vinyl alcohol)

Poly(vinyl alcohol) is a versatile polymer with many industrial applications, and it may be the only polymer with an all carbon–carbon bond backbone that is truly biodegradable. The monomer is vinyl acetate, which is readily polymerized with free radical initiators to form poly(vinyl acetate), and the latter can be hydrolyzed either partially to prepare vinyl alcohol-vinyl acetate copolymers, or fully to prepare poly(vinyl alcohol). At high vinyl alcohol contents, the copolymers, and the vinyl alcohol homopolymer, are water soluble, and the aqueous solutions of these polymers find many applications; for example, as adhesives or for coatings. The homopolymer and the very high vinyl alcohol copolymers are crystalline with melt transitions between 180 and 230 °C and glass transitions between 58 and 85 °C, depending on the vinyl alcohol content. The polymers can be melt processed to form molded plastics, fibers and films [36].

At least 55 species of microorganisms have been shown to participate in the degradation of poly(vinyl alcohol), including bacteria, molds, yeast and fungi. Hence, this polymer can be degraded in a variety of environments with active microbial populations, including activated sludge, settling ponds, anaerobic digesters, septic systems, aquatic systems and activated landfills. In such environments poly(vinyl alcohol) samples with number average molecular weights as high as 100 000 have been reported to be completely degraded by soil bacteria, especially by *Pseudomonads* [37].

The basic mechanism involved in the degradation of this water-soluble polymer is believed to be a random chain cleavage process on the polymer backbone resulting from two types of enzyme-catalyzed oxidation processes [38, 39]. The first of these processes converts the 1,3-glycol structure of two successive repeating units to a β-diketone by random oxidative dehydrogenation reactions catalyzed by extracellular enzymes that attack secondary hydroxyl groups. The second reaction is a carbon–carbon bond breaking reaction, which converts one of the ketone groups to a carboxylic group with concomitant chain rupture. The enzymatic carbon–carbon bond breaking reaction may be essentially the same as that which occurs in the final step of the reaction sequence above in which an alkane is degraded to release acetic acid:

$$\underset{\underset{\displaystyle \sim\!\!\sim\!\!\sim CH_2CHCH_2CHCH_2\sim\!\!\sim\!\!\sim}{}}{\overset{OH \quad\quad OH}{\big| \quad\quad\quad \big|}} \xrightarrow{E} \underset{}{\overset{O \quad\quad O}{\sim\!\!\sim\!\!\sim CH_2\overset{\|}{C}CH_2\overset{\|}{C}CH_2\sim\!\!\sim\!\!\sim}}$$

$$\xrightarrow{E} \sim\!\!\sim\!\!\sim CH_2COOH + CH_3\overset{\overset{\displaystyle O}{\|}}{C}CH_2\sim\!\!\sim\!\!\sim$$

The final reaction is essentially the reverse reaction of a Claisen condensation in organic chemistry, and acetic acid can eventually be formed by that reaction and utilized by the microorganism.

A similar degradation process could also occur by the reactions of hydroxyl radicals, which are generated by peroxidase enzymes, with the secondary alcohol groups in poly(vinyl alcohol) and with the ketone groups formed either by this type of chemical reaction or by the enzymatic reactions described above. The mechanisms of the chemical reactions are shown below:

$$\underset{\displaystyle \sim\!\!\sim\!\!\sim CH_2CHCH_2CHCH_2\sim\!\!\sim\!\!\sim}{\overset{OH \quad\quad OH}{\big| \quad\quad\quad \big|}} + HO\cdot \;\rightarrow\; \sim\!\!\sim CH_2\overset{\overset{\displaystyle O}{\|}}{C}CH_2\underset{\displaystyle CHCH_2\sim\!\!\sim}{\overset{OH}{\big|}}$$

$$\sim\!\!\sim \xrightarrow{OH\cdot} \sim\!\!\sim CH_2\overset{\overset{\displaystyle O}{\|}}{C}CH_2\overset{\overset{\displaystyle O}{\|}}{C}CH_2\sim\!\!\sim \xrightarrow{OH\cdot} \sim\!\!\sim CH_2COOH + CH_3\overset{\overset{\displaystyle O}{\|}}{C}CH_2\sim\!\!\sim$$

In addition, if poly(vinyl alcohol) is first chemically oxidized to convert a portion of the alcohol groups to ketone groups, the resulting polymer is much more susceptible to biodegradation [24].

The internal enzymatic degradation of high molecular weight poly(vinyl alcohol) by the processes shown above would, of course, be greatly facilitated by the solubility of this polymer in water and the accessibility of the internal units to the enzyme in the aqueous solution. In this regard, it should be noted that poly(vinyl alcohol) cannot be prepared directly, but instead, as described above, it is obtained by the hydrolysis of poly(vinyl acetate). If the latter polymer is only partially hydrolyzed, to less than approximately 70% conversion of acetate to alcohol groups, the resulting copolymer is reportedly not biodegradable by the enzymatic processes shown above [40].

6.5 Poly(acrylic acid)

Poly(acrylic acid), which is also readily soluble in water, can also be degraded in natural environments [41]. However, much less is known about the mechanism of degradation of this polymer, and samples of only relative low molecular weights (below about 4000) seem to be biologically active.

6.6 Polymer Modifications to Facilitate Biodegradation

Because oligomers and polymers with main chains containing only carbon–carbon bonds, except for those with large numbers of polar groups on the main chain, such as poly(vinyl alcohol), show little or no susceptibility to enzyme-catalyzed degradation reactions, especially at higher molecular weights, several approaches have been used to insert "weak links" within or immediately attached to the backbones of such polymers [42]. These "weak links" are designed to permit the controlled degradation of an initially high molecular weight, hydrophobic polymer to low molecular weight oligomers, which could then be utilized and consumed by microorganisms through biodegradation processes. Particular emphasis in this approach to create useful biodegradable polymers has been placed on two types of polymer modifications; namely by (1) the insertion of functional groups in the main chain, especially ester groups, which can be cleaved by chemical hydrolysis, and (2) the insertion of functional groups in or on the main chain that can undergo photochemical chain-cleavage reactions, especially carbonyl groups.

An exceedingly clever method for inserting main chain ester groups into vinyl-type polymers, including polystyrene and polyethylene, is to carry out a free radical copolymerization reaction on the appropriate vinyl monomer (e.g. styrene or ethylene) with a special monomer that undergoes a free radical, ring-opening reaction to generate a main chain ester group [43, 44]. Methylene-substituted, cyclic acetal and $ortho$-ester monomers can participate in such a reaction by free radical copolymerization according to the mechanism shown below for one of these types of comonomers, 2-methyleneoxepane ($R = H$ for polyethylene or $R = C_6H_5$ for polystyrene-based copolymers):

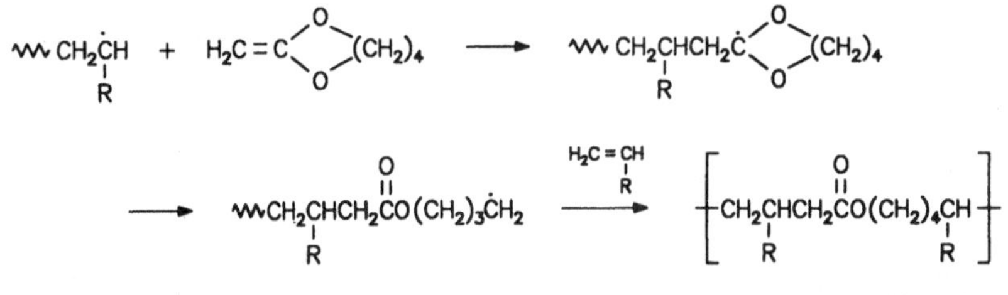

By using this procedure several different types of otherwise all-carbon chain polymers have been prepared, and after hydrolysis to their low molecular weight carboxy- and hydroxyl-terminated oligomers, the latter can be degraded by fungal attack.

The other principal approach, the preparation of a photodegradable copolymer, also utilizes a free radical copolymerization reaction, but in this case the comonomer is one which will create a ketone group either in the main chain or immediately attached to the main chain. Both carbon monoxide, Reaction I below, and vinyl ketones, Reaction II, will form such "weak links", and both of these comonomers have been used effectively in small amounts to prepare useful copolymers with a variety of vinyl-type monomers (again, especially for polystyrene, $R = C_6H_5$, and polyethylene, $R = H$) as shown below [45, 46]:

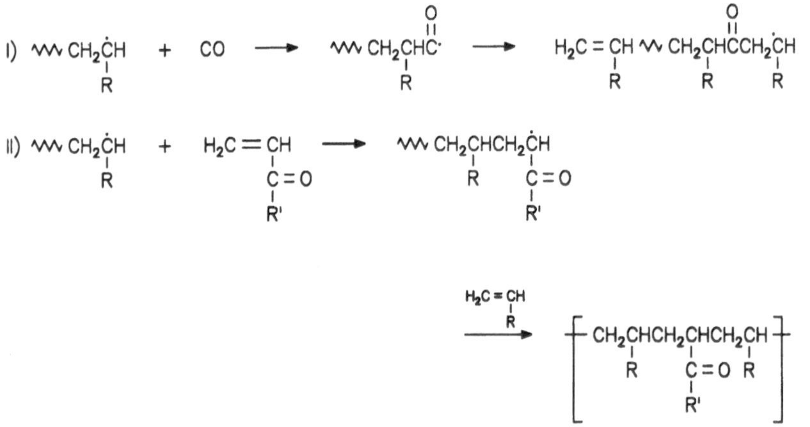

I) $\text{CH}_2\dot{\text{C}}\text{H}$ + CO $\longrightarrow$ $\text{CH}_2\text{CHC}\overset{O}{\overset{\|}{\text{C}}}\cdot$ $\longrightarrow$ $\text{H}_2\text{C}=\text{CH}$ $\text{CH}_2\text{CHCCH}_2\text{CH}$
(with R substituents)

II) $\text{CH}_2\dot{\text{C}}\text{H}$ + $\text{H}_2\text{C}=\text{CH}$ $\longrightarrow$ $\text{CH}_2\text{CHCH}_2\dot{\text{C}}\text{H}$

$\longrightarrow$ $\left[\text{CH}_2\text{CHCH}_2\text{CHCH}_2\text{CH}\right]$

On irradiation with ultraviolet light, the activated ketone groups present can take part in two different types of free radical, bond-breaking reactions. In organic photochemistry, these two reactions are referred to as Norrish I and Norrish II Reactions, and their mechanisms are shown below for the degradation of copolymers of ethylene and carbon monoxide [46, 47]:

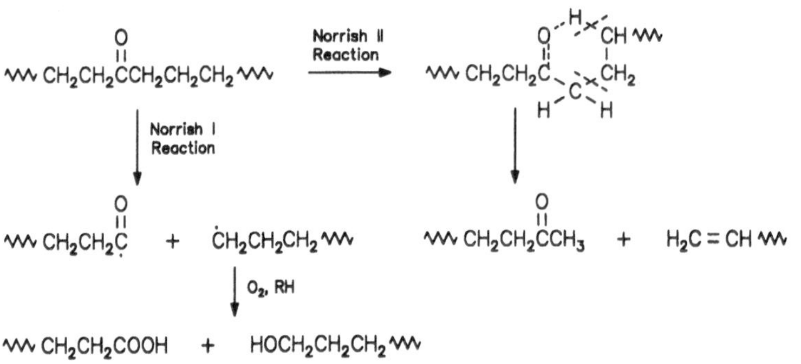

$\text{CH}_2\text{CH}_2\overset{O}{\overset{\|}{\text{C}}}\text{CH}_2\text{CH}_2\text{CH}_2$ $\xrightarrow{\text{Norrish II Reaction}}$ (Norrish II product)

$\downarrow$ Norrish I Reaction

$\text{CH}_2\text{CH}_2\overset{O}{\overset{\|}{\text{C}}}$ + $\dot{\text{C}}\text{H}_2\text{CH}_2\text{CH}_2$ $\text{CH}_2\text{CH}_2\overset{O}{\overset{\|}{\text{C}}}\text{CH}_3$ + $\text{H}_2\text{C}=\text{CH}$

$\downarrow$ O$_2$, RH

$\text{CH}_2\text{CH}_2\text{COOH}$ + $\text{HOCH}_2\text{CH}_2\text{CH}_2$

The Norrish I Reaction fragments the polymer to generate both carbonyl and alkyl radicals, and in the presence of oxygen, as shown above, these reactive fragments form carboxylic acid-terminated and hydroxyl-terminated lower molecular weight polymers. Further reactions of this type generate oligomers that can presumably interact with enzymes by the processes described above for the biodegradation of alkanes. The Norrish II Reaction creates fragments with vinyl and methyl ketone end groups, and the latter can undergo further photochemical and oxidative reactions to form carboxylic acid end groups, which, again, should be susceptible to enzymatic interaction and oxidation if the fragments are of sufficiently low molecular weight.

Another method for photodegrading polyethylene is to include metal salts, which catalyze photooxidation reactions, in the solid polymer. The compounds most generally used for that purpose are divalent transition-metal salts of higher aliphatic acids, such as stearic acid or dithiocarbonates or acetoacetic acid. The photochemical reaction is an oxidation-reduction reaction that forms free radicals capable of reacting with polyethylene, RH, to initiate an autooxidation chain reaction, as follows:

$$M^{+2}(X)_n \xrightarrow{U.V.} M^{+1}(X)_{n-1} + X\cdot$$

$$X\cdot + R\text{-}H \rightarrow XH + R\cdot$$

$$R\cdot + O_2 \rightarrow ROO\cdot \xrightarrow{RH} ROOH + R\cdot$$
$$\qquad\qquad\qquad\qquad\quad \longrightarrow RO\cdot + \cdot OH$$

and so on.

6.7 Blends of Biodegradable and Non-degradable Polymers

The blending of biodegradable polymers, such as starch, with inert polymers, such as polyethylene, has received a considerable amount of attention for possible application in the waste disposal of plastics. The reasoning behind this approach is that, in principal, if the biodegradable component is present in sufficient amounts and if it is removed by microorganisms in the waste disposal environment, the plastic or film containing the remaining inert component should lose its integrity, disintegrate and disappear. This concept has found its principal application in blends of minor amounts of starch with polyethylene in which the latter constitutes the continuous phase so that the blend can be melt processed to form films or plastics with polyethylene-like properties.

Granular starch, either in virgin form or chemically modified on the granule surface to increase its compatibility with the matrix polymer, has been used to form the type of blends described above [48–51]. In a biologically-active environment containing microorganisms that secrete amylases, the exposed starch granules on the surfer of the sample and those granules within the sample which are in direct contact with the surface granules, can be enzymatically hydrolyzed and completely removed to create pits or voids. When a sufficient amount of the starch present in the blend is degraded and removed in that manner, the sample should lose its strength and/or continuity and disintegrate. However, this described effect occurs only for samples containing fairly large amounts of starch, in the order of 30% by volume, and polyethylene plastics and films containing so much granular starch have substantially decreased tensile, tear and impact strengths. That is, the effective connectivity and accessibility of the starch granules, which is required for extensive enzymatic hydrolysis and removal, is achieved only at relatively high starch contents. At lower starch contents, with blend compositions much below the threshold level for the connectivity of granules, very little effect on mechanical properties results from the biodegradation of the accessible starch component [52].

The melt blending of non-biodegradable thermoplastics, such as polystyrene, poly(vinyl chloride) and polyethylene, with biodegradable thermoplastics is receiving increasing attention. The biodegradable polymers that have melting temperatures which permit melt blending in a reasonable temperature range are primarily polycaprolactone [56], which has a melt transition at 60 °C, and the bacterial copolyesters containing 3-hydroxybutyrate and 3-hydroxyvalerate units [57], which melt between 80 and 180 °C depending on their repeating unit compositions. The rates of biodegradation of such blends can vary widely with variations in compatibility of the two components and in blending procedure. These parameters greatly effect the morphology of the resulting blend, which, in turn, controls the accessibility of the biodegradable component and, therefore, its biodegradability [57].

6.8 Starch-Based Plastics

Starch can be made thermoplastic by plasticization with water at elevated temperatures and pressures and by this means, starch can be melt processed and melt blended with other thermoplastics [53, 54]. That is, the melt transition of the crystalline component of starch, amylose, can be strongly depressed with good solvents such as water, and in the presence of such solvents, starch can be melted and processed at temperatures well below its thermal degradation temperature. For example, with 10 weight % of water dissolved in starch at elevated temperatures under pressure, the glass transition and melting temperatures of this heterogeneous, normally infusible polymer are reduced to approximately 60 °C and 180 °C, respectively, and with 20 weight % water these values are reduced to approximately 40 and 50 °C, respectively.

Other effective plasticizers for starch for imparting melt processibility include a variety of low molecular weight compounds, such as glycerol and diethylene glycol, and also polymers such as poly(ethylene-co-vinyl alcohol) [55]. Furthermore, starch plasticized in that manner can be melt blended with minor amounts of hydrophobic thermoplastics, such as polyethylene and poly(methyl methacrylate), to obtain "biodisintegratable" molded articles with good mechanical properties.

7 Rate and Mechanism of Biodegradation of Polymers in the Solid State

The enzymatic degradation of polymers has been investigated primarily by biochemists and microbiologists, and to some extent by biochemical engineers, but polymer chemists, with a few notable exceptions [24], have shown little interest in this subject. The biochemists/microbiologists have been concerned with the detailed roles of the microorganisms and their enzymes on the manner and rate of polymer degradation, but most studies of that type have been concerned only superficially with the role of the polymer substrate. That is, for the biodegradation of solid polymers, they have given little attention to the effect of such important polymer properties and solid state characteristics as solid state morphology (especially the size and distribution of crystalline regions), degree of crystallinity, polymer mobility and glass transition temperature effects, molecular weight and molecular weight distribution, surface composition (particularly end group concentration on the surface), and so on. Similarly, biochemical engineers have been primarily concerned with developing rate equations and mass balances for cell growth and substrate consumption based on surface area effects, cell substrate affinity, and solubility and the diffusion of degradation products, not with the role of the polymer on the degradation rate.

The polymer chemistry aspects of biodegradation of natural and synthetic polymers would address such important questions as the following: (1) for exoenzymes, how does the enzyme locate the end groups of the polymer molecule; (2) for a semi-crystalline polymer, how are the polymer molecules, which are within the crystalline state, attacked and degraded by enzymes; (3) for a degradation process in which a specific enzyme molecule remains attached to and causes multiple bond scissions along a single polymer molecule, how is that contact maintained and how do the enzyme and polymer molecules move in a relative and absolute sense; (4) if the degradation reaction involves a free, extracellular enzyme, what determines the course by which the enzyme penetrates into the polymer matrix as the degradation process proceeds; (5) if the degradation reaction involves an enzyme associated with the outer cell wall of a microorganism, how does the microorganism penetrate into the polymer matrix as the degradation process proceeds; and similar concepts and concerns on the specific role of the polymer.

As the discussion above implies, there are several different types of enzymatic attack on solid polymers. The simplest case in concept is attack on the polymer by a soluble, freely diffusing, extracellular enzyme which degrades the polymer to form water soluble, low molecular weight fragments that dissolve and diffuse through the aqueous environment until they are taken up by the microorganisms that produced and secreted the depolymerase enzyme. A more complex case is that in which the enzyme is associated with or bound to the outer wall or cell surface of the microorganism, which must attach itself to the polymer surface to permit the immobilized enzyme to seek out the section of the polymer (that is, the terminal units for an exoenzyme and the internal repeat units for an endoenzyme), which it recognizes and can bind to in order to bring the catalytic site of the enzyme into play. In either case, the enzyme molecules can only complex with the polymer molecules on the surface of the solid polymer because enzymes are high molecular weight proteins, and unless the polymer is highly swollen, even the soluble enzymes cannot penetrate into the bulk of the polymer in advance of the degrading polymer front.

7.1 Kinetics of Biodegradation

The actual degradation process for a specific enzyme-polymer system can vary according to the reversibility of formation of the enzyme-polymer substrate complex and on the number of bond scissions (one or more) which a given enzyme will cause to occur on a given polymer molecule before they dissociate from each other. A general treatment of the latter, which covers both single scission and multiple scission processes has been derived by Lotan and colleagues [58]. Their treatment assumes that the initial attack can occur at random points along the polymer chain, but that after initial cleavage the enzyme can either dissociate from the chain, which would constitute a single scission process, or the enzyme remains attached and "slides down" the chain causing successive or multiple bond scissions as it moves along; that is, the process begins by an endo-type degradation, then continues for multiple scissions by an exo-type degradation. The principal variables in this treatment are the molecular weight of the substrate polymer (which determines the probability of the location along the chain for the site of initial attack), the size of the fragments formed by the initial attack, and the number of repetitive scissions which occur while the enzyme remains complexed with one of the two fragments formed from the initial polymer molecule (zero for a single scission process).

Huang and Chou have developed a kinetic model for microbial attack on, and utilization of solid substrates [59]. Their model is based on observations of the anaerobic digestion of water insoluble substrates such as long chain fatty acids, cellulose and biological sludge. They assume that degradation does not occur by the attack of soluble extracellular enzymes, but that the microorganisms involved must attach themselves to the insoluble substrate surface, and the transfer of the substrate, or of molecular fragments from the substrate, into the cell occurs at or near the point of cell adhesion and enzyme attack. The important variables in this type of process are the cell-substrate affinity, the concentration of available sites on the solid substrate for complex formation with, and attack by, the enzyme, the substrate particle size and total surface area (including the accessible area within macroscopic pores), cell concentration and cell-substrate loading.

7.2 Effect of Polymer Morphology

The effect of solid state morphology on the course of the biodegradation of semi-crystalline polymers has been shown to be of controlling importance by Huang and coworkers in their study of the biodegradation of polycaprolactone [24]. We have recently observed very similar effects in our investigations of the degradation of bacterial polyesters by purified enzymes, and Fig. 2 is a photo-micrograph of the appearance of the surface of a film of an HB/HV copolymer (see Sect. 5.4) before and after it has been exposed to a very active hydrolase enzyme in an aqueous medium [60]. The surface of the film before degradation (on the left in Fig. 2) is essentially smooth with some undulations most likely resulting from the presence or absence of spherulitic crystalline regions just below a disordered surface layer. After partial degradation of the polymer by enzymatic hydrolysis (as shown on the right of Fig. 2), the surface has been deeply etched or eroded, and the exposed regions surrounded by pits and channels are most likely those spherulites referred to in the previous sentence. If this interpretation is correct then, as previously suggested by Huang and coworkers [24], the enzymatic hydrolysis is apparently much faster on the polymer molecules which are present in the amorphous regions as compared to those embedded in macroscopic crystalline regions (in spherulites and lamellar

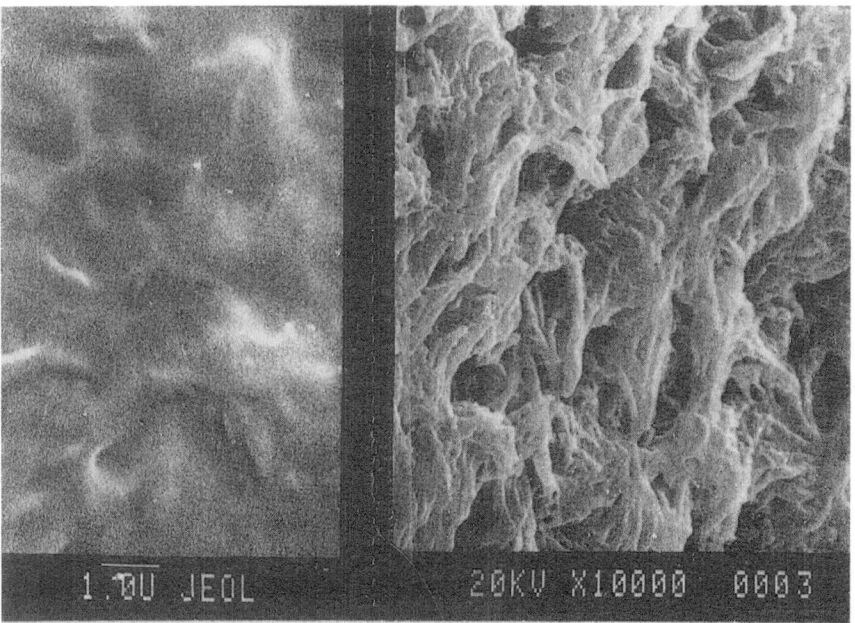

Fig. 2. Electron photomicrograph of surface of HB/HV copolymer: left – before degradation; right – after exposure for 20 hours to an aqueous solution of a purified exocellular enzyme produced by *Pseudomonas lemoignei*

crystalline regions). Eventually, of course, the latter regions are also degraded and removed, but by then the degradation process has proceeded well into the amorphous and defect crystalline regions which are well below the remnants of the original surface.

Other physical restraints and rate controlling effects on the process involved in the microbial degradation of solid polymers have been considered and analyzed by Cole for starch-polyethylene blends [61]. His analysis is concerned with the limitations imposed on the invasion of an aerobic microorganism and the enzymatic degradation of the starch phase by low rates of diffusion of oxygen and hydrolytic enzymes into, and diffusion of the nutrients released out of, the pores created in the degradation process. It is assumed that deep intrusion of the degrading microbes into the degradable polymer is not possible until extensive erosion has occurred so that the primary mechanism of attack on the starch phase is by amylase production at the entrance to or near a pore, followed by diffusion of the enzyme into the pore, and simultaneous diffusion of the soluble hydrolysis products back to the surface, where they are metabolized. That is, the microbial cell cannot enter into the small pores, those with diameters less than approximately 5 μm or so.

8 Plastic Waste Management

At the time of writing, the applications of biodegradable polymers are confined mostly to the field of agriculture, where they are used in products with limited lifetimes, such as mulch films and pellets for the controlled release of herbicides. The synthetic polyesters used in medical applications, principally polylactide and poly(lactide-co-glycolide), while claimed to be biodegradable, are degraded in the body mainly, if not entirely, by chemical hydrolysis. There is little evidence that the hydrolysis of these polyesters of α-hydroxyacids can be catalyzed by hydrolase or depolymerase enzymes.

The potential application of biodegradable polymers, which has created the greatest amount of interest, is in the waste management of plastics. Waste plastics are currently disposed of by either incineration, burying in landfills, or recycling. By far, most waste plastics are disposed of in landfills, most of which are biologically inactive, so in such an environment even those polymers which are biodegradable will remain untouched and entombed indefinitely. However, small but increasing numbers of landfills are now maintained biologically active by supplying water well below the surface of the landfill to permit the growth of anaerobic microorganisms, which can utilize waste organic matter. The principal, ultimate by-products of the anaerobic biodegradation of organic matter are water: CO_2, and methane, and by using gas collection systems within the landfill, the latter can be recovered for energy use.

8.1 Composting

Composting could be an effective procedure for the biological utilization of biodegradable polymers by microorganisms. Composting is, principally an aerobic biological process in which plant waste and other organic wastes are decomposed under more or less controlled conditions by a highly diverse population of microorganisms, principally fungi [58]. The solid product of this type of degradation, termed "compost" or "humus", is a heterogeneous mixture of biomass and other organic matter that can be used in land reclamation and in agriculture as a topsoil replacement, although it is not effective as a fertilizer.

Composting normally requires a digestion period of several weeks to months depending on how well the degrading waste material is kept aereated and moist. The temperature within the compost pile can reach as high as 60 °C, and the microbial population throughout the pile can include hundreds of different types of bacteria and fungi. Alternatively, procedures and equipment are being developed to accelerate the composting process and reduce the time required to days or even hours.

8.2 Standard Tests for Biodegradation

Because of the increasing industrial interest in, and applications for, biodegradable polymers, as well as the growing importance of plastic waste management and the confusion over the basic concepts of biodegradation, standard test methods are being developed for plastics that can determine if enzymatic biodegradation occurs in a given environment, how it occurs, and how it can be accurately measured. A committee of the American Society of Testing and Materials [62], ASTM Committee D20.96, has undertaken the task of setting specific conditions and developing specific test procedures for evaluating both the aerobic and anaerobic biodegradation of plastics in a variety of environments, including simulated landfill, composting, marine, waste water treatment plant, and sewage sludge systems.

The Agency of Industrial Science and Technology in Japan (MITI) is sponsoring the development of an analytical method for determining biodegradability of organic polymers which can be decomposed in activated sludge [63]. In the "modified MITI method" the "Degree of Biodegradability" in % (DB) is given by the following equation:

$$DB = \frac{BOD - B}{TOD} \times 100$$

in which BOD is the biochemical oxygen demand of the sample (in mg), B is the oxygen consumption of the culture medium containing a standard activated sludge selected from the region in Japan in which the measurement is made (in mg), and TOD is the theoretical oxygen demand as calculated for the

complete oxidation of all organic carbon in the sample (in mg). An electrochemi-
cal analyzer is used to determine the oxygen requirement in a closed system at
specific sample and sludge concentrations at 25 °C over a standard test period
(usually 14 or 28 days.)

8.3 Fate Analysis

The biodegradation of biopolymers in natural environments provides nutrients
for the growth and survival of microorganisms and plants. For the former, the
low molecular weight products so formed are taken up and metabolized by the
surrounding microorganisms and ultimately converted into water, carbon dixo-
ide, and/or methane. Hence, the biodegradation of natural polymers in nature
presumably leaves no byproducts which can pose a threat to plants or living
species. However, is that also true for natural polymers in non-natural environments (e.g. in compost, activated sludge, etc.), and is that true for the biodegradation of synthetic polymers in all environments? That is, what happens to
the lower molecular weight products of such enzymatic degradation processes;
what is their fate?

The fate analysis of the biodegradation of water soluble polymers has long
been a concern of the detergent industry [64]. The biodegradation of such
compounds in an aerobic environment is assessed by a carbon balance of the
process using the following equation [65]:

$$C_e = CO_2 + C_c + C_r$$

in which C_e is the carbon content of the sample of detergent which is being
degraded, CO_2 is the carbon dioxide evolved, C_c is the carbon content of the
detergent that becomes incorporated into cellular matter, and C_r is residual
carbon from the detergent that remains in the environment. All of these terms
can be measured, and the effectiveness of biodegradation in a given period of
time can be described as follows:

(1) when $C_r = 0$ the detergent was totally biodegraded and $C_e = CO_2 + C_c$;
(2) when $C_r > 0$, incomplete biodegradation has occurred; and
(3) when $C_r = C_e$ no biodegradation has occurred

If $C_r > 0$, the fate of the incompletely biodegraded residue is important,
especially the toxicity of the residue. For an anaerobic process, the equivalent
analysis would be described by the following equation:

$$C_e = CO_2 + CH_4 + C_c + C_r$$

Acknowledgement. The author is grateful to the following organizations which provided financial
support for his research programs on bacterial polyesters and on the biodegradation of polymers
– the Office of Naval Research (Contract number N00014-K-89), the DuPont Co., Procter and
Gamble Co., Rohm and Haas Co., and the Eastman Chemical Co. The author is also grateful to his
coworkers, Prof. R. Clinton Fuller, Dr. David Gilmore and Mark Timmins, and to Dr. Graham
Swift of the Rohm and Haas Co for their help and advice in the preparation of this manuscript.

9 References

1. Narayan R (1990) In: Barenberg SA, Brash JL, Narayan R, Redpath AE (eds) Degradable materials: perspectives, issues and opportunities. CRC, Boca Raton
2. Lehninger AL (1982) Principles of biochemistry. Worth, New York
3. Cooke TF (1990) J Polym Eng 9: 171
4. Sui RGH (1951) Microbial decomposition of cellulose. Reinhold, New York; Cowlind EB (1963) In: Enzymatic hydrolysis of cellulose and related materials. Reese EW (ed). Macmillan, New York
5. Sutherland IW (1990) Biotechnology of microbial exopolysaccharides. Cambridge University Press, New York
6. Husenann E, Fritz B, Lippert R, Pfanneumuller B, Schupp E (1958) Makromol Chem 26: 181; Husemann E, Lippert R, Pfannemuller B (1956) Makromol Chem 26: 214; Pfannemuller B, Burchard W (1969) Makromol Chem 121: 1
7. Doi Y (1990) Microbial polyester. VCH, New York
8. Anderson AJ, Dawes EA (1990) Microbiol Rev 54: 450
9. Brandl H, Gross RA, Lenz RW, Fuller RC (1990) Adv Biochem Engineering/Biotech 41: 77
10. Marchessault RH, Blulim TL, Deslandes Y, Hamer GK, Orts WJ, Gundararajan PR, Taylor MG, Bloembergen S, Holden DA (1988) Makromol Chem Makromol Symp 19: 235; Owen AJ, Bauer H, Owen AJ (1988) Colloid Polymer Sci 226: 241
11. Nagodawithana TW, Steinkraus KH (1976) Applied Environ Microbiol 31: 158
12. Ljungdahl LG, Eriskson K-E (1985) In: Marshall KC (ed) Adv Microbiol Ecology, vol 8. Plenum, New York, p 237
13. Aubert J-P, Beguin P, Millet J (eds) (1988) Biochemistry and genetics of cellulose degradation. Academic, New York
14. (a) Robyt JF, (1984) In: Whistler RL, Benuller JN, Paschall EP (eds) Starch; Chemistry and technology. Academic Press, New York, pp 87ff; (b) (1980) Marshall JJ (ed) Mechanisms of saccharide polymerization and depolymerization. Academic, p 55ff
15. Berkeley RCW, Gooday GW, Ellwood DC (1979) Microbial polysaccharides and polysaccharases. Academic, New York
16. Chowdhury AA (19B) Arch Microbiol 47: 167
17. (a) Lusty CJ, Doudoroff M (1966) Proc Natl Acad Sci USA 56: 960; (b) Nakayama K, Saito T, Fukui T, Shirakura Y, Tonuta K (1985) Biochem Biophysics Acta 827: 63; (c) Stinson MW, Merrick JM (1974) J Bacteriology 119: 152; (d) Delafield FP, Cooksey KE, Duodorff M (1965) J Biol Chem 240: 4023
18. (a) Tanio T, Fukui T, Shirakura Y, Saito T, Tomita K, Kaiho T, Masamune S (1982) Dur J Biochem 124: 71; (b) Shirakura Y, Fukui T, Tanio T, Nakayama K, Matsuno R, Tomita K (1983) Biochem Biophys Acta 748: 331
19. Doi Y, Kanesawa Y, Kunioka M, Saito T (1990) Macromolcules 23: 26
20. Glass JE, Swift G (eds) (1990) Agricultural and synthetic polymers: biodegradability and utilization. American Chemical Society, Washington, DC
21. Huang SJ (1989) In: Eastmond GC, Ledwith A, Russo S, Sigwalt P (eds) Comprehensive polymer science, vol 6. Pergamon, New York, p 597
22. Huang SJ (1985) In: Encyclopedia of polymer science and engineering, 2nd edn vol 2. John Wiley, New York, p 221
23. Potts JE (1984) In: Encyclopedia of chemical technology, Suppol Vol. Wiley Interscience, New York, p 626
24. Huang J-C, Shetty AS, Wang M-S (1990) Adv Polymer Tech 10: 23
25. Diamond MJ, Freedman B, Garibaldi JA (1975) Int Biodetn Bull 11: 127
26. Tokiwa Y, Audo T, Suzuki T, Takeda K in Ref 19, pp 136–148
27. Union Carbide Chemicals and Plastics Co Inc, Bound Brook NJ. Product Brochure for Tone Polymers
28. Jarrett P, Benedict CV, Bell JP, Cameron JA, Huang SJ (1985) In: Shalaby SW, Hoffman AS, Ratner BD, Horbett TA (eds) Polymers as biomaterials. Plenum Publishers, New York, p 181ff; Benedict CV, Cameron JA, Huang SJ (1983) J Appl Polym Sci 28: 335
29. Kawai F (1987) In: CRC critical reviews in biotechnology. CRC, vol 6, Issue 3
30. Potts JE (1978) In: Jellinck HHG (eds) Aspects of degradation and stabilization of polymers. Elsevier, New York, p 617ff

31. Jen-Hao L, Schwartz A (1961) Kumstoffe 51: 317
32. Albertsson A-C, Karlsson S in Ref 19, pp 60–64; Albertsson A-C, Andersson SO, Karlsson S (1987) Polymer Deg Stability 18: 73; Albertsson A-C, Karlsson S (1988) J Appld Polymers Sci 35: 1289
33. Gage P (1990) Tappi J 161
34. Mills J, Eggins HOW (1970) Int Biodeterior Bull 6: 13; Jones PH, Prasad D, Heskins M, Morgan MH, Guillet JE (1974) Environ Sci Technol 8: 919
35. Ferguson GM (1991) Chem Australia Jan-Feb: 70
36. Axelrod RJ, Phillips JH (1991) Air Products and Chemicals Inc, Allentown PA presentation at the Plastics Waste Management Workshop. New Orleans, LA
37. Shimao M, Kato N (1990) In: International Symposium on Biodegradable Polymers. Program and Abstract. Biodegradable Plastics Society, Tokyo, Japan, p 80
38. Suzuki T, Ichihara Y, Yamada M, Tonomura K (1975) Agric Biol Chem 39: 747; (1979) J Appl Polymer Sci Appl Polymer Symp 35: 431
39. Watanabe Y, Morita M, Hamada M, Towomura K (1975) Agric Biol Chem 39: 2447; (1976) Arch Biochem Biophys 174: 575
40. Matsumura S, Maeda S, Takahashi J, Yoshikawa S (1988) Kobunshi Ronbunshu 45: 317
41. Matsumura S, Yoshikawah S in Ref 19, pp 124–135; Lehman HJ (1973) Chemie Unsurer Zeit 3: 82; Schefer W (1980) Seifen-ole-Fette-Wachse 109: 399
42. Cole MA (1990) Preprints of Proceedings of ACS Division of Polymeric Materials: Science and Engineering 63: 877
43. Bailey WJ, Kuruganti VK, Angle JS in Ref 19, pp 149–160; Bailey WJ, Gu J, Lin Y, Zheng Z (1991) Makromol Chem Macromol Symp 42/43: 195
44. Bailey WJ, Gapud B (1985) Polymer Stabil and Deg 280: 423
45. Scott G, Mellar DC, Moir AB (1973) Eur Polymer J 9: 219; Colin G, Cooney JD, Carlsson DJ, Wiles DM (1981) J Appld Polymer Sci 26: 509
46. Guillet JE in Ref 1, pp 55–97
47. Scott G (1989) Polymer News 14: 169
48. Gage P (1990) TAPPI J 161
49. Griffin GJL (1974) Adv Chem Ser 134: 159; (1977) US Patent 4,016,117; (1977) UK Patent 1,485,833; (1977) UK Patent 1,487,050
50. Otey FH, Westhoff RP, Doane WM (1987) Ind Eng Chem Res 26: 1659
51. Roper H, Koch H (1990) Starch/Starke 42: 123
52. Peanasky JS, Long JM, Wool RP (1991) J Polymer Sci, Part B: Polymer Phys 29: 565; Goheen SM, Wool RP (1991) J Appld Polymer Sci 42: 2691
53. Zobel HF (1984) In: Whistler RL, Bemiller JN, Paschall EF (eds) Starch: Chemistry and technology 2nd (edn). Academic, p 285
54. Stepto RFT, Tomka I (1987) Chimia 41; Wittwer F, Tomka I (1987) US Patent 4,673,438
55. Lay G, Rehm J, Stepto RFT, Thomas M (1989) European Patent Appln 89810078.9
56. Koleske J (1978) In: Paul DR, Newman S (eds) Polymer blends. Academic, New York, vol 2, p 369
57. Juni K, Nakano M, Kubota M (1986) J Controlled Release 4: 25; Koenig W, Seidel HR, Sandow JK (1985) European Patent Appln 133988 AZ; Kumagai Y, Doi Y (1992). Polymer Degradation and Stability, in press; Dave P, Ashar N, Gross R, McCarthy SP (1990) Polymer Preprints 33: No 1
58. Azhari R, Sideman S, Lotan N (1991) Polymer Degradation Stability 33: 35
59. Huang S-Y, Chou M-S (1990) Biotech Bioeng 35: 547
60. Timmins M (1992) PhD research progress report, University of Massachusetts, Amherst
61. Cole MA in Ref 20, p 76ff
62. (a) Narayan R (1989) Kunstoffe German Plastics 79: 92; (b) Cole MA, Leonas KK (1991) Biocycle 56; (c) Morgan D (1991) Scientist 6: 1
63. Masuda T (1991) Techno Japan 24: 56
64. Swisher RD (1987) Surfactant biodegradation: 2nd edn. Marcel Dekker, New York
65. Swift G (1991) In: Program and Abstracts of the Internationl Symposium on Biodegradable Polymers. October, Tokyo, Japan, Biodegradable Polymer Society, p 61

Received: August 12, 1992

Poly (Ortho Esters)

J. Heller

Controlled Release and Biomedical Polymers Department, SRI International, Menlo Park, CA 94025, USA

The synthesis and characterization of four distinct families of poly (ortho esters) are described and designated as poly (ortho esters) I, II, III and IV. Poly (ortho ester) I is prepared by the transesterification of diethoxytetrahydrofuran with diols. Poly (ortho ester) II is prepared by the condensation of 3,9-bis (ethylidene 2,4,8,10-tetraoxaspiro [5, 5] undecane) with diols to produce a linear polymer or with a triol to produce a crosslinked polymer. Poly (ortho ester) III is prepared by the condensation of a flexible triol with and alkyl orthoacetate to produce ointment-like materials. Poly (ortho ester) IV is prepared by the condensation of a rigid triol with and alkyl orthoacetate to produce solid materials. The detailed mechanism of hydrolysis of these polymers has been determined and drug release data for a number of therapeutic agents are presented.

1 Introduction

Drug release from bioerodible polymers with therapeutic agents homogeneously dispersed in the polymer matrix can be classified into devices that undergo a bulk hydrolysis process where drug release is dominated by Fickian diffusion and those that undergo a surface hydrolysis process where drug release is dominated by polymer solubilization in the outer layers of the device [1]. This latter process, also known as surface erosion, is clearly preferable because an ability to control kinetics of polymer hydrolysis will also control kinetics of drug release.

Thus, considerable effort has been devoted to the development of polymers that are able to erode by a well controlled surface hydrolysis process. Dominant among these are polyanhydrides [2] and poly (ortho esters) [3].

2 Surface Hydrolysis

The first example of a surface eroding polymer was a partially esterified copolymer of methyl vinyl ether and maleic anhydride, published in 1978 [1]. These polymers solubilize by ionization of carboxylic acid groups as shown in Scheme 1 and because at low pH the carboxyl groups are unionized and hence the polymer is insoluble, these materials are useful as enteric coatings to protect oral dosage forms from dissolution in the stomach [4–6].

When a water-insoluble drug such as hydrocortisone is physically dispersed in the polymer and the polymer fabricated into thin disks which are then placed in a buffer at pH 7.4, release kinetics shown in Fig. 1 are obtained. Clearly, this system exhibits excellent zero order drug release kinetics with concomitant linear rate of polymer weight loss. The simultaneous polymer erosion and drug release is a clear indication that erosion occurs at the surface of the device and that drug release is controlled by erosion of the polymer.

Provided that diffusional release of the drug is minimal so that release is truly dominated by polymer erosion, drug release from surface eroding polymers is characterized by the following properties: (a) rate of drug release is directly proportional to drug loading, (b) lifetime of the device is directly proportional to device dimensions and (c) rate of drug release is directly proportional to the total area of the device. Because total surface area decreases

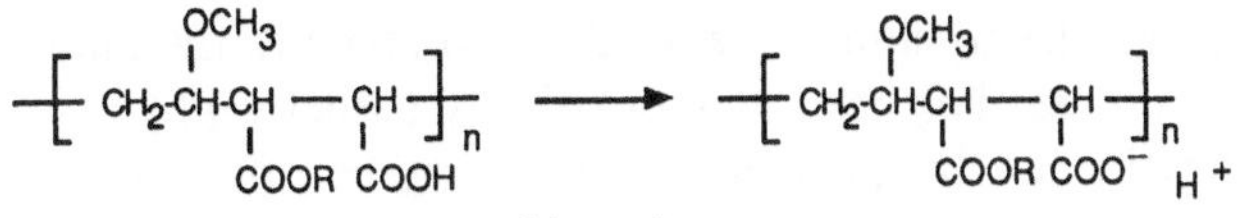

Scheme 1

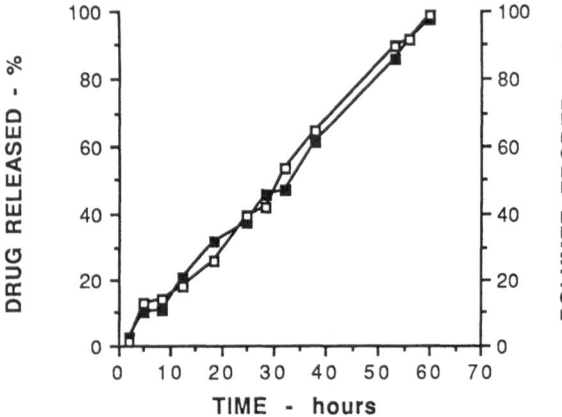

Fig. 1. Rate of polymer erosion (■) and rate of hydrocortisone release (□) from thin disks of an *n*-butyl half ester of methyl vinyl ether and maleic anhydride at pH 7.4 and 37 °C. Drug loading 10 wt% [1]. Reprinted with permission

as the erosion process proceeds, kinetics of release will remain constant only for devices that maintain constant surface area.

Kinetics of drug release from various geometrical shapes undergoing surface erosion can be calculated from the following equation [7]:

$$\frac{M_t}{M_\infty} = 1 - [1 - k_0 t/C_0 a]^n$$

where k_0 is the erosion constant, t is the time, C_0 is the initial uniform concentration of the drug in the matrix, a is the radius of a sphere or cylinder or half-thickness of a slab and n = 3 for a sphere, n = 2 for a cylinder and n = 1 for a slab. Thus, constant release rate is only provided for a slab-shaped device and because thin disks were used, release rates shown in Fig. 1 are zero order.

Clearly, a device will only undergo surface erosion if the rate of the reaction that leads to polymer solubilization occurs exclusively at the polymer-water interface, or if the rate of such a process is significantly higher at the interface relative to the bulk. In this respect, a partially esterified copolymer of methyl vinyl ether and maleic anhydride represents a unique situation because solubilization of the polymer occurs by ionization of carboxylic acid groups and surface erosion is due to a difference in pK_a values between carboxylic acid groups at the polymer-water interface and those in the interior of the matrix. Thus, carboxylic acid groups at the polymer surface are adjacent to an aqueous phase of high dielectric constant so that the acid dissociation constant is that normally associated with an aqueous medium. However, carboxylic acid groups in the interior of the matrix are in a medium of much lower dielectric constant and therefore the degree of ionization is well below that observed at the polymer-water interface. Therefore, only carboxylic acid groups at the polymer-water interface will ionize and the erosion process will be confined to the outer surface of the device. This process has been mathematically modeled [1].

However, the solubilization of this particular polymer system produces a high molecular weight, nondegradable, water-soluble polymer. Therefore, it is only suitable for topical applications where elimination of such a polymer can take place with no difficulty, but is not suitable for use as an implant.

To be useful as an implant, the polymer must hydrolyze to small, water soluble and toxicologically safe molecules and to be useful as a surface-eroding system, the hydrolysis must occur at much higher rates in the outer layers than it does in the bulk. Therefore, the successful development of such devices requires the selection of bonds that are capable of undergoing rapid hydrolysis. Two such bonds are anhydrides which are rapidly hydrolyzed to diacids even at the physiological pH of 7.4 and ortho esters which at pH 7.4 are slow to hydrolyze but which hydrolyze at increasingly rapid rates as the pH is lowered. Polymers based on both of these linkages are under intensive development and this chapter will cover, in depth, the development and current status of poly (ortho esters).

3 Ortho Ester Hydrolysis Mechanism

The hydrolysis of acetals, ketals and ortho esters proceeds as is illustrated in Scheme 2 for ortho esters [8].

The rate determining step is the formation of a carbonium ion and as expected, substituents that are able to stabilize the carbonium ion will strongly accelerate the hydrolysis reaction. This effect is shown in Table 1. However, as also shown in Table 1, this is not true for ortho esters where substituent effects are much smaller than would be predicted by analogy with acetal or ketal hydrolysis, and in some cases are in the opposite direction. These data are consistent with the hypothesis that in ortho ester hydrolysis, there is very little carbonium ion character in the transition state and that in the hydrolysis of ortho esters, addition of a proton is concerted with the breaking of a C-O bond.

Ortho esters were shown to undergo general acid catalysis as early as 1929 [9]. This type of catalysis is experimentally observed by noting an increase in reaction rate with increasing buffer concentrations at constant pH. In this type of catalysis the actual catalytic species is not the hydronium ion but instead is

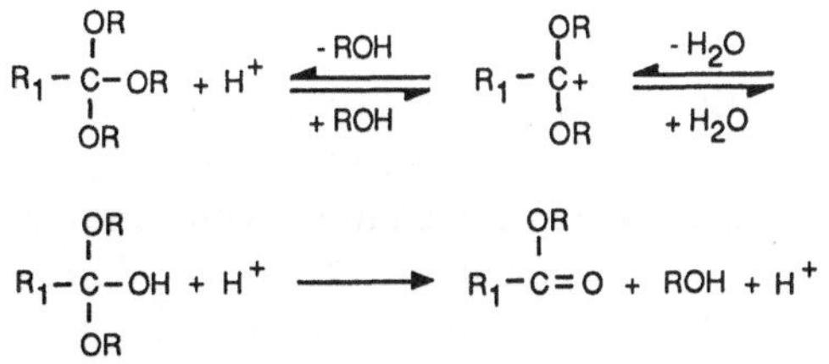

Scheme 2

Table 1. Relative reactivities of certain Acetals and Ortho Esters

Substrate	Relative reactivity
Acetals	
$CH_2(OEt)_2$	1.00^a
$CH_3CH(OEt)_2$	6×10^3
$C_6H_5CH(OEt)_2$	1.71×10^5
$(CH_3)_2CH(OEt)_2$	1.83×10^7
Ortho Esters	
$HC(OEt)_3$	1.00^b
$CH_3C(OEt)_3$	38.5
$CH_3CH_2C(OEt)_3$	24.3
$CH_3CH_2OC(OEt)_3$	0.17

[a] $k_2 = 4.13 \times 10^5$ mole^{-1} sec^{-1} l
[b] $k_2 = 5.38 \times 10^2$ mole^{-1} sec^{-1} l
Adapted from [59]

the actual buffer molecules. Thus, rate of hydrolysis is not only affected by buffer concentration but also by the chemical nature of the buffer. This is an important consideration and will be further discussed in Sect. 4.2.5, Self-regulated Insulin Device.

4 Development of Poly (Ortho Esters)

Up till now, four families of poly (ortho esters) have been prepared. These will be designated as poly (ortho ester) I, poly (ortho ester) II, poly (ortho ester) III and poly (ortho ester) IV.

4.1 Poly (Ortho Ester) I

4.1.1 Polymer Synthesis

The first example of a family of poly (ortho esters) was described in a series of patents in 1978–79 [10–14]. The polymer is prepared as shown in Scheme 3.

This scheme was evolved after considerable experimentation demonstrated that linear poly (ortho esters) can only be prepared if the reactivity of one alkoxy on the ortho ester group is greatly diminished relative to the other two. If this is not done, only crosslinked products are isolated. Because alkoxy groups in an

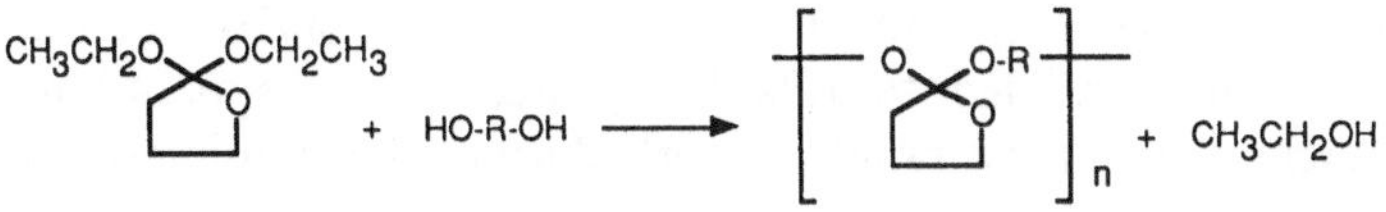

Scheme 3

ortho ester linkage rapidly equilibrate, such a decreased reactivity can only be realized by using a cyclic structure such as diethoxytetrahydrofuran shown in Scheme 3.

4.1.2 Typical Experimental Procedure

To 45 g (0.312 mol) of anhydrous *trans*-1,4-cyclohexanedimethanol and 0.05 g polyphosphoric acid in a commercially available polymerization reactor is added, with constant stirring under an inert nitrogen environment and normal atmospheric pressure, 50 g (0.312 mol) of anhydrous 2,2-diethoxytetrahydrofuran. Next, the mixture is heated to 110–115 °C and held at that temperature for 1.5 to 2 h with slow distillation of ethanol. Then, while maintaining the temperature, the pressure is gradually reduced to 0.01 mm of mercury and at this reduced pressure the temperature is slowly increased to 180 °C. The reaction is continued at this temperature for 24 h. The polymer is isolated by extrusion from the reactor.

4.1.3 Polymer Hydrolysis

When this poly (ortho ester) is placed in an aqueous environment, an initial hydrolysis to a diol and γ-butyrolactone takes place. The γ-butyrolactone then rapidly hydrolyzes to γ-hydroxybutyric acid. This reaction path is shown in Scheme 4.

The hydrolysis is an autocatalytic process because the γ-hydroxybutyric acid hydrolysis product accelerates hydrolysis of the acid-sensitive ortho ester linkages. Therefore, in order to prevent autoacceleration of the hydrolysis, a base must be incorporated into the polymer to neutralize the γ-hydroxybutyric acid.

The metabolic fate of the degradation products from a polymer prepared from diethoxy tetrahydrofuran and *cis/trans*-cyclohexanedimethanol has been

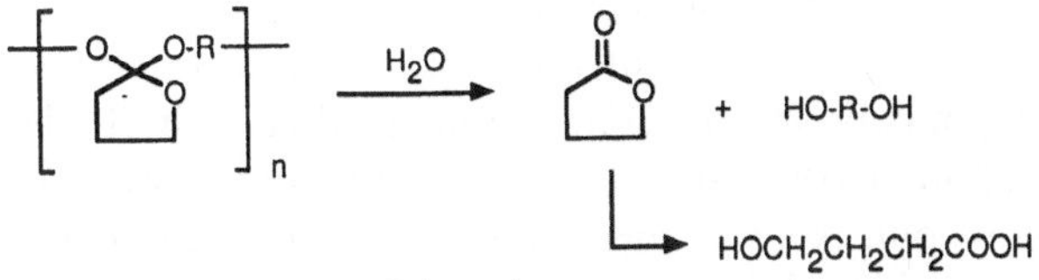

Scheme 4

studied. It was found that when an aqueous solution of $[^{14}C]$ 30/70 *cis/trans* 1,4-cyclohexanedimethanol was fed to rats by gavage in doses of 40 to 400 mg/kg body weight, the diol was rapidly absorbed from the gastrointestinal tract and after 48 hours, 95% of the dose was excreted in the urine, 2.5% in the feces, 0.03% respired as $^{14}CO_2$ and 0.4% remained in the carcass [15]. ^{14}C-carboxyl-labelled γ-hydroxybutyrate and γ-butyrolactone were found to be metabolized very rapidly to $^{14}CO_2$ in the intact rat and the major pathway of metabolism does not appear to involve formation of succinic acid [16].

4.1.4. Development of Specific Delivery Systems

This poly (ortho ester) was originally denoted by the Alza Corporation as Chronomer and is currently known as Alzamer. Its use has been described in a number of publications. Unfortunately, these do not identify the material beyond the code names of C111 and C101ct. However, it can be inferred that C101ct refers to a polymer prepared from diethoxytetrahydrofuran and *cis/trans*-cyclohexanedimethanol and C111 refers to a polymer prepared from diethoxytetrahydrofuran and 1,6-hexanediol.

Release of naltrexone. The C101ct polymer has been investigated as a bioerodible naltrexone delivery system [17]. In this study 20 wt% naltrexone was dispersed into the polymer, pressed into films and $1 \times 1 \times 14$ mm devices punched from the films using a heated punch. Although not stated, the devices very likely also contained 10 wt% Na_2CO_3. Sterilization was achieved by ^{60}Co 2.5 mrad irradiation.

Figure 2 shows in vitro release of ^3H-naltrexone and urinary ^3H release in rats. The pattern of urinary excretion follows closely the in vitro data for the first 10 days. Then, despite the accelerated in vitro release, urinary excretion remains relatively constant until day 25 where it drops rapidly. Only 2% of the initial ^3H was found in devices retrieved from the rats at the termination of the experiment at 34 days. However, no erosion data were presented so that it is not clear how much naltrexone is released by diffusion and how much by erosion of the matrix. Recovery of ^3H in the urine amounted to 43% and it was presumed that the remainder was excreted in the feces.

Release of contraceptive steroids. The C101ct polymer was also used in the development of a bioerodible contraceptive implant under the sponsorship of the Contraceptive Development Branch of NIH and the World Health Organization [18]. Initial toxicological evaluations of devices with incorporated norethindrone and Na_2CO_3 in dogs, rats and primates was somewhat equivocal in that there was evidence of swelling in dogs after a latent period of about two weeks. There was no swelling in baboons, but some swelling did occur in rhesus monkeys. Following these preliminary studies, a small local irritation study in human volunteers was carried out. None of the volunteers experienced itching with the placebo devices, but itching did occur with some volunteers that had

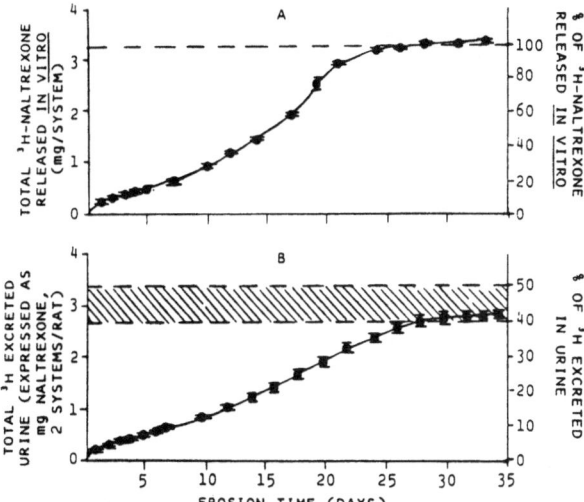

Fig. 2. Cumulative release of ^{3}H-naltrexone in vivo (**A**) and urinary excretion of ^{3}H in rats. (**B**) polymer/^{3}H-naltrexone (20 w/w) system, rods $1 \times 1 \times 14$ mm. (**A**) Each value is the mean $\pm$ standard deviation for 5 systems eroded in vitro. The *dashed line* represents the total naltrexone content; (**B**) each value is the mean $\pm$ standard deviation for 5 rats, 2 systems in the subcutaneous tissue of each rat. The *hatched area* represents the range of expected total ^{3}H-excretion in urine [17]. Reprinted with permission

active devices implanted [19]. Serum concentrations of norethindrone of four human volunteers is shown in Fig. 3 [18].

Because the only adverse effect was itching and some redness which rapidly resolved when the device was removed, another human study using levonorgestrel was carried out [20]. During that study, itching and redness was again observed and further work with this system was discontinued.

Treatment of burns. The C111 polymer, which is assumed to be a polymer prepared from diethoxy tetrahydrofuran and 1,6-hexanediol, has been investigated as a bioerodible ointment for the controlled release of 4-hosulfanilamide

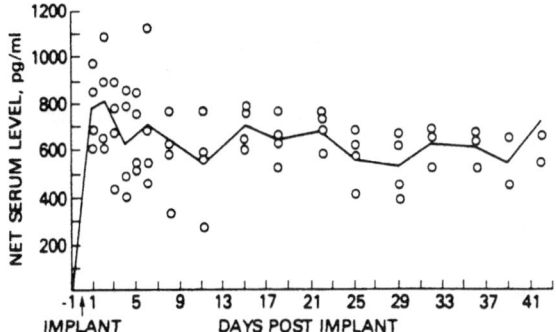

Fig. 3. Serum concentration of norethindrone in four women receiving two poly (ortho ester) norethindrone devices each. The *solid line* represents the mean [18]. Reprinted with permission

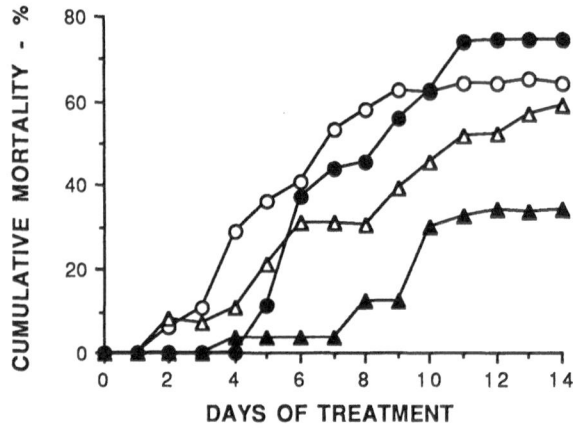

Fig. 4. Cumulative mortality of rats (○) C111 polymer (●) polyester fabric (△) Sulfamylon cream containing homosulfanilamide acetate (▲) C111 polymer with homosulfanilamide free base impregnated in polyester fabric [21]. Reprinted with permission

in the management of *Pseudomonas aeruginosa* burn wound sepsis [21]. In this work, 4-homosulfanilamide was mixed into the C111 polymer and the mixture impregnated into a polyester double knit material. The impregnated fabric was then placed on rat burns infected with a strain of *Pseudomonas aeruginosa* and changed daily. This treatment was compared to that using a commercially available Sulfamylon hydrophilic cream and controls consisting of C111 polymer alone and polyester fabric alone. Results of these studies are shown in Fig. 4. Clearly, survival time with the C111 polymer containing 4-homosulfanilamide is significantly better than that achieved with the commercial preparation.

4.2 Poly (Ortho Ester) II

4.2.1 Polymer Synthesis

Ortho esters can also be prepared by the reaction between a ketene acetal and an alcohol as shown in Scheme 4.

Clearly, formation of polymers requires the availability of a bifunctional ketene acetal. One diketene acetal that has been described is 1,1,4,4-tetramethoxy-1,3-butadiene which can be prepared by the multistep synthesis shown in Scheme 5 [22].

Reaction between this diketene acetal and a diol should proceed as shown in Scheme 6.

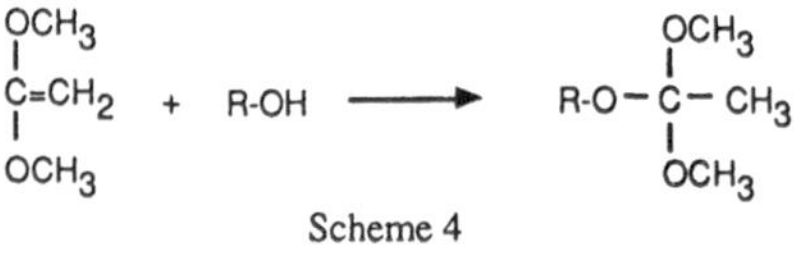

$$
\begin{array}{ccc}
\underset{\substack{| \\ \text{OCH}_3}}{\overset{\text{OCH}_3}{\underset{|}{\text{C}=\text{CH}_2}}} \; + \; \text{R-OH} & \longrightarrow & \text{R-O}-\underset{\substack{| \\ \text{OCH}_3}}{\overset{\overset{\text{OCH}_3}{|}}{\text{C}}}-\text{CH}_3
\end{array}
$$

Scheme 4

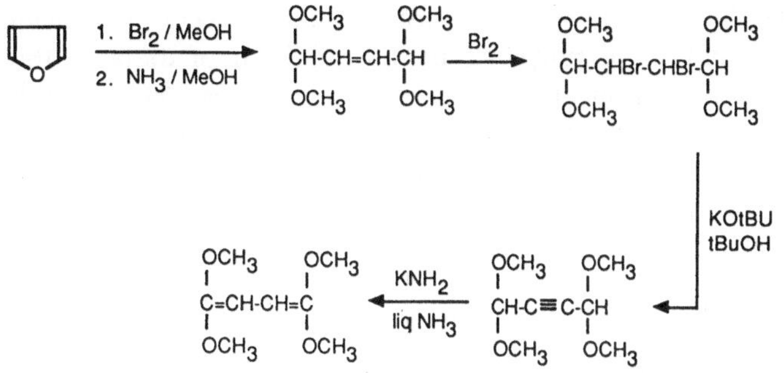

$$
\begin{array}{c}
\text{(furan)} \xrightarrow[\text{2. NH}_3\text{/ MeOH}]{\text{1. Br}_2\text{/ MeOH}}
\underset{\overset{|}{\text{OCH}_3}}{\overset{\overset{\text{OCH}_3}{|}}{\text{CH-CH=CH-CH}}}\underset{\overset{|}{\text{OCH}_3}}{\overset{\overset{\text{OCH}_3}{|}}{}}
\xrightarrow{\text{Br}_2}
\underset{\overset{|}{\text{OCH}_3}}{\overset{\overset{\text{OCH}_3}{|}}{\text{CH-CHBr-CHBr-CH}}}\underset{\overset{|}{\text{OCH}_3}}{\overset{\overset{\text{OCH}_3}{|}}{}}
\end{array}
$$

KOtBU
tBuOH

$$
\underset{\overset{|}{\text{OCH}_3}}{\overset{\overset{\text{OCH}_3}{|}}{\text{C=CH-CH=C}}}\underset{\overset{|}{\text{OCH}_3}}{\overset{\overset{\text{OCH}_3}{|}}{}}
\xleftarrow[\text{liq NH}_3]{\text{KNH}_2}
\underset{\overset{|}{\text{OCH}_3}}{\overset{\overset{\text{OCH}_3}{|}}{\text{CH-C}\equiv\text{C-CH}}}\underset{\overset{|}{\text{OCH}_3}}{\overset{\overset{\text{OCH}_3}{|}}{}}
$$

Scheme 5

$$
\underset{\overset{|}{\text{OCH}_3}}{\overset{\overset{\text{OCH}_3}{|}}{\text{C=CH-CH=C}}}\underset{\overset{|}{\text{OCH}_3}}{\overset{\overset{\text{OCH}_3}{|}}{}}
+ \text{HO-R-OH} \longrightarrow
\left[\!\!\!\!-\text{O-}\underset{\overset{|}{\text{OCH}_3}}{\overset{\overset{\text{OCH}_3}{|}}{\text{C}}}\text{-CH}_2\text{-CH}_2\text{-}\underset{\overset{|}{\text{OCH}_3}}{\overset{\overset{\text{OCH}_3}{|}}{\text{C}}}\text{-O-R}-\!\!\!\!\right]_n
$$

Scheme 6

However, attempts to produce linear polymers using this approach were futile and it was not possible to suppress the competing crosslinking reaction shown in Scheme 7.

This facile crosslinking reaction is due to the extreme ease with which alkoxy groups on an ortho ester linkage transesterify. To prevent this transesterification, a scheme was used where the alkoxy groups were made part of a cyclic structure. In this particular case, two alkoxy groups in the final polymer were made part of a cyclic structure. To prepare such a polymer, the cyclic diketene acetal 3,9-bis (methylene 2,4,8,10-tetraoxaspiro [5,5] undecane) was used. It was prepared as shown in Scheme 8 [23].

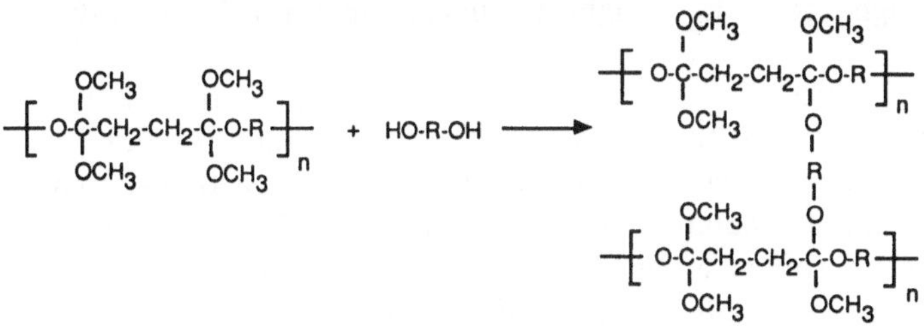

$$
\left[\!\!\!\!-\text{O-}\underset{\overset{|}{\text{OCH}_3}}{\overset{\overset{\text{OCH}_3}{|}}{\text{C}}}\text{-CH}_2\text{-CH}_2\text{-}\underset{\overset{|}{\text{OCH}_3}}{\overset{\overset{\text{OCH}_3}{|}}{\text{C}}}\text{-O-R}-\!\!\!\!\right]_n
+ \text{HO-R-OH} \longrightarrow
$$

Scheme 7

$$
\underset{\overset{|}{\text{Cl}}}{\overset{}{\text{CH}_2\text{-CH}}}\!\!\overset{\text{OCH}_3}{\underset{\text{OCH}_3}{\diagdown}}
+
\text{HOCH}_2\!\!\underset{\text{HOCH}_2}{\overset{\text{CH}_2\text{OH}}{\diagup\diagdown\text{C}\diagup\diagdown}}\!\!\text{CH}_2\text{OH}
\longrightarrow
$$

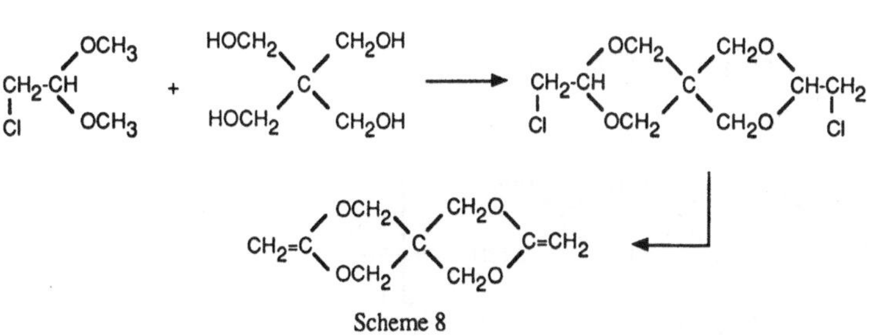

Scheme 8

However, because the addition of an alcohol to a ketene acetal is an acid-catalyzed reaction, formation of polymers by the addition of diols to this diketene acetal is greatly complicated by the extreme susceptibility of this monomer towards a competing cationic polymerization. Nevertheless, linear polymers could be prepared by using iodine in pyridine catalysis [24]. This polymerization, illustrated for 1,6-hexanediol, is shown in Scheme 9. The polymer was characterized by ^{13}C-NMR spectroscopy shown in Fig. 5 [24]. The band assignments are shown in the figure.

The facile cationic polymerization of the diketene acetal is due to the activation of the double bond by the two alkoxy electron donor groups. To convert this monomer to a more useful one, it is necessary to block this facile cationic polymerization and yet retain the reactivity of the ketene acetal group towards additions of alcohols. This was achieved by the introduction of steric hindrance about the double bond by replacing a hydrogen by a methyl group. The structure of the two compounds is compared in Scheme 10.

The second diketene acetal, 3,9-bis (ethylidene 2,4,8,10-tetraoxaspiro [5, 5] undecane) can be prepared as shown in Scheme 11.

The increase in steric hindrance by the introduction of the methyl group was sufficient to prevent the facile cationic polymerization and linear polymers could be prepared by using acid catalysts such as p-toluenesulfonic acid. Further, this diketene acetal could be easily purified and analyzed and the synthesis readily scaled up.

Formation of poly (ortho esters) using the diketene acetal 3,9-bis (ethylidene 2,4,8,10-tetraoxaspiro [5, 5] undecane) is shown in Scheme 12 [25].

The reaction proceeds readily at room temperature and to prepare polymers it is merely necessary to dissolve the monomers in a polar solvent such a tetrahydrofuran and to add a trace of an acid catalyst. Polymerization is exothermic and high molecular weight polymers are formed virtually instantaneously.

Mechanical properties of the polymer can be readily controlled by an appropriate choice of the diols that are used in the condensation reaction [26]. Use of the rigid diol trans-cyclohexanedimethanol produces a rigid polymer having a glass transition temperature of 120 °C. while use of the flexible diol 1,6-hexanediol produces a soft material having a glass transition temperature of 20 °C. Mixtures of the two diols produce polymers that have glass transition

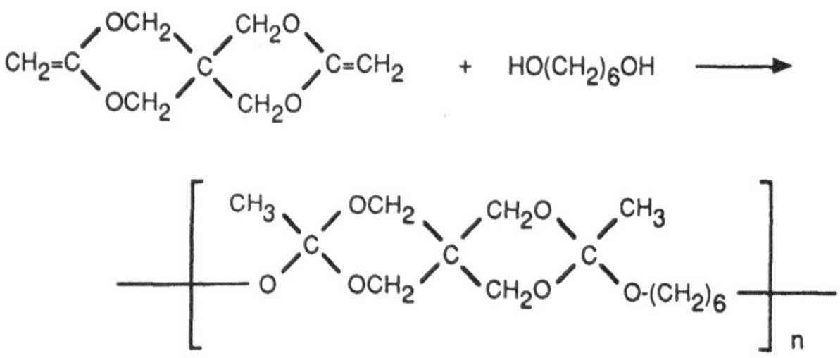

Scheme 9

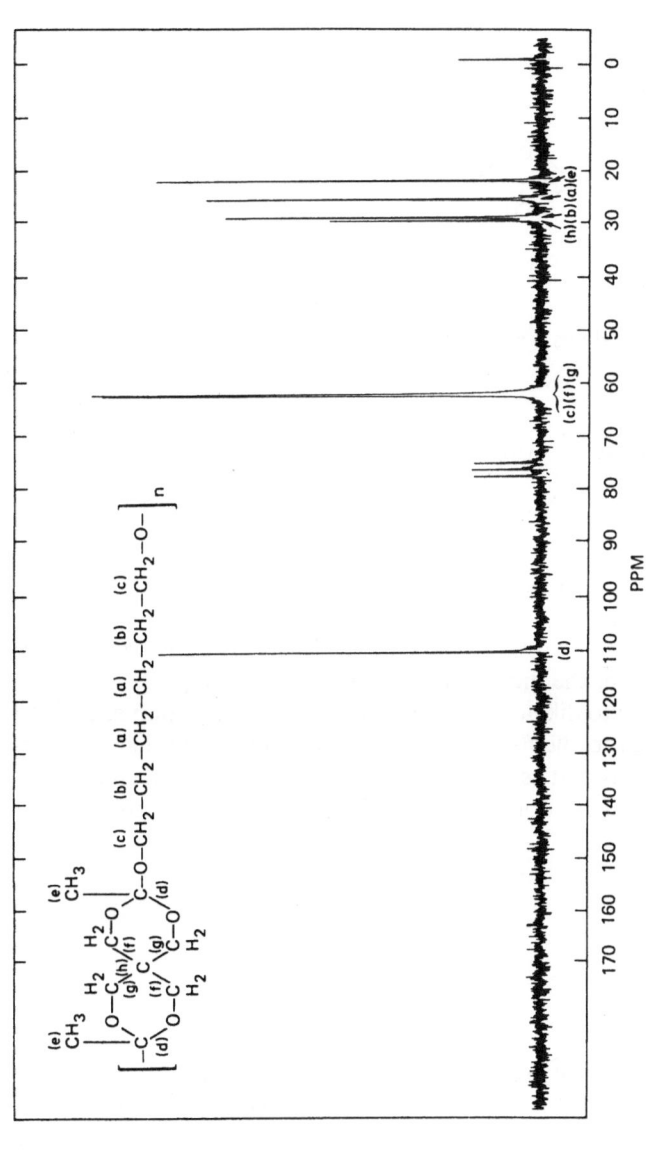

Fig. 5. 25.2 MHz ^{13}C-NMR spectrum of a polymer prepared from 3,9-bis (methylene-2,4,8,10-tetraoxaspiro [5, 5] undecane) and 1,6-hexanediol in CDCl$_3$ at room temperature [24]. Reprinted with permission

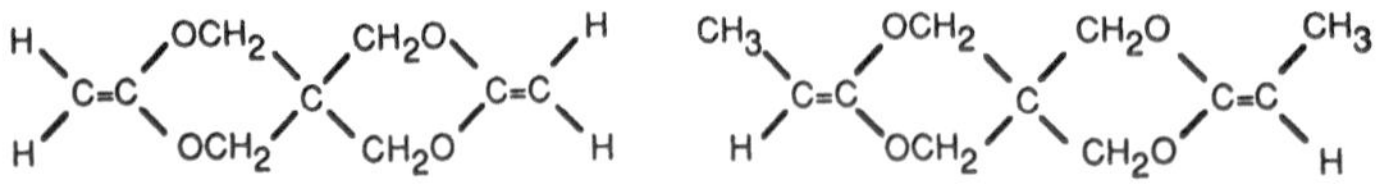

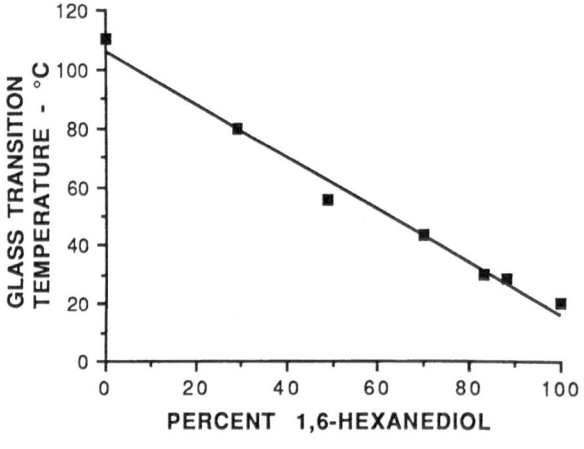

Scheme 10

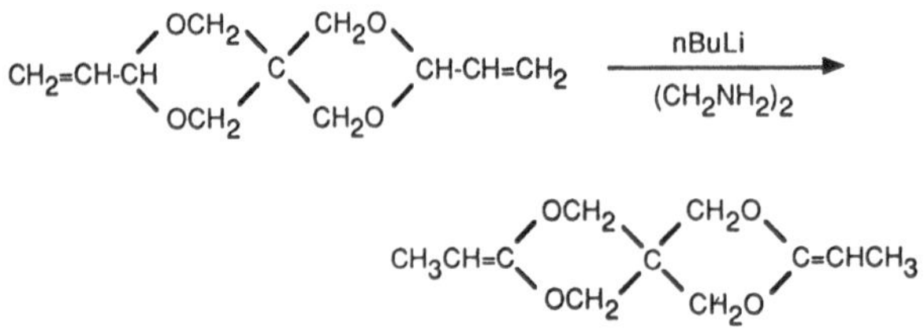

Scheme 11

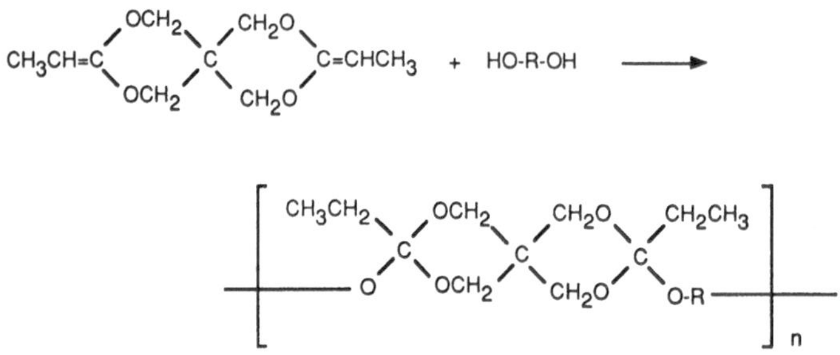

Scheme 12

temperatures between these two values. Variation of the glass transition temperature with composition of the diol mixture is shown in Fig. 6 [26].

The addition of diols to diketene acetals is similar to the addition of diols to diisocyanates that leads to the formation of polyurethanes. And, like in polyurethanes, mechanical properties can be widely varied by using different diols. Further, because the condensation between a diketene acetal and a diol, just like

Fig. 6. Glass transition temperature of 3,9-bis (ethylidene-2,4,8,10-tetraoxaspiro [5, 5] undecane) /*trans*- cyclohexanedimethanol/ 1,6-hexanediol polymer as a function of mol % 1,6-hexanediol [26]. Reprinted with permission

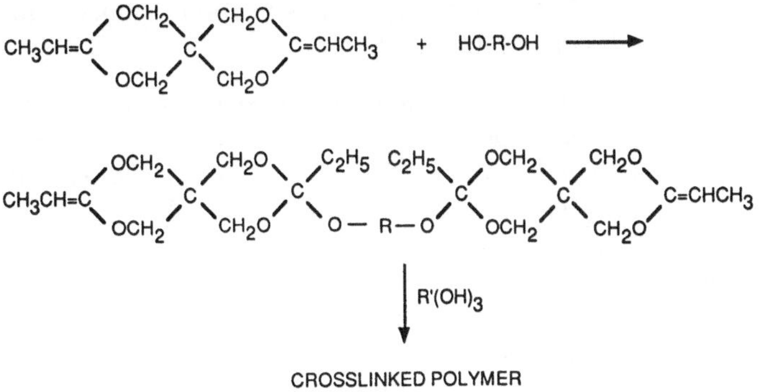

CH₃CH=C(OCH₂)(OCH₂)C(CH₂O)(CH₂O)C=CHCH₃ + HO-R-OH ⟶

$$CH_3CH=C \underset{OCH_2}{\overset{OCH_2}{\diamond}} C \underset{CH_2O}{\overset{CH_2O}{\diamond}} C \underset{O-R-O}{\overset{C_2H_5 \quad C_2H_5}{\diamond}} C \underset{OCH_2}{\overset{OCH_2}{\diamond}} C \underset{CH_2O}{\overset{CH_2O}{\diamond}} C=CHCH_3$$

↓ R'(OH)₃

CROSSLINKED POLYMER

Scheme 13

that between a diisocyanate and a diol, proceeds without the evolution of volatile by-products, dense, crosslinked materials can be produced by using reagents having a functionality greater than two [27].

To prepare crosslinked materials, a molar excess of the diketene acetal is used and the resulting prepolymer with ketene acetal end-groups is reacted with a triol or a mixture of diols and triols. This synthesis is shown in Scheme 13.

4.2.2 Typical Experimental Procedures

Preparation of 3,9-bis (ethylidene-2,4,8,10-tetraoxaspiro [5,5] undecane). In a 3 L three necked flask fitted with a mechanical stirrer, argon inlet tube, thermometer and rubber septum is placed 1.2 L ethylene diamine. The flask is cooled with ice water and the contents kept at about 8 °C under an argon atmosphere. A hexane solution of 130 g (2 mol) on n-butyllithium is added via a stainless steel hypodermic U-tube pushed through the rubber septum using carefully controlled argon pressure over a period of 1 h.

Next, a mixture of 530 g (2.5 mol) 3,9-bis (vinyl-2,4,8,10-tetraoxaspiro [5,5] undecane) and 0.5 L ethylenediamine is cooled to 8 °C and added to the three necked flask. After stirring at 8 °C for 3 h, the reaction mixture is poured into 3 L ice-water with vigorous stirring. The aqueous mixture is extracted twice with 1 L portions of hexane. The combined hexane extracts are washed three times with 1 L portions of waters, dried over anhydrous magnesium sulfate and filtered under suction. The filtrate is evaporated to dryness on a rotary evaporator to give 413 g (78%) crude material containing 90% of 3,9-bis (ethylidene 2,4,8,10-tetraoxaspiro [5,5] undecane).

The crude product is dissolved in 2 L of hexane containing 10 ml of triethylamine and the solution placed in a 4 L filter flask, sealed and stored in a freezer at − 20 °C for 2 days. The crystals thus formed are collected by basket centrifugation at − 5 °C under an argon atmosphere. Distillation of the brownish product through a 12 -inch vigreaux column at reduced pressure gives 313 g (61%) 3,9-bis (ethylidene 2,4,8,10-tetraoxaspiro [5,5] undecane) as a colorless liquid, b.p. 82 °C (13.33 Pa) which crystallizes at room temperature, m.p. 30 °C; characteristic IR band at 1700 cm^{-1}.

Preparation of linear polymers. Into a 5 L, three-necked flask equipped with an overhead stirrer, an argon inlet tube and a condenser are placed 89.57 g (0.621 mole) *trans*-cyclohexanedimethanol, 39.52 g (0.334 mol) 1,6-hexanediol, and 1.8 L distilled tetrahydrofuran. The mixture is stirred until all solids have dissolved; then, 200 g (0.942 mole) 3,9-bis (ethylidene 2,4,8,10-tetraoxaspiro [5,5] undecane) is added. The polymerization is initiated by the addition of 2 mL of a solution of *p*-toluenesulfonic acid (20 mg/mL) in tetrahydrofuran.

The polymerization temperature rapidly rises to the boiling point of tetrahydrofuran and then gradually decreases. Stirring is continued for about 2 h, 10 mL of triethylamine stabilizer added, and the reaction mixture then very slowly poured with vigorous stirring into about 15 gallons of methanol containing 100 mL of triethylamine. The precipitated polymer is collected by vacuum filtration and dried in a vacuum oven at 60 °C for 24 h. The weight of the dried polymer is 325 g (98.8% yield).

Preparation of crosslinked polymers. To a solution of 31.84 g (0.159 mol) 3,9-bis (ethylidene 2,4,8,10-tetraoxaspiro [5,5] undecane) in 200 mL distilled tetrahydrofuran is added 10.42 g (0.100 mole) 2-methyl-1,4-butanediol. The solution is stirred under argon and 0.5 mL *p*-toluenesulfonic acid solution in tetrahydrofuran (20 mg/ml) is added to initiate the reaction. After the heat of reaction has subsided, the solution is stirred until the temperature is once more ambient and then concentrated on a rotary evaporator followed by heating in a vacuum oven at 40 °C to remove residual solvent.

Devices are then prepared by mixing into the prepolymer an excess of 1,2,6-hexanetriol and the desired excipients and curing the mixture in a mold at 75 °C for 5 h. Best results are obtained when the mole ratio of hydroxyl to ketene acetal is about 1.3.

4.2.3 Polymer Hydrolysis

When these poly (ortho esters) are placed in an aqueous environment they hydrolyze as shown in Scheme 14 [26].

Even though the hydrolysis eventually produces an acid, polymer erosion rate is controlled by hydrolysis of the ortho ester bonds. The subsequent hydrolysis of the ester bonds takes place at a much slower rate so that the neutral, low molecular weight reaction products can diffuse away from the implant before hydrolysis to an acid takes place. Thus, unlike the poly (ortho ester) system I, no autocatalysis is observed and it is not necessary to use basic excipients to neutralize the acidic hydrolysis products.

The exact path of the hydrolytic process depends on whether initial protonation occurs on the exocyclic or endocyclic alkoxy groups [28]. The two possible paths are shown in Schemes 15 and 16.

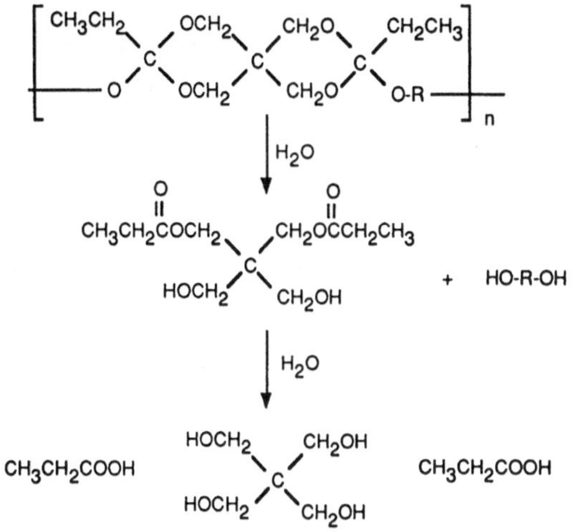

$$\left[\begin{array}{c} CH_3CH_2 \quad OCH_2 \quad CH_2O \quad CH_2CH_3 \\ \diagdown C \diagup \diagdown C \diagup \diagdown C \diagup \\ O \quad OCH_2 \quad CH_2O \quad O\text{-}R \end{array} \right]_n$$

$\downarrow H_2O$

$$\begin{array}{c} O \qquad\qquad O \\ \parallel \qquad\qquad \parallel \\ CH_3CH_2COCH_2 \qquad CH_2OCCH_2CH_3 \\ \diagdown \diagup \\ C \\ \diagup \diagdown \\ HOCH_2 \qquad CH_2OH \end{array} \qquad + \qquad HO\text{-}R\text{-}OH$$

$\downarrow H_2O$

$$CH_3CH_2COOH \qquad \begin{array}{c} HOCH_2 \quad CH_2OH \\ \diagdown C \diagup \\ \diagup \diagdown \\ HOCH_2 \quad CH_2OH \end{array} \qquad CH_3CH_2COOH$$

Scheme 14

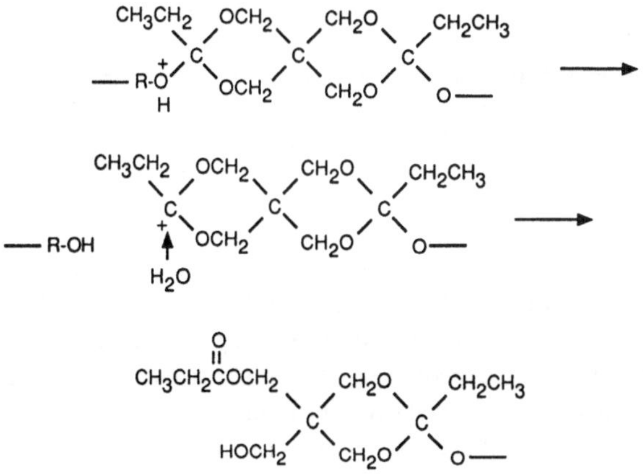

$$\begin{array}{c} CH_3CH_2 \quad OCH_2 \quad CH_2O \quad CH_2CH_3 \\ \diagdown C \diagup \diagdown C \diagup \diagdown C \diagup \\ \text{—}R\text{-}O^+ \quad OCH_2 \quad CH_2O \quad O\text{—} \\ \mid \\ H \end{array} \qquad \longrightarrow$$

$$\begin{array}{c} CH_3CH_2 \quad OCH_2 \quad CH_2O \quad CH_2CH_3 \\ \diagdown C \diagup \diagdown C \diagup \diagdown C \diagup \\ \text{—}R\text{-}OH \quad \uparrow^+ OCH_2 \quad CH_2O \quad O\text{—} \\ H_2O \end{array} \qquad \longrightarrow$$

$$\begin{array}{c} O \\ \parallel \\ CH_3CH_2COCH_2 \qquad CH_2O \quad CH_2CH_3 \\ \diagdown C \diagup \diagdown C \diagup \\ HOCH_2 \quad CH_2O \quad O\text{—} \end{array}$$

Scheme 15

$$\begin{array}{c} CH_3CH_2 \quad OCH_2 \quad CH_2O \quad CH_2CH_3 \\ \diagup C \diagdown \diagup C \diagdown \diagup C \diagdown \\ \text{—}R\text{-}O \quad {}^+OCH_2 \quad CH_2O \quad O\text{—} \\ \mid \\ H \end{array} \qquad \longrightarrow$$

$$\begin{array}{c} CH_3CH_2 \quad OCH_2 \quad CH_2O \quad CH_2CH_3 \\ \diagup C \diagdown \diagup C \diagdown \diagup C \diagdown \\ \text{—}R\text{-}O \quad ^+\uparrow HOCH_2 \quad CH_2O \quad O\text{—} \\ H_2O \end{array} \qquad \longrightarrow$$

$$\begin{array}{c} O \\ \parallel \\ \text{—}R\text{-}O\text{-}C\text{-}CH_2CH_3 \qquad \begin{array}{c} HOCH_2 \quad CH_2O \quad CH_2CH_3 \\ \diagdown C \diagup \diagdown C \diagup \\ HOCH_2 \quad CH_2O \quad O\text{—} \end{array} \end{array}$$

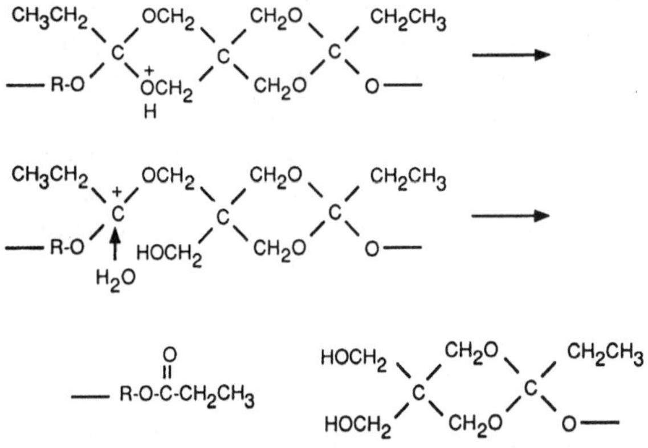

Scheme 16

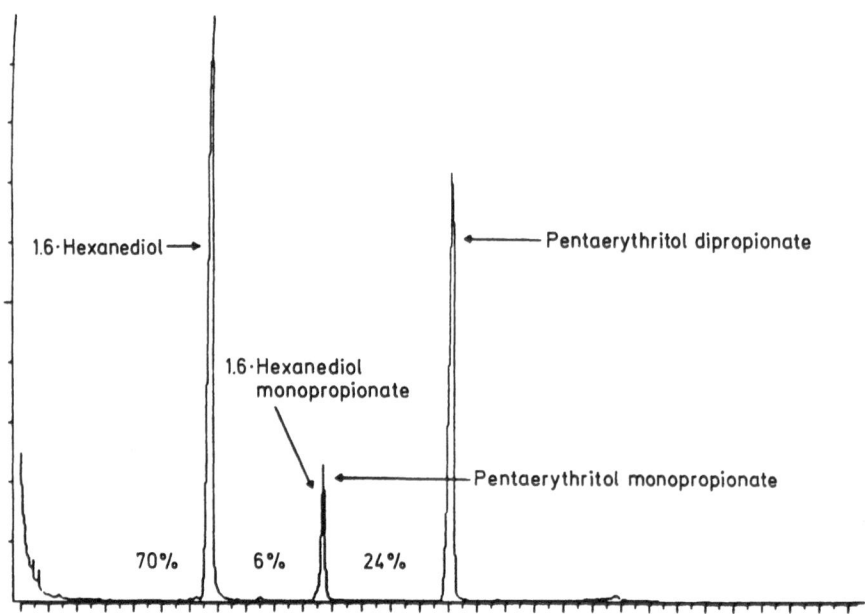

Fig. 7. Gas chromatogram of hydrolysis products of a linear polymer prepared from 3,9-bis (ethylidene-2,4,8,10-tetraoxaspiro [5, 5] undecane) and 1,6-hexanediol [28]. Reprinted with permission

In order to ascertain the relative importance of these two reaction paths, a careful analysis of the hydrolysis products from a linear polymer based on 3,9-bis (ethylidene 2,4,8,10-tetraoxaspiro [5, 5] undecane) and 1,6-hexanediol was carried out [28]. If the hydrolysis proceeds via initial protonation of the exocyclic alkoxy group, the hydrolysis proceeds to yield pentaerythritol dipropionate and the free diol. If, on the other hand, the hydrolysis proceeds via initial protonation of the endocyclic alkoxy group, then the hydrolysis proceeds to yield pentaerythritol and hexanediol propionate.

Results of gas chromatographic analysis of the hydrolysis products, presented in Fig. 7, show two major peaks due to the diol and pentaerythritol dipropionate and two minor peaks, identified as the monoesters of 1,6-hexanediol and pentaerythritol. When the area under the peaks is corrected for unequal detector response, the area corresponding to the monopropionate of 1,6-hexanediol corresponds to about 4.5% of the total area. According to these results, 95.5% of the hydrolysis proceeds via initial protonation and cleavage of the exocyclic alkoxy group.

4.2.4 Control of Erosion Rate

While ortho ester linkages are quite labile in solution, when they are incorporated into a highly hydrophobic polymer matrix, rate of hydrolysis is very slow.

Thus, to accelerate polymer hydrolysis and concomitant release of incorporated therapeutic agents, the hydrolysis needs to be accelerated. Such acceleration can be achieved by the addition of acidic excipients, by increasing the hydrophilicity of the polymer matrix, or both. It is also possible to retard polymer hydrolysis by using basic excipients which stabilize ortho ester linkages.

Use of acidic excipients. When a hydrophobic polymer with a physically dispersed acidic excipient is placed into an aqueous environment, water will diffuse into the polymer, dissolve the acidic excipient in the surface layers and the lowered pH will accelerate hydrolysis of the ortho ester bonds.

This process is schematically shown in Fig. 8 where it has been analyzed in terms of the movement of two fronts, V_1, the movement of a hydrating front and V_2, the movement of an erosion front [29]. Clearly, the ultimate behavior of a device will be determined by the relative movement of these two fronts. If $V_1 > V_2$, the thickness of the reaction zone will gradually increase and at some time, the matrix will be completely permeated by water. At that point, all ortho ester linkages will hydrolyze at comparable rates and bulk hydrolysis will take place. However, if $V_1 = V_2$, then hydrolysis is confined to the surface layers and only surface hydrolysis will take place. In this latter case, rate of polymer erosion will be completely determined by the rate at which water intrudes into the polymer.

The sorption of water by poly (ortho esters) has been found to be relatively small, about 0.30 to 0.75% with a diffusion coefficient ranging from a high of 4.07×10^{-8} cm^2 s^{-1} for a polymer based on 1,6-hexanediol (T_g 22 °C) to a low of 2.11×10^{-8} cm^2 s^{-1} for a polymer based on *trans*-cyclohexanedimethanol (T_g 122 °C) [30]. Thus, assuming a disk thickness of about 2 mm and using the lowest diffusion coefficient, the disk would be completely permeated by water in about 20 days if water penetrates only from one side of the disk and about 10 days if water penetrates from both sides of the disk. Thus, the use of acidic excipients limits the design of surface eroding devices to lifetimes that do not exceed two to four weeks, depending on the actual size.

A number of acidic excipients have been found to usefully accelerate polymer hydrolysis. Among these are latent acids such as various anhydrides, lactides or

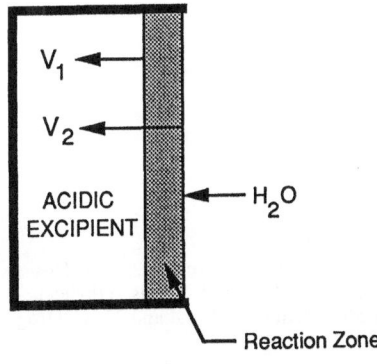

Fig. 8. Schematic representation of water intrusion and erosion for one side of a bioerodible device containing dispersed acidic excipient [29]. Reprinted with permission

glycolides, or free acids. When rapid erosion is desired, free acids that are soluble in the polymer, have a low water solubility and a pKa of about 3 must be used to yield devices that combine long term stability (6 months at 25 °C) and rapid erosion (12 to 24 hours).

The effect of the latent acidic excipient maleic anhydride on the rate of release of a marker drug is shown in Fig. 9 [31]. These data clearly show the effect on polymer erosion rate and the consequent ability to vary erosion rates by relatively minor adjustments in the concentration of the acidic excipient.

The ability of acidic excipients to induce surface erosion is further demonstrated in Figs. 10–12 [32]. Figure 10 shows the concomitant release of the incorporated marker, methylene blue, release of the anhydride excipient hydrolysis product succinic acid, and total weight loss of the device. It is clear that release of both materials coincides with weight loss of the device, as expected for a surface erosion process.

The release kinetics are characterized by an initial lag-phase, a zero release phase and a depletion phase. During the lag-phase water intrudes into the polymer matrix and activates the latent catalyst. During the zero-order release phase, an equilibrium between water intrusion and polymer erosion is established and an eroding front, (V_2), that penetrates the device is established. Because thin disks were used, device geometry remains essentially constant and zero order release uncomplicated by a decrease in total surface area is observed. The depletion phase characterizes a decrease in device and depletion of the incorporated acidic excipient.

Because drug release in a surface-eroding polymer occurs as a consequence of an eroding front that moves through the device, rate of drug release depends

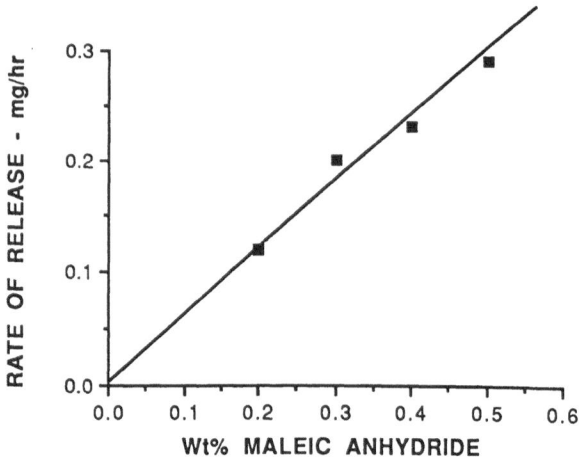

Fig. 9. Release rate of timolol maleate from a 7:3 blend of polymers prepared from 3,9-bis (ethylidene-2,4,8,10-tetraoxaspiro [5,5] undecane) and 1,6-hexanediol and 3,9-bis (ethylidene-2,4,8,10-tetraoxaspiro 5,5] undecane and *trans*-cyclohexanedimethanol at pH 7.4 and 37 °C. Drug loading 2 wt% [31]. Reprinted with permission

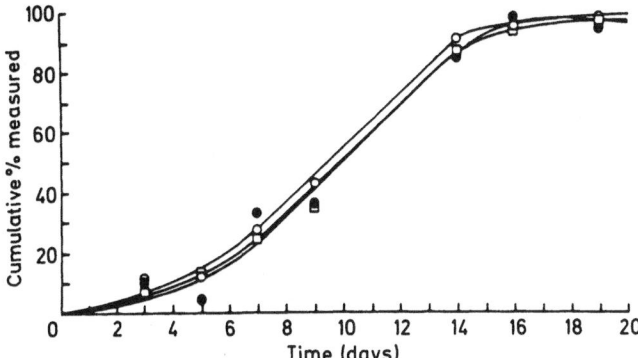

Fig. 10. Cumulative release of methylene blue (○) [1,4-^{14}C] succinic acid (□), and weight loss (●) from a polymer disk prepared from 3,9-bis (ethylidene-2,4,8,10-tetraoxaspiro [5, 5] undecane) and a 50/50 mol ratio of *trans*-cyclohexane dimethanol and 1,6-hexanediol at pH 7.4 and 37 °C. Polymer contains 0.1 wt% [1,4-^{14}C] succinic acid and 0.3 wt% methylene blue [32]. Reprinted with permission

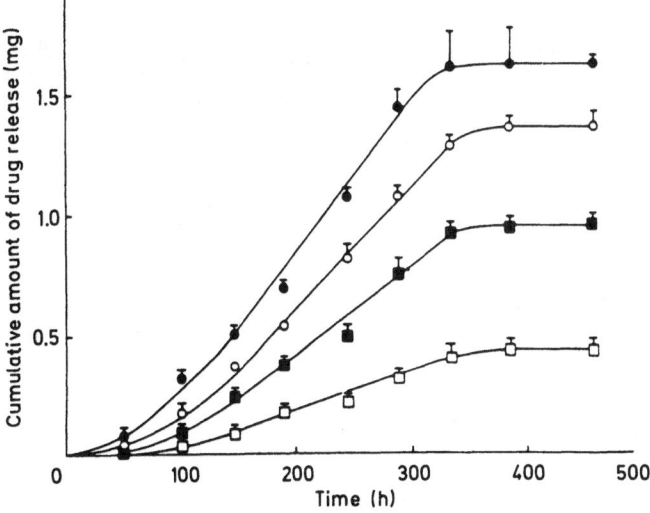

Fig. 11. Effect of drug loading on cumulative drug release from a polymer disk prepared from 3,9-bis (ethylidene-2,4,8,10-tetraoxaspiro [5, 5] undecane) and a 50/50 mol ratio of *trans*-cyclohexane dimethanol and 1,6-hexanediol at pH 7.4 and 37 °C. Drug loading (●) 8 wt%, (○) 6 wt%, (■) 4 wt% and (□) 2 wt% [32]. Reprinted with permission

on the rate of movement of the eroding front and on the concentration of the drug in the polymer. This effect is shown in Fig. 11 which shows rate of release of a drug at four different loadings. As expected, rate of drug release is directly proportional to loading.

Surface erosion has been further demonstrated by noting a linear relationship between surface area and drug release rate shown in Fig. 12 and a linear relationship between device thickness and lifetime, shown in Fig. 13.

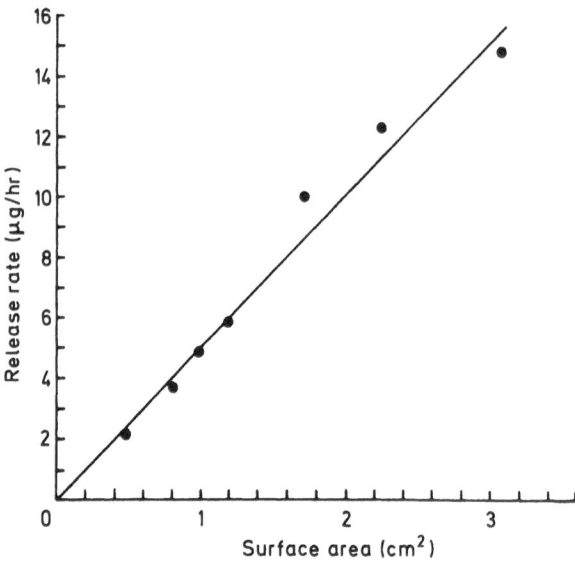

Fig. 12. Effect of surface area on rate of drug release from a polymer disk prepared from 3,9-bis (ethylidene-2,4,8,10-tetraoxaspiro [5,5] undecane) and a 50/50 mol ratio of *trans*-cyclohexane dimethanol and 1,6-hexanediol at pH 7.4 and 37 °C. Polymer contains 4 wt% drug and 0.2 wt% poly (sebasic anhydride) [32]. Reprinted with permission

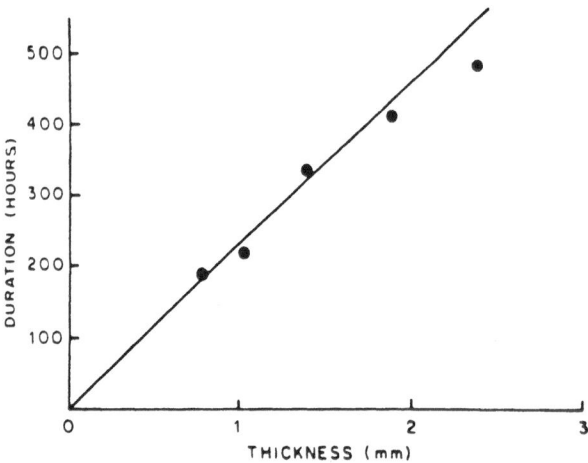

Fig. 13. Effect of thickness on duration of drug release from polymer disks prepared from 3,9-bis (ethylidene-2,4,8,10-tetraoxaspiro [5,5] undecane) and a 50/50 mol ratio of *trans*-cyclohexane dimethanol and 1,6-hexanediol at pH 7.4 and 37 °C. Polymer contains 4 wt% drug and 0.2 wt% poly (sebasic anhydride) [32]. Reprinted with permission

The release kinetics have been mathematically modelled and the formation of residual shells corresponding to uneroded polymer which occurs as a consequence of depletion of acidic excipient has been described [33, 34].

Control of polymer hydrophilicity. As already mentioned, ortho ester linkages are labile even at neutral pH, but when they are incorporated into a highly hydrophobic matrix, they become very unreactive. However, their hydrolysis rate can be accelerated by using more hydrophilic monomers which increases polymer water uptake.

A useful nontoxic, hydrophilic monomer is triethylene glycol. Acceleration of hydrolysis rate and concomitant release of incorporated 5-fluorouracil has been demonstrated by preparing crosslinked polymers based on a prepolymer formed from 3,9-bis (ethylidene 2,4,8,10-tetraoxaspiro [5,5] undecane) and varying proportions of the hydrophobic diol 1,2-propylene glycol and the hydrophilic diol, triethylene glycol, crosslinked with 1,2,6-hexanetriol. Results of this study are shown in Fig. 14. Clearly, the rate of release of 5FU is very sensitive to the amount of incorporated triethylene glycol and release rate can be readily varied by changing the ratio of the two diols. Release rate can be further varied as shown in Fig. 15 by choosing one composition and varying drug loading. These devices are currently under investigation for control of L1012 leukemia tumors in rats [35].

The hydrolysis rate of poly (ortho esters) can be further manipulated by a combination of hydrophilicity and the use of diols that contain a pendant carboxylic acid, such as 9,10-dihydroxystearic acid [36]. Figure 16 shows the release of a marker compound *p*-nitroacetanilide from a hydrophobic matrix prepared from a 60/40 mole ratio of *trans*-cyclohexanedimethanol and 1,6-hexanediol containing varying amounts of 9,10-dihydroxystearic acid. While there is a noticeable effect, it is relatively minor. However, when the hydrophilicity of the matrix is increased by using a 60/10/30 mole ratio of *trans*-cyclohexanedimethanol, 1,6-hexanediol and triethylene glycol, the effect of 9,10-dihydroxystearic acid is greatly magnified, as shown in Fig. 17.

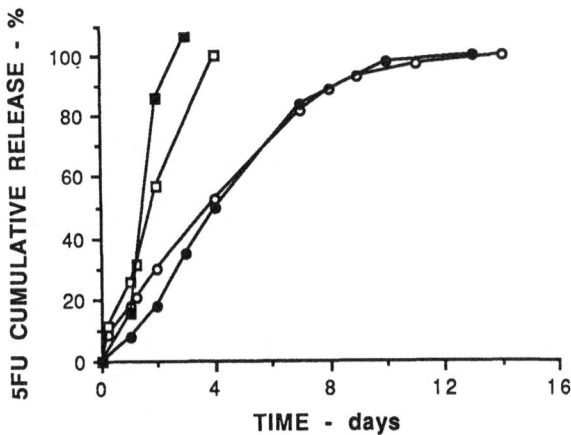

Fig. 14. Cumulative release of 5 fluorouracil (5FU) at pH 7.4 and 37 °C from polymer disks prepared by crosslinking a 3,9-bis (ethylidene-2,4,8,10-tetraoxaspiro [5, 5] undecane), 1,2-propylene glycol (PG) and triethylene glycol (TEG) prepolymer with 1,2,6-hexanetriol. (■) TEG/PG 60/40, (□) TEG/PG 45/55, (○) TEG/PG 30/70, (●) TEG/PG 15/85

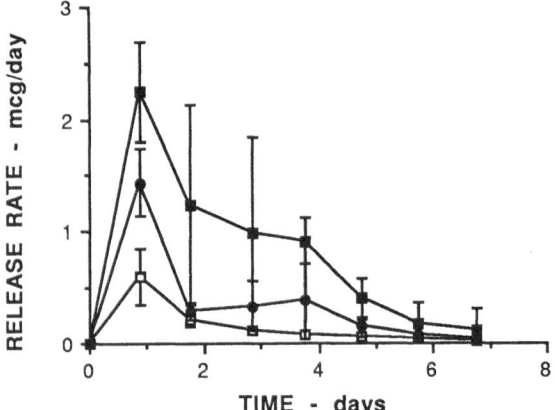

Fig. 15. Daily release of 5 fluorouracil (5FU) at pH 7.4 and 37 °C from polymer disks prepared by crosslinking a 3,9-bis (ethylidene-2,4,8,10-tetraoxaspiro [5, 5] undecane, 1, 2-propylene glycol (PG) and triethylene glycol (TEG) prepolymer with 1,2,6-hexanetriol. TEG/PG 45/55. (□) 5 wt%, (●) 10 wt%, (■) 20 wt%

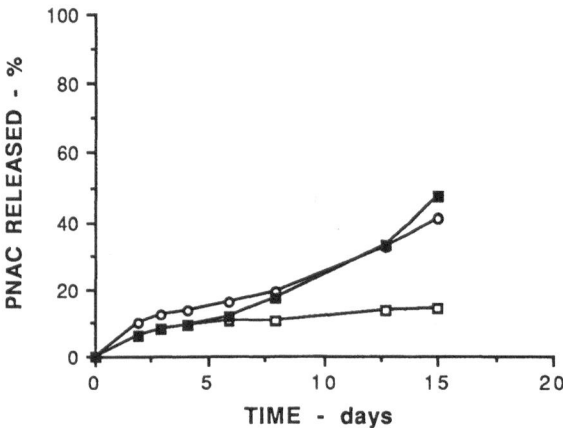

Fig. 16. Release of *p*-nitroacetanilide (PNAC) as a function of time from polymer disks prepared from 3,9-bis (ethylidene-2,4,8,10-tetraoxaspiro [5, 5] undecane) and a 60/40 mol ratio of *trans*-cyclohexane dimethanol and 1,6-hexanediol at pH 7.4 and 37 °C. Polymer contains copolymerized 9,10-dihydroxystearic acid (DHSA) and 2 wt% PNAC. (□) no DHSA, (○) 0.25 mol% DHSA, (■) 0.50 mol% DHSA [36]. Reprinted with permission

Use of basic excipients. As discussed in Sect. 4.2.4, a typical poly(ortho ester) device will be completely permeated by water in a matter of a few weeks so that the use of acidic excipients is limited to delivery systems having a maximum life-time of about one month. However, ortho ester linkages are stable in devices containing a base, and very long erosion times are possible if the polymer is stabilized with a base which prevents hydrolysis even though the matrix is completely permeated by water. A plausible mechanism for erosion of devices that contain the base $Mg(OH)_2$ is shown in Fig. 18. According to this mech-

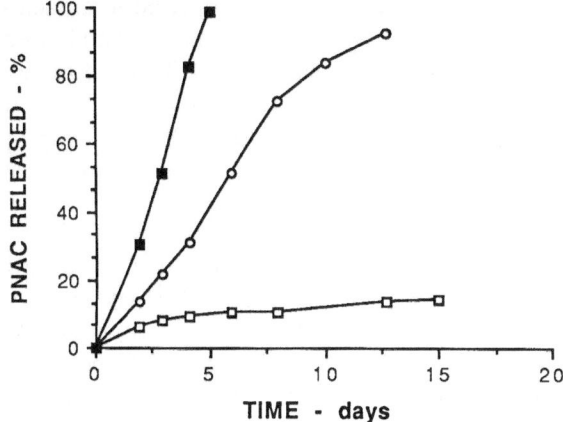

Fig. 17. Release of *p*-nitroacetanilide (PNAC) as a function of time from polymer disks prepared from 3,9-bis (ethylidene-2,4,8,10-tetraoxaspiro [5,5] undecane) and a 60/10/30 mol ratio of *trans*-cyclohexane dimethanol, 1,6-hexanediol and triethylene glycol at pH 7.4 and 37 °C. Polymer contains copolymerized 9,10-dihydroxystearic acid (DHSA) and 2 wt% PNAC. (□) no DHSA, (○) 0.25 mol% DHSA, (■) 0.50 mol% DHSA [36]. Reprinted with permission

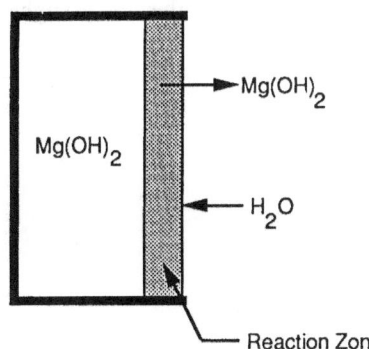

Fig. 18. Schematic representation of water intrusion, $Mg(OH)_2$ diffusion and erosion for one side of a bioerodible device containing dispersed $Mg(OH)_2$

anism, $Mg(OH)_2$ stabilizes the interior of the device and erosion can only occur in the surface layers where the base has been eluted or neutralized. This is believed to occur by water intrusion into the matrix and diffusion of the slightly water-soluble $Mg(OH)_2$ out of the device where it is neutralized by the external buffer. Polymer erosion then occurs in the $Mg(OH)_2$-depleted layer.

4.2.5 Development of Specific Delivery Systems

Self-regulated insulin device. Because replacement of insulin in diabetic patients by simple injection is not able to prevent the serious consequences of the disease [37], considerable effort is currently being devoted to the development of devices that can release insulin in response to external glucose concentration.

One such device is shown in Fig. 19 where insulin is dispersed in an acid-sensitive polymer which is surrounded by a hydrogel containing immobilized glucose oxidase [38]. When glucose diffuses into the hydrogel, it is oxidized by the enzyme glucose oxidase to gluconic acid and the consequent pH will accelerate polymer erosion and concomitant insulin release. Clearly, the key to such a device is a bioerodible polymer that can reproducibly and reversibly change erosion rate in response to small changes in the external pH.

While poly (ortho esters) are acid-sensitive, their sensitivity is not sufficient for use in a self-regulated insulin delivery system. However, when a diol contains a tertiary amine in the backbone is used, the pH-sensitivity is enormously increased. The structure of a polymer from 3,9-bis (ethylidene 2,4,8,10-tetraoxaspiro [5,5] undecane) and N-methyldiethanolamine is shown in Scheme 17.

When disks containing insulin dispersed in the polymer were subjected to well controlled low pH pulses and the cumulative amount of insulin released during each pulse determined by radioimmunoassay, the results shown in Fig. 20 were obtained. The progressive decrease with repeated stimulation represents a gradual depletion of insulin from the device.

It is clear that control over insulin release is excellent and it is important to note that response of the polymer to a decrease in pH is virtually instantaneous. Further, release stops as soon as the pH increases, although there is a slight

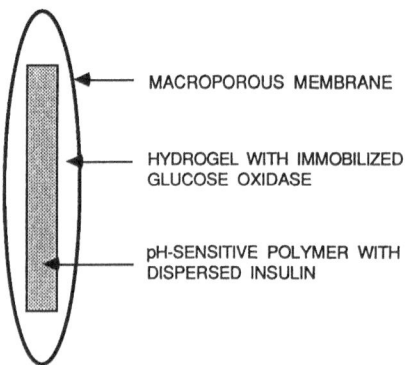

MACROPOROUS MEMBRANE

HYDROGEL WITH IMMOBILIZED
GLUCOSE OXIDASE

pH-SENSITIVE POLYMER WITH
DISPERSED INSULIN

Fig. 19. Schematic representation of proposed insulin delivery system [38]. Reprinted with permission

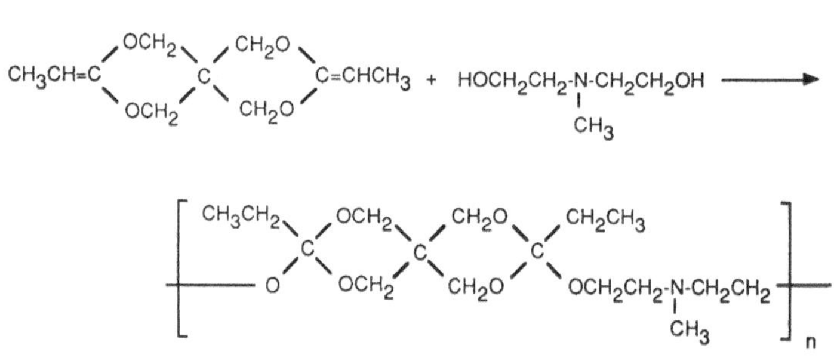

Scheme 17

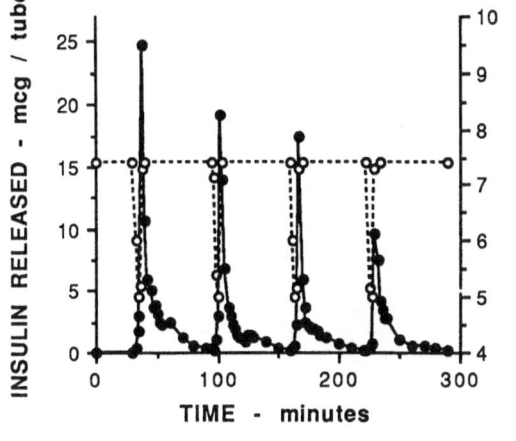

Fig. 20. Release of insulin from a linear polymer prepared from 3,9-bis (ethylidene-2,4,8,10-tetraoxaspiro [5,5] undecane) and N-methyldiethanolamine as a function of external pH variations between pH 7.4 and 5.0 at 37 °C. Buffer was continuously perifused at a flow rate of 2 ml/min and total effluent collected at 1–10 min intervals. (○) buffer pH, (●) insulin release [38]. Reprinted with permission

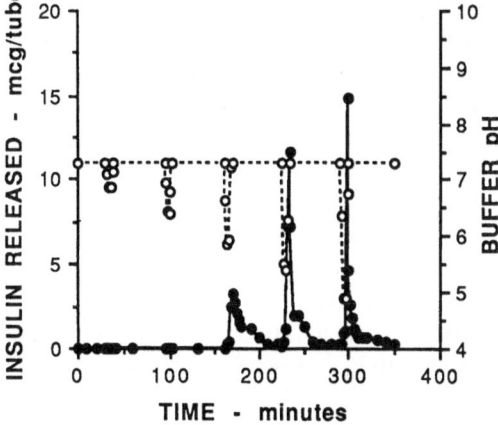

Fig. 21. Release of insulin from a linear polymer prepared from 3,9-bis (ethylidene-2,4,8,10-tetraoxaspiro [5,5] undecane) and N-methyldiethanolamine as a function of external pulses of decreasing pH. Buffer was continuously perifused at a flow rate of 2 ml/min and total effluent collected at 1–10 min intervals. (○) buffer pH, (●) insulin release [38]. Reprinted with permission

tailing off. The rapid response to an increase in pH is important if hypoglycemia is to be avoided.

Figure 21 shows the response of the device to pulses of progressively decreasing pH. Insulin release is first detected at pH 6.0 and release increases as the pH decreases, a highly desirable property for an eventually therapeutically useful device.

However, these in vitro studies were carried out in a citrate buffer. When the studies were repeated in a physiologic buffer, response of the device was only minimal, even at very low pH pulses. A more detailed study using two different buffers at various concentrations and at constant pH is shown in Fig. 22 [39]. These data clearly show general acid catalysis and also show that the desired behavior can only be achieved in a citrate buffer. Thus, further work with this polymer was discontinued and a search for another bioerodible polymer that will undergo specific hydronium ion catalysis is currently underway.

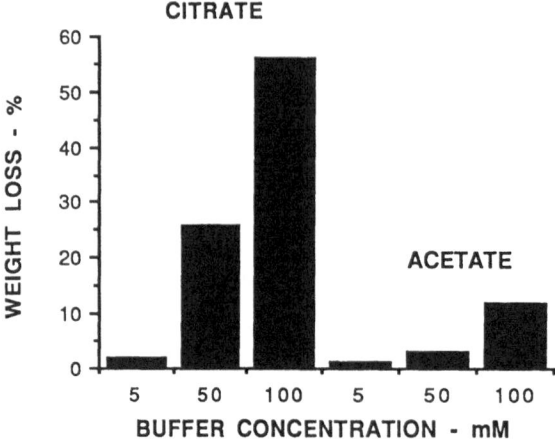

Fig. 22. Weight loss of a polymer prepared from 3,9-bis (ethylidene-2,4,8,10-tetraoxaspiro [5, 5] undecane) and N-methyldiethanolamine in two buffers at pH 5.0 at differing buffer concentrations at 37 °C. Time in buffer was 30 min

Release of 5-Fluorouracil. 5-Fluorouracil (5-FU) is a well known antineoplastic agent that finds applications in cancer chemotherapy [40] and in the prevention of fibroblast proliferation following glaucoma filtration surgery [14].

After investigating a number of linear poly (ortho esters) a material prepared from 3,9-(bis ethylidene 2,4,8,10-tetraoxaspiro [5, 5] undecane) and 1,6-hexanediol was selected as the best material [42]. Figure 23 shows results of a study where both 5FU release and weight loss were determined. The data show that with this particular system, concomitant drug release and polymer erosion has been achieved. Further, because the molecular weight of the residual polymer remains unchanged, the hydrolysis process is confined to the outer surface of the device and surface erosion has been achieved.

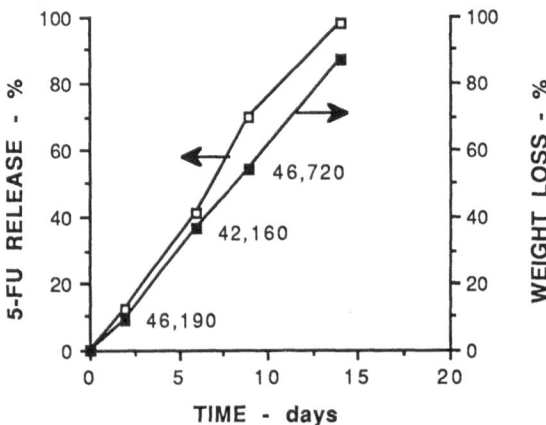

Fig. 23. Cumulative weight loss (■) and cumulative release of 5-fluorouracil (5FU) (□) from polymer disks prepared from 3,9-bis (ethylidene-2,4,8,10-tetraoxaspiro [5, 5] undecane) and 1,6-hexanediol at pH 7.4 and 37 °C. Devices contain 10 wt% 5FU and 0.15 wt% suberic acid. Numbers indicate weight average molecular weight of residual polymer [42]. Reprinted with permission

An investigation of the effectiveness of these devices in the treatment of L1210 leukemia in mice has been carried out and preliminary data are shown in Fig. 24 [43]. The data show survival times of DBA$_2$ mice inoculated intraperitoneally with L1210 cells (10^5) and subsequently treated intraperitoneally on day 1 with polymer alone and a polymer containing 10 wt% 5-FU and 0.15 wt% suberic acid.

As shown in Fig. 24, untreated animals, or those treated with polymer alone, died within the first 20 days but those treated with a 5-FU implant led to a mean increase in survival time of 143%. These data are very encouraging because L1210 leukemia is not particularly sensitive to 5-FU, a drug often used clinically to treat colon cancer.

However, this particular polymer has a very low glass transition temperature and when thin disks are placed in an aqueous environment, they undergo deformation and the consequent irreproducible change in surface area results in changes in drug delivery rates. For this reason, work with the linear polymer has been discontinued and work is currently in progress using crosslinked polymers which maintain their shape. This work has already been described in Sect. 4.2.4, Control of polymer hydrophilicity.

Release of Naltrexone. Naltrexone is a narcotic antagonist that occupies the same receptors as morphine but produces no euphoric effects. Therefore, a patient on naltrexone therapy experiences no euphoria upon intake of heroin which is rapidly metabolized in the body to morphine. Naltrexone therapy is currently a method of choice in the rehabilitation of opiate dependent individuals because it provides an enforced opiate free life [44].

However, continuing a maintenance of oral naltrexone therapy requires very strong motivation because discontinuing therapy produces no withdrawal effects. For this reason, it is desirable to develop a naltrexone-releasing implant which could be implanted by a physician so that removal by the patient would be impossible, or at least very difficult.

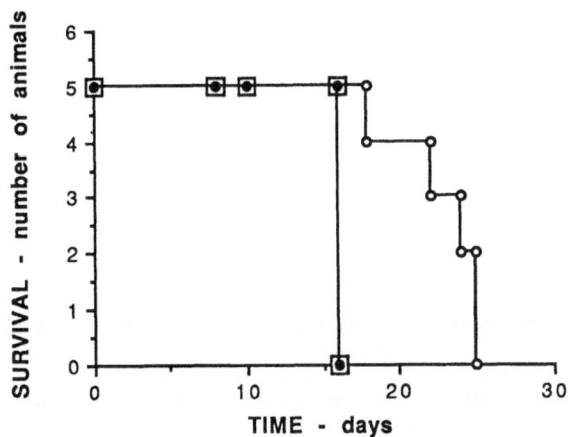

Fig. 24. Survival times of DBA$_2$ mice inoculated with L1210 tumor. (●) control, (□) polymer alone, (○) polymer with 5FU [43]. Reprinted with permission

Because poly (ortho esters) are stable in base and because naltrexone is a base with a pKa of 8.13 which produces a saturated aqueous solution having a pH of about 10, it can not be used as such and a neutral or slightly acidic salt must be used. Therefore, the salt naltrexone pamoate was used. Use of pamoates to decrease water solubility of various drugs has been described.

Previous studies have shown that in order to maintain a therapeutically effective naltrexone blood plasma levels of 2 ng/ml, the device must release naltrexone at about 3–4 mg/day [45]. Because of this high rate, preparation of devices that contain loadings of at least 50 wt% is essential. For this reason, the soft, low glass transition temperature 100% 1,6-HD polymer was used. Figure 25 shows release of naltrexone pamoate as a function of amount of incorporated suberic acid [46]. Because naltrexone pamoate is slightly acidic due to the unreacted phenolic hydrogens, naltrexone pamoate is released from the uncatalyzed matrix with concomitant polymer erosion. The data also show that despite a loading of 50 wt%, good linear release has been achieved and that drug depletion and polymer erosion coincide, indicating that naltrexone pamoate is released by an erosion controlled process. The data plotted as daily release for a device containing 1 wt% suberic acid is shown in Fig. 26. Aside from an initial burst which subsides after day 2, excellent constant release has been achieved. Rod-shaped devices 20 × 2.4 mm have been fabricated by a transfer-molding process and the devices implanted in rabbits. However, during fabrication, even at moderate temperatures, a naltrexone-induced decomposition of the polymer takes place with a significant decrease in molecular weight. Thus, despite excellent in vitro release, the in vivo behavior was not satisfactory. This approach has, however, been resumed with another poly (ortho ester), described in Sect. 4.3.4, Release of naltrexone.

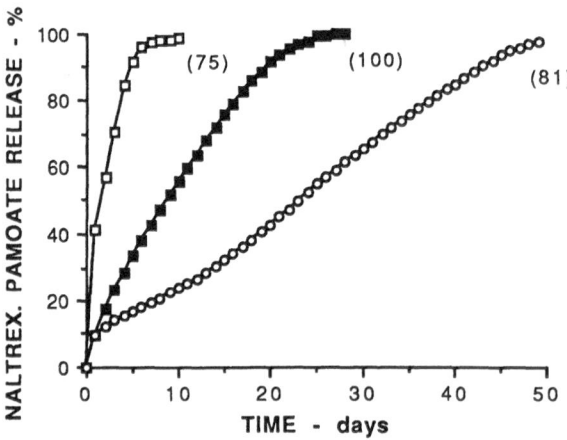

Fig. 25. Cumulative release of naltrexone pamoate from polymer slabs (25 × 4 × 1.25 mm) prepared from 3,9-bis (ethylidene-2,4,8,10-tetraoxaspiro [5,5] undecane) and 1,6-hexanediol at pH 7.4 and 37 °C. Numbers in parentheses indicate percent weight loss. Devices contain 50 wt% drug and varying amounts of suberic acid (SA). (□) 3 wt% SA, (■) 1 wt% SA, (○) no SA [46]. Reprinted with permission

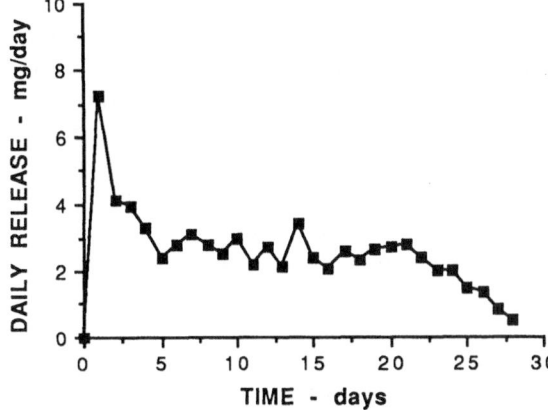

Fig. 26. Daily release of naltrexone pamoate from polymer slabs ($25 \times 4 \times 1.25$ mm) prepared from 3,9-bis (ethylidene-2,4,8,10-tetraoxaspiro [5,5] undecane) and 1,6-hexanediol at pH 7.4 and 37 °C. Devices contain 50 wt% drug and 1 wt% suberic acid [46]. Reprinted with permission

Release of levonorgestrel. Considerable work has been expended on the development of a poly (ortho ester) implant that can deliver a contraceptive steroid from a subcutaneous or intramuscular implant for about one year at close to zero order kinetics and where drug depletion and polymer erosion occur at about the same time. Initial funding for the development for the poly (ortho ester) II system was provided by the Contraceptive Development Branch of NIH with the objective of developing such a contraceptive system.

In the initial approach, and before the action of acidic excipients was clearly understood, the slightly acidic and relatively water insoluble salt, calcium lactate, was used with the hopes of catalyzing long term surface erosion of linear poly (ortho esters). In this work, rod-shaped devices were prepared by incorporating 30 wt% levonorgestrel and 2 wt% calcium lactate into a polymer prepared from 3,9-bis (ethylidene-2,4,8,10-tetraoxaspiro [5,5] undecane) and a 60/40 mol mixture of *trans*-cyclohexane dimethanol and 1,6-hexanediol and the devices were implanted into rabbits. The devices were then explanted at various time intervals and examined by scanning electron microscopy. A device explanted after 10 weeks is shown in Fig. 27 [25].

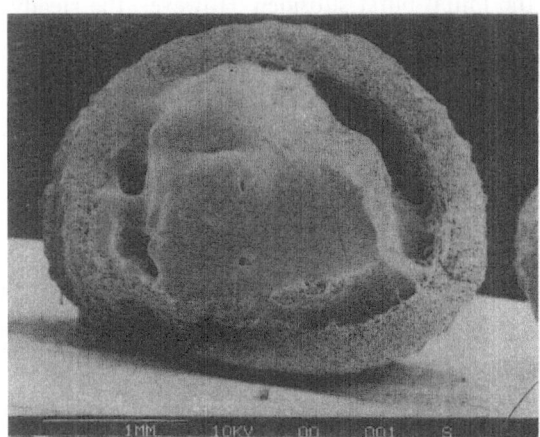

Fig. 27. Scanning electron micrograph of polymer prepared from 3,9-bis (ethylidene-2,4,8,10-tetraoxaspiro [5,5] undecane) and a 60/40 mol ratio of *trans*-cyclohexane dimethanol and 1,6-hexanediol. Polymer rods 2.4×20 mm containing 30 wt% levonorgestrel and 2 wt% calcium lactate. Device implanted subcutaneously in rabbit for 10 weeks, 30X [25]. Reprinted with permission

The appearance of the device can be rationalized as follows: when it is placed in an aqueous environment, a calcium lactate-induced surface erosion takes place and levonorgestrel is released into the surrounding environment. However, because rate of polymer erosion exceeds the rate at which levonorgestrel can solubilize, erosion will take place between the dispersed levonorgestrel particles producing the foam-like layer evident in Fig. 27. As polymer erosion continues deeper into the device, degradation products accumulate underneath this layer and act as plasticizers which dissolve the polymer and create the observed voids. Even though details of this process have not been completely elucidated, it is clear that the use of acidic excipients to catalyze long-term surface erosion is not possible.

For this reason, and as already discussed in Sect. 4.2.4, Use of basic excipients, the development of devices that have long erosion times requires the use of a base which will stabilize the interior of a device and only allow erosion to take place in the outer layers from which the base has been depleted by diffusion. Because in early studies we have found that use of water soluble bases such as Na_2CO_3 leads to osmotic imbibing of water with consequent swelling of the device [47], $Mg(OH)_2$ which has a water-solubility of only 0.8 mg/100 ml was selected. No swelling of devices containing $Mg(OH)_2$ was observed. Thus, this basic salt was used in these studies.

In the development of these devices a crosslinked polymer was used. Devices were fabricated by first preparing a ketene acetal terminated prepolymer derived from two equivalents of the diketene acetal 3,9-bis (ethylidene 2,4,8,10 tetraoxaspiro [5, 5] undecane) and one equivalent of the diol 3-methyl-1,5-pentanediol. Then, 30 wt% levonorgestrel, 7 wt% $Mg(OH)_2$ and a 30 mol% excess of 1,2,6-hexanetriol were mixed into the prepolymer and this mixture then extruded into rods and cured. Erosion and drug release from these devices was studied by implanting the rod shaped devices subcutaneously into rabbits, explanting at various time intervals and measuring weight loss and residual drug [29].

· Levonorgestrel blood plasma levels determined by radioimmunoassay are shown in Fig. 28 [48]. These data were encouraging since the blood plasma level was reasonably constant once the initial burst subsided. However, the steady state plasma level was too low and thus a more rapidly eroding polymer was needed. To achieve a more rapid erosion, a material containing 7 wt% $Mg(OH)_2$ and 1 mol% copolymerized 9,10-dihydroxystearic acid was prepared. The devices were again implanted into rabbits and levonorgestrel blood plasma level determined. Results of these studies are shown in Fig. 29 [3]. Clearly, the more rapidly eroding polymer produces a much higher drug plasma level.

The explanted devices were also examined by scanning electron microscopy and the results shown in Fig. 30 [29]. The pictures clearly show a progressive diminution of a central uneroded zone and the development of voids around the periphery of the rod-shaped device. The presence of voids suggest that once erosion starts, generation of hydrophilic degradation products at that location accelerate further polymer hydrolysis.

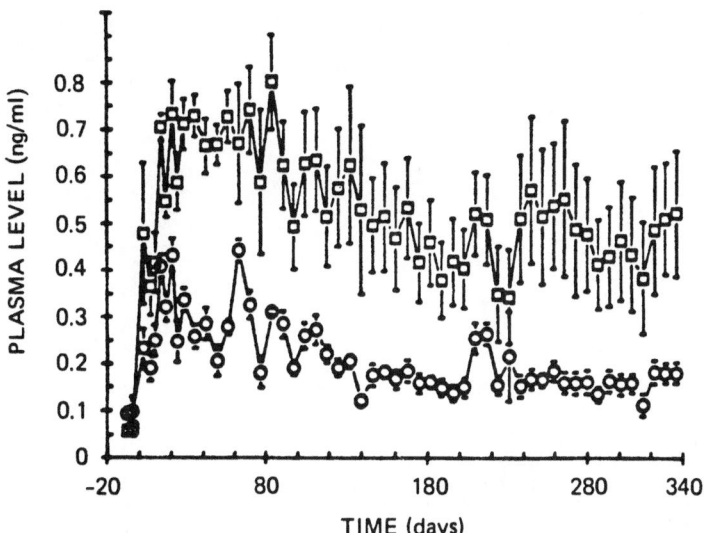

Fig. 28. Daily rabbit blood plasma levels of levonorgestrel from a crosslinked polymer prepared from 3,9-bis (ethylidene-2,4,8,10-tetraoxaspiro [5,5] undecane)/3-methyl-1,5-pentanenediol prepolymer crosslinked with 1,2,6-hexanetriol. Polymer rods 2.4 × 20 mm containing 30 wt% levonorgestrel and 7.1 mol% $Mg(OH)_2$. Devices implanted subcutaneously in rabbits (○) 1 device/rabbit, (□) 2 devices/rabbit. [48]. Reprinted with permission

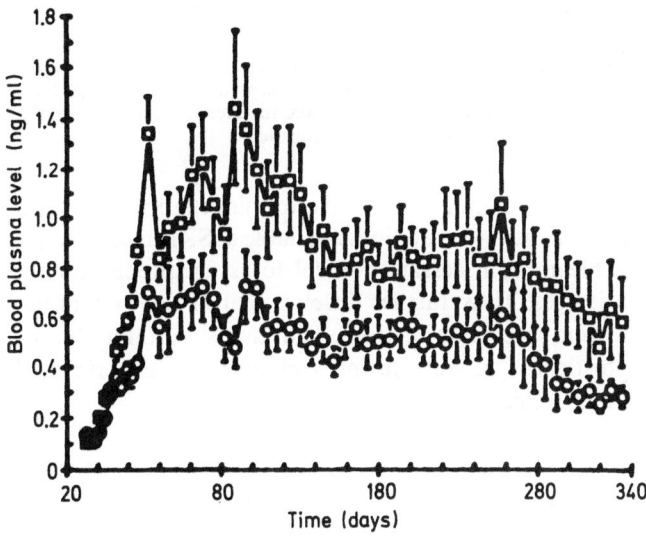

Fig. 29. Daily rabbit blood plasma levels of levonorgestrel from a crosslinked polymer prepared from 3,9-bis (ethylidene-2,4,8,20-tetraoxaspiro [5,5] undecane)/3-methyl-1,5-pentanenediol prepolymer crosslinked with 1,2,6-hexanetriol. Prepolymer contains 1 mol% copolymerized 9,10-dihydroxystearic acid. Polymer rods 2.4 × 20 mm containing 30 wt% levonorgestrel and 7.1 mol% $Mg(OH)_2$. Devices implanted subcutaneously in rabbits. (○) 1 device/rabbit, (□) 2 devices/rabbit. [3]. Reprinted with permission

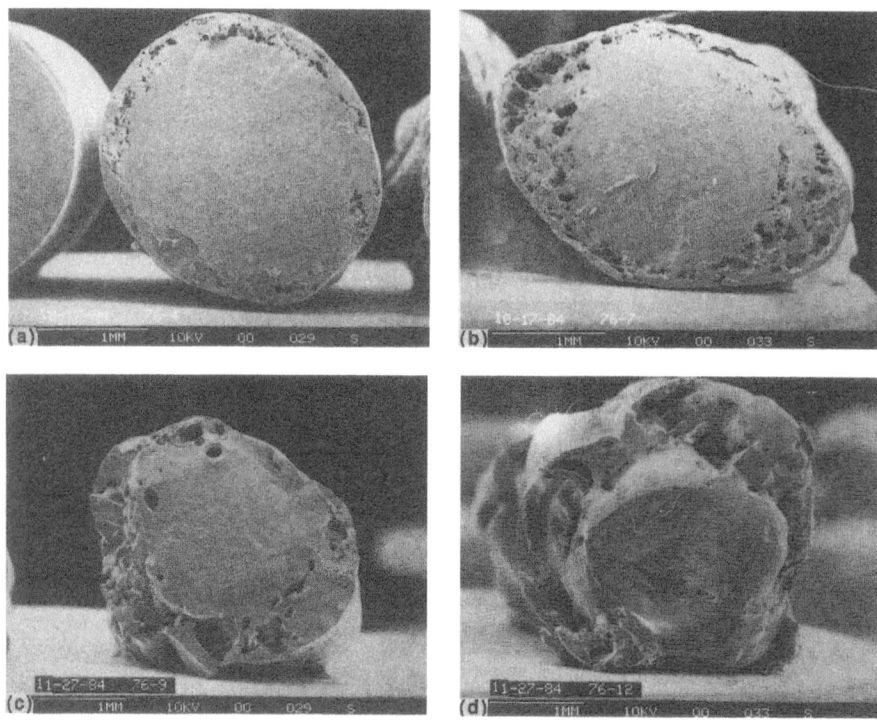

Fig. 30a–c. Scanning electron micrographs of crosslinked polymer prepared from 3,9-bis (ethylidene-2,4,8,20-tetraoxaspiro [5, 5] undecane)/3-methyl-1,5-pentanenediol prepolymer cross-linked with 1,2,6-hexanetriol. Prepolymer contains 1 mol% copolymerized 9,10-dihydroxystearic acid. Polymer rods 2.4×20 mm containing 30 wt% levonorgestrel and 7.1 mol% Mg(OH)$_2$. Devices implanted subcutaneously in rabbits. (**a**) after 6 weeks, 30X, (**b**) after 9 weeks, 30X, (**c**) after 12 weeks, 25X, (**d**) after 16 weeks, 25X [29]. Reprinted with permission

Unfortunately, this system is extremely sensitive to minor fabrication variables and with the limited fabrication capabilities available at SRI International, it was not possible to achieve the necessary control to assure preparation of reproducible batches of devices. Thus, further development of this system has been discontinued. However, successful devices could almost certainly be developed at a more sophisticated manufacturing facility where experimental variables can be closely controlled.

4.3 Poly (Ortho Ester) III

4.3.1 Polymer Synthesis

This family of poly (ortho esters) can be prepared as shown in Scheme 18 by reacting a triol with two vicinal hydroxyl groups and one removed by at least three methylene groups with an alkyl orthoacetate [49].

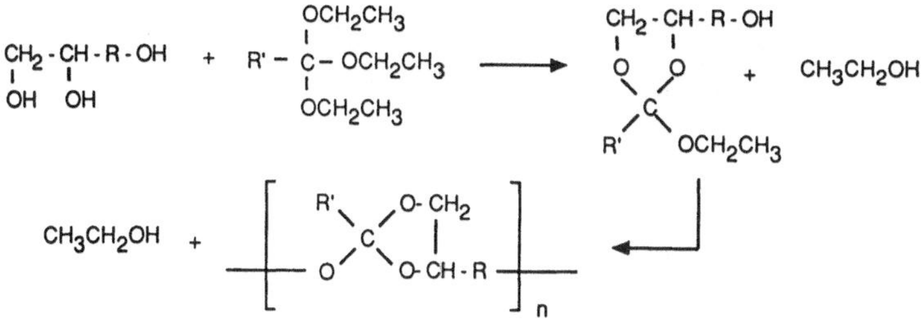

Scheme 18

The intermediate does not have to be isolated and continuous reaction produces a polymer. The use of flexible triols such as 1,2,6-hexanetriol produces highly flexible polymers that have ointment-like properties even at relatively high molecular weights. Properties such as viscosity and hydrophobicity can be readily varied by controlling molecular weight and the size of the alkyl group R'.

4.3.2 Typical Experimental Procedure

Under anhydrous conditions 48.67 g (0.30 mol) triethyl orthoacetate, 40.25 g (0.30 mol) 1,2,6-hexanetriol and 20 mg p-toluenesulfonic acid were weighed into a 500-mL round bottom flask equipped with a magnetic stirring bar. Next, 300 mL cyclohexane was added and the flask adapted to a 60 cm spinning band column. The reaction flask was heated to 100 °C with vigorous stirring and the ethanol-cyclohexane azeotrope was rapidly removed at 55 °C. Throughout the procedure, strictly anhydrous conditions were maintained. When the boiling point began to climb above 55 °C, the take-off ratio of the column was reduced to 1/20 distillation/reflux ratio until the boiling point reached 81 °C where the take-off was set for total reflux. After heating for an additional 4 h, the solution was cooled to room temperature. Five drops of triethylamine were then added to stabilize the product and the solvent removed by distillation. The product was a viscous, ointment-like material with an average molecular weight of 29 000 as determined by gel permeation chromatography using a Waters 150-C instrument with Waters ultrastyrogel 10^3 and 10^4 columns, with tetrahydrofuran solvent at 30 °C with a small amount of triethylamine stabilizer. Polystyrene was used for calibration.

4.3.3 Polymer Hydrolysis

Polymer hydrolysis occurs as shown in Scheme 20 for a polymer prepared from 1,2,6-hexanetriol. Initial hydrolysis occurs at the labile ortho ester bonds to generate one or more isomeric monoesters of the triol. This initial hydrolysis is followed by a much slower hydrolysis of the monoesters to produce a carboxylic

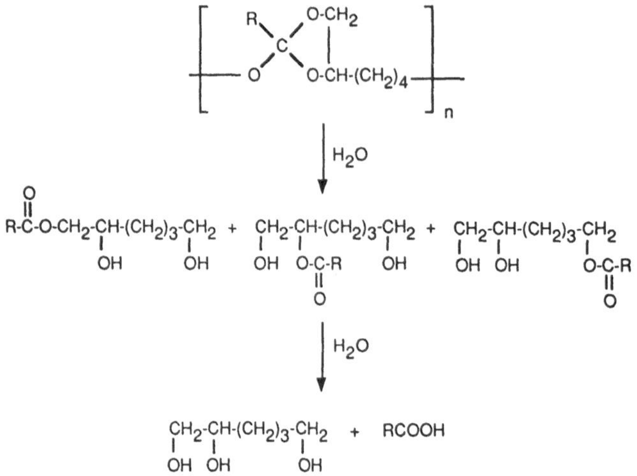

Scheme 20

acid and a triol. Thus, as with the poly (ortho ester) II described in Sect. 4.2.1, no autocatalysis is observed.

Because a knowledge of all degradation products is important for regulatory approval, studies aimed at establishing the exact hydrolysis path were carried out [50, 51]. As already described in Sect. 4.2.3, the acid-catalyzed hydrolysis of the cyclic ortho ester bond can proceed via two paths, depending on whether cleavage occurs at the exocyclic alkoxy or endocyclic alkoxy group.

If protonation of the exocyclic alkoxy group takes place and bond cleavage occurs as shown in Scheme 21,1,2,6-hexanetriol esterified in the 1 and 2 position is obtained.

If protonation of the endocyclic alkoxy group takes place and bond cleavage occurs as shown in Scheme 22,1,2,6-hexanetriol esterified in the 1 and 6 position is obtained.

To establish the identity of the hydrolysis products, the acetate polymer was hydrolyzed and the hydrolysis products analyzed by capilary gas chromatography [51]. Figure. 31a shows that two hydrolysis products have been isolated.

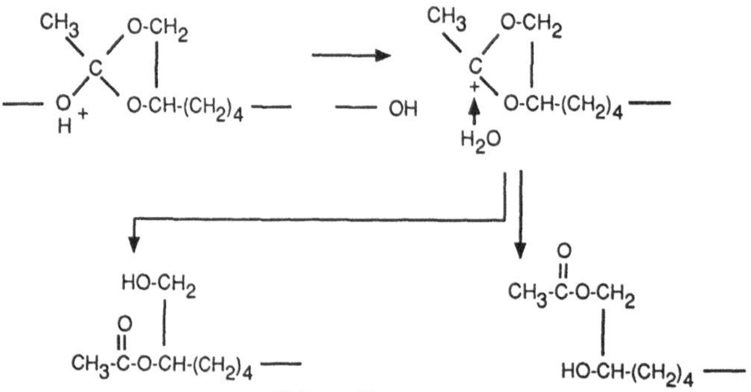

Scheme 21

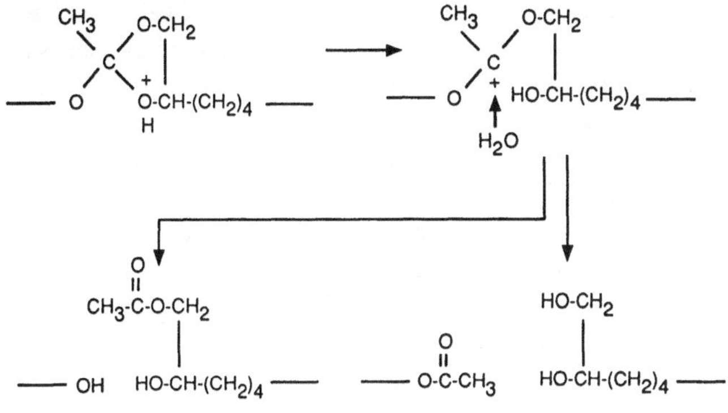

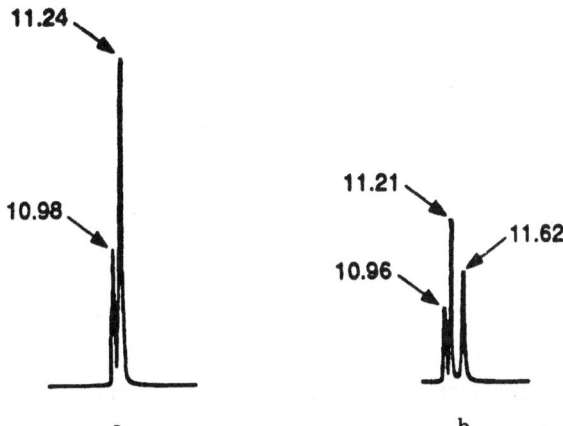

Scheme 22

Fig. 31a, b. Chromatographic trace of hydrolysis products of polymer prepared from 1,2,6-hexanetriol and triethylorthoacetate (**a**) polymer hydrolysis products, (**b**) products from esterification of 1,2,6-hexanetriol with acetic acid. [50]. Reprinted with permission

When retention times are compared to Fig. 31b which shows the three reaction products from a reaction between equimolar amounts of 1,2,6-hexanetriol and acetic acid, it is clear that the polymer hydrolysis products are two of the three monoacetates of 1,2,6-hexanetriol.

To establish the identity of these products, the 6-monoester of 1,2,6-hexanetriol was prepared as shown in Scheme 23 [50].

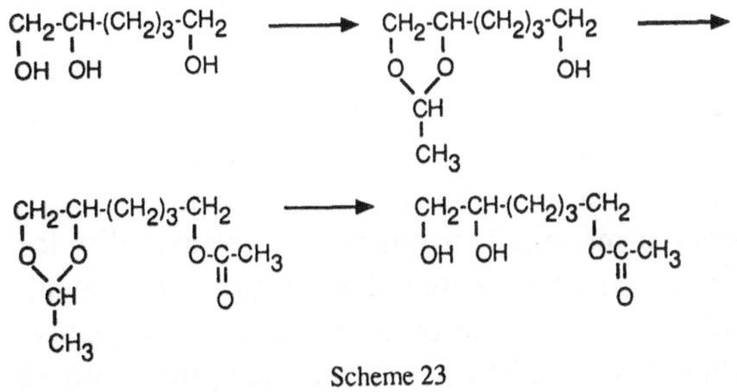

Scheme 23

When this 6-isomer was added to the reaction product of 1,2,6-hexanetriol and acetic acid, the 11.62 peak was enhanced. Thus, the 10.98 and 11.24 peaks in chromatogram 31a are the 1 and 2-isomers. Because the sensitivity of the gas chromatographic assay is such that as little as 0.01% of the 6-isomer can be detected, hydrolysis of this particular polymer proceeds exclusively by the exocyclic cleavage path shown in Scheme 20. This is in very good agreement with a reported hydrolysis study of the closely related model compounds 2-methoxy-1,3-dioxalane (1) and 2-phenyl, 2-methoxy-1,3-dioxalane (2) which have been reported to undergo respectively about 2.5% and less than 0.1% endocyclic cleavage [52].

4.3.4 Development of Specific Delivery Systems

Ointment-like materials allow the incorporation of therapeutic agents by a simple mixing procedure without the need to use solvents or elevated temperatures. Thus, these materials are of interest for the delivery of sensitive materials such as proteins that can undergo loss of tertiary structure when elevated temperatures or solvents are used.

Release of hydrocortisone. Initial exploratory in vitro studies designed to ascertain the usefulness of this polymer system have been carried out with a polymer based on 1,2,6-hexanetriol and various alkyl ortho esters. In these studies 2 wt% hydrocortisone was physically mixed into the ointment and the mixture placed into an erosion cell. A pH 7.4 buffer solution was then pumped across the cell at 9.5 ml/h, samples collected using an automatic fraction collector and analyzed for hydrocortisone by HPLC. The same study was also carried out by mixing into the ointment 2 wt% hydrocortisone and 2 wt% adipic acid. Results of these studies are shown in Fig. 32 [49]. The data clearly shows that without the incorporation of an acidic excipient erosion rate of the polymer is so slow that no hydrocortisone is released for two days. However, when an acidic excipient is mixed into the polymer, a fairly constant release is obtained.

Release of 4-homosulfanilamide. The usefulness of controlled delivery of 4-homosulfanilamide in the management of *Pseudomonas aeruginosa* burn wound sepsis has already been discussed in Sect. 4.1.4, Treatment of burns.

Because this poly (ortho ester) forms excellent ointments, its suitability for the release of 4-homosulfanilamide was investigated. This compound is available as both the acidic hydrochloride and as the basic free drug. As expected, and as shown in Fig. 33 [49], when 10 wt% of the hydrochloride is incorporated

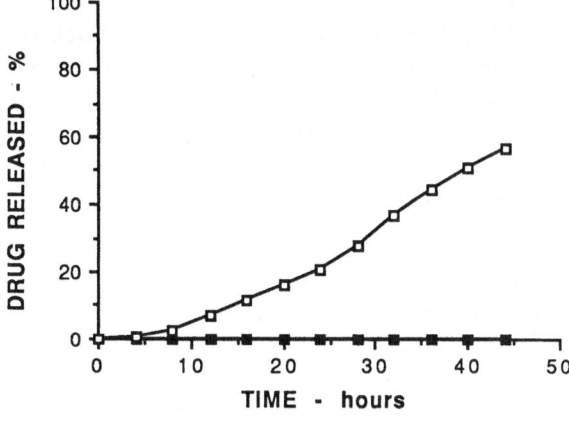

Fig. 32. Release of hydrocortisone from a polymer prepared from 1,2,6-hexanetriol and triethyl-orthopropionate at pH 7.4 and 37 °C. Polymer contains 2 wt% hydrocortisone and the indicated amounts of adipic acid (AA). (□) 2 wt% AA, (■) no AA [49]. Reprinted with permission

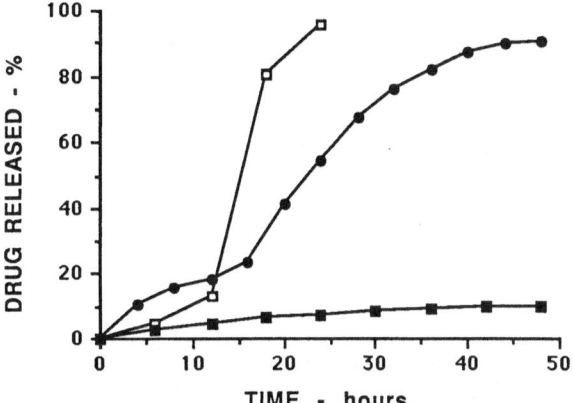

Fig. 33. Release of 4-homosulfanilamide from a polymer prepared from 1,2,6-hexanetriol and triethylorthoacetate at pH 7.4 and 37 °C. Polymer contains 10 wt% drug. (■) 4-homosulfanilamide, (□) 4-homosulfanilamide hydrochloride, (○) 50/50 mixture of free base and hydrochloride

into the polymer, rapid release occurs while virtually no release takes place when the free base is used. These data suggest that a blend of the two forms of 4-homosulfanilamide should produce a formulation where the drug is released at an intermediate rate. As also shown in Fig. 33 when a 50/50 mixture is used, intermediate release rates can be achieved [49]. The possible application of this delivery system for the treatment of burns is now under consideration.

Release of ganciclovir. 9-(1,3-dihydroxy-2-propoxymethyl) guanine (Ganciclovir) is an acyclic nucleoside analogue of the guanine base which has shown promising therapeutic effects in the treatment of cytomegalovirus infections common in AIDS patients [53]. Current therapy involves either massive doses of the antiviral drug ganciclovir through a catheter inserted into the chest with consequent side-effects, or a daily injection of ganciclovir into the eye, an unpleasant and painful procedure.

80 — J. Heller

Preliminary studies were conducted on the release of ganciclovir from an acetate polymer shown in Fig. 34 [54]. The data show a reasonably liner release for about two days when the neat drug is incorporated into the polymer. When 1 wt% NaHCO$_3$ is incorporated into the polymer, the release is extended to about one week. With proper formulation, therapeutically useful doses of ganciclovir lasting a few weeks can almost certainly be produced, thus significantly reducing the time between treatment.

Release of 5-fluorouracil. The release of 5-fluorouracil from the ointment-like material is currently under investigation as an adjunct for glaucoma filtration surgery. A preliminary study showing release of 5FU for about one week from an acetate polymer containing 2.5 wt% Mg(OH)$_2$ is shown in Fig. 35 [55]. As with the ganciclovir study described under 4.3.4, suitable formulation and polymer selection should produce devices that release 5FU for the desired length of time.

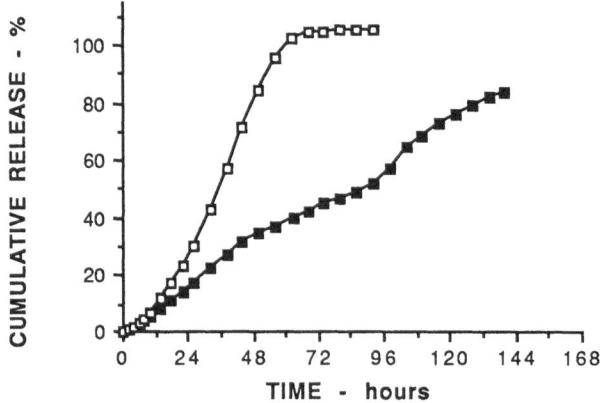

Fig. 34. Release of ganciclovir from a polymer prepared from 1,2,6-hexanetriol and triethylorthoacetate at pH 7.4 and 37 °C. Polymer contains 10 wt% drug and the indicated amount of excipient. (■) 1 wt% NaHCO$_3$, (□) no excipient

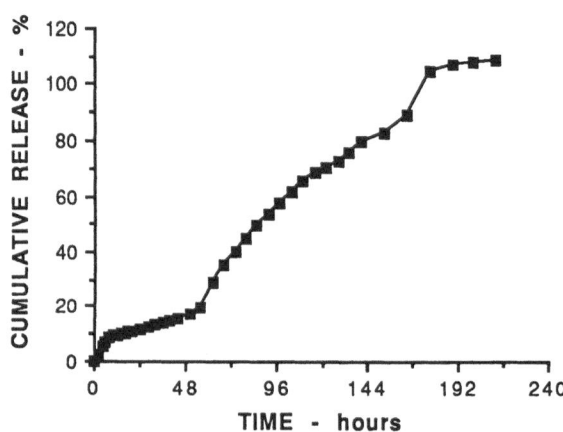

Fig. 35. Release of 5-fluorouracil from a polymer prepared from 1,2,6-hexanetriol and triethylorthoacetate at pH 7.4 and 37 °C. Polymer contains 10 wt% drug and 2.5 wt% Mg(OH)$_2$

Release of naltrexone. The release of naltrexone using poly (ortho ester) II has already been discussed in Sect. 4.2.5. Since it was not possible to form highly loaded devices that had useful mechanical properties due to an interaction between naltrexone and the polymer at the device fabrication temperatures, it was decided to use the ductile polymer since naltrexone could be mixed into the polymer at room temperature and with appropriate mechanical mixing, high drug loadings could be achieved.

Preliminary results are shown in Figs. 36–38 [56]. As with 4-homosulfanilamide, the pH of the drug was found to have a significant effect on kinetics of release. Thus, naltrexone pamoate is acidic by virtue of its unreacted phenolic hydrogens and as shown in Fig. 36, release rate is relatively fast. On the other hand, naltrexone is basic and release rate as shown in Fig. 37 is slower. As shown in Fig.38, when a 50/50 mixture of the two forms is used, good linear release kinetics have been achieved.

Release of polypeptides. Because therapeutic agents can be incorporated into the ointment-like polymer at room temperature and without the use of solvents, it is

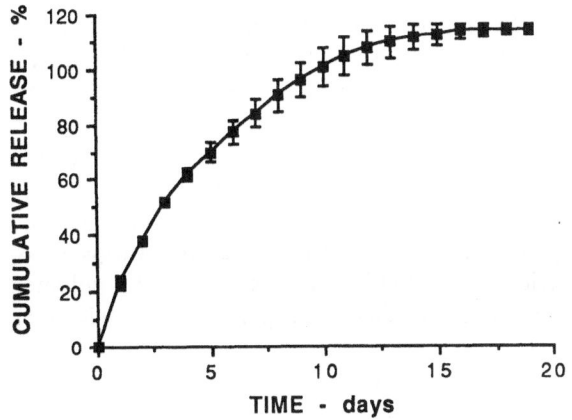

Fig. 36. Release of naltrexone pamoate from a polymer prepared from 1,2,6-hexanetriol and triethylorthoacetate at pH 7.4 and 37 °C. Polymer contains 50 wt% drug

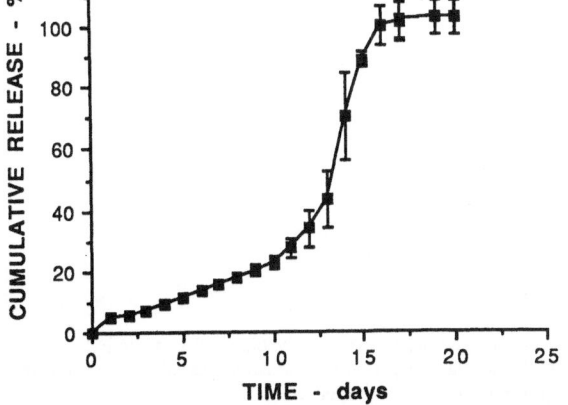

Fig. 37. Release of naltrexone from a polymer prepared from 1,2,6-hexanetriol and triethylorthoacetate at pH 7.4 and 37 °C. Polymer contains 50 wt% drug

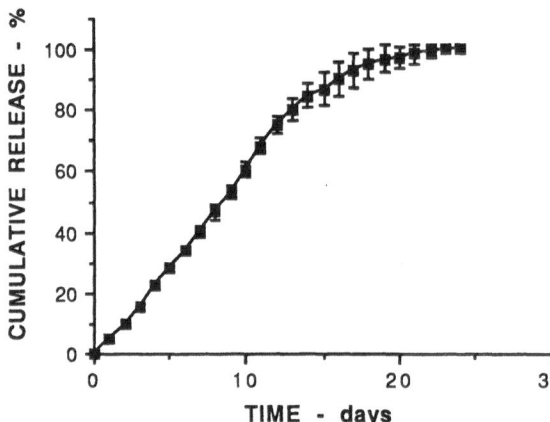

Fig. 38. Release of 50/50 mol mixture of naltrexone pamoate and naltrexone from a polymer prepared from 1,2,6-hexanetriol and triethylorthoacetate at pH 7.4 and 37 °C. Polymer contains 50 wt% drug

of particular interest in the delivery of proteins which are easily denatured by loss of tertiary structure. Initial work was carried out with the enzyme lysozyme because any denaturation can be readily detected by noting changes in the rate of lysis of the enzyme substrate *Micrococcus lysodeikticus* [57].

Figure 39 shows release of the enzyme from an acetate polymer of three different molecular weights [50]. These data show that the enzyme is released from the matrix in about 24 hours but that release occurs after an induction period that varies from about six hours for the lowest molecular weight polymer to about 3.5 days for the highest molecular weight polymer. The same trend is shown in Fig. 40 which shows release of lysozyme from a propionate polymer of three different molecular weights. In this particular case, the lowest molecular weight polymer releases lysozyme at a fairly constant rate for about three days with only a slight induction period while the highest molecular weight polymer releases lysozyme after an induction period of 4 days. Figure 41 shows release of

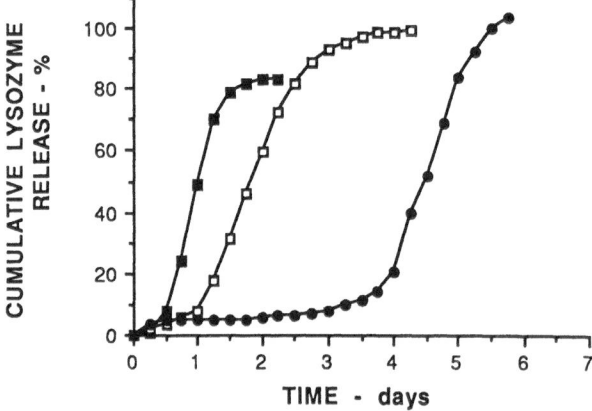

Fig. 39. Release of lysozyme from a polymer prepared from 1,2,6-hexanetriol and triethyl orthoacetate at pH 7.4 and room temperature. Molecular weights are as indicated. (■) 5350 (□) 6800 (●) 12000. Lysozyme loading 5 wt% [50]. Reprinted with permission

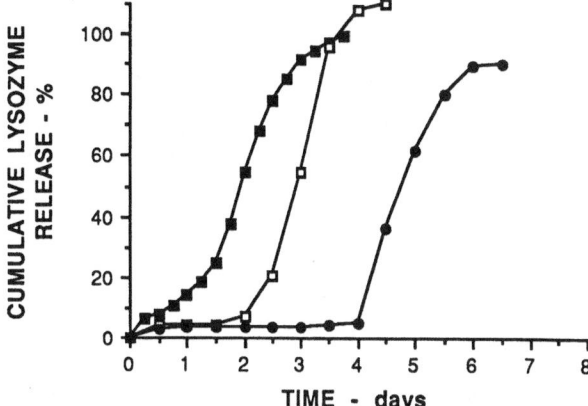

Fig. 40. Release of lysozyme from a polymer prepared from 1,2,6-hexanetriol and triethyl orthopropionate at pH 7.4 and room temperature. Molecular weights are as indicated. (■) 3200 (□) 4600 (●) 6600. Lysozyme loading 5 wt% [50]. Reprinted with permission

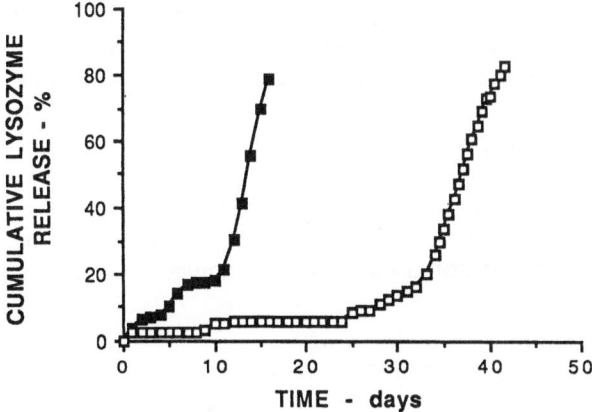

Fig. 41. Release of lysozyme from a polymer prepared from 1,2,6-hexanetriol and triethyl orthovalerate at pH 7.4 and room temperature. Molecular weights are as indicated. (■) 5500 (□) 9300. Lysozyme loading 5 wt% [50]. Reprinted with permission

lysozyme from a valerate polymer of two different molecular weights. In this particular case the lowest molecular weight polymer shows an induction period of about 10 days while the highest molecular weight polymer shows an induction period of over one month.

These data clearly show that release of lysozyme is affected by both polymer molecular weight and by the nature of the alkyl substituent R'. The magnitude of the dependency on R' is shown in Fig. 42 where rate of release of lysozyme from the acetate, propionate and valerate polymers of about the same molecular weights is compared.

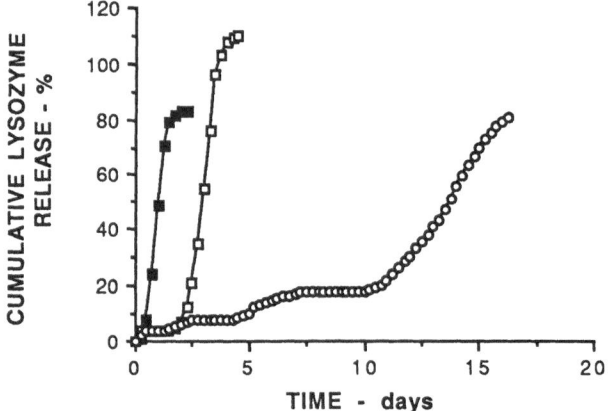

Fig. 42. Effect of alkyl group on release of lysozyme at pH 7.4 and room temperature. Molecular weights and R'-groups are as indicated. (■) methyl, 5350 (□) propyl, 4600 (○) pentyl, 5500. Lysozyme loading 5 wt% [50]. Reprinted with permission

Stability studies of lysozyme immobilized in a polymer where R' is methyl are shown in Fig. 43 [50]. The activity of the released lysozyme was assayed by noting rate of lysis of *Micrococcus lysodeikticus* [57]. The control was a solution made up at the start of the experiment and stored at room temperature. This solution was sampled at each time point of the release experiment.

As shown, virtually complete retention of enzyme activity was achieved indicating that the mixing procedure and release from the polymer does not lead to enzyme deactivation. However, whether more sensitive proteins can also be incorporated without loss of activity still needs to be determined.

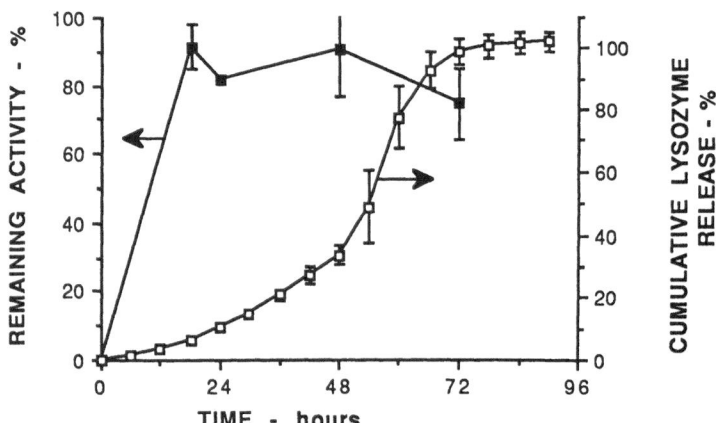

Fig. 43. Assay of lysozyme activity following release from a polymer prepared from 1,2,6-hexanetriol and triethyl orthoacetate at pH. 7.4 and room temperature. Activity assayed by noting change in rate of lysis of *Micrococcus lysodeikticus*. Lysozyme loading 10 wt% [50]. Reprinted with permission

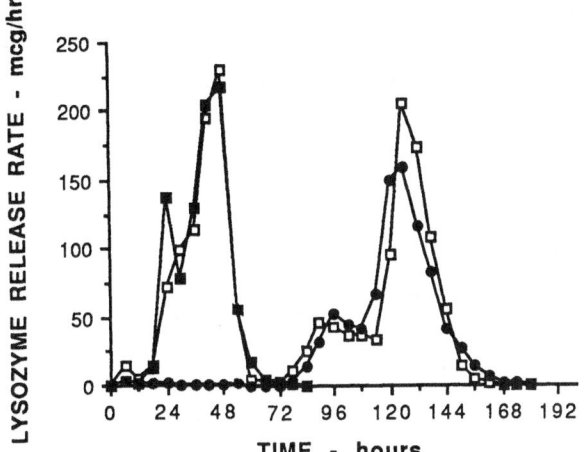

Fig. 44. Release of lysozyme from two cells, connected in series and in parallel. One cell contains a 6000 molecular weight polymer prepared from 1,2,6-hexanetriol and triethyl orthoacetate at pH 7.4 and room temperature and the other cell contains a 12 000 molecular weight polymer prepared from 1,2,6-hexanetriol and triethyl orthoacetate also at pH. 7.4 and room temperature. Lysozyme loading 5 wt%. (■, ●) cells connected in parallel (□O20) cells connected in series [50]. Reprinted with permission

The availability of polymers having different delay times makes possible the construction of devices that can release proteins in well defined and well spaced pulses. To do so, it is only necessary to use a device that contains two or more different polymer formulations in separate domains. This can be achieved by placing these formulations in a thin, bioerodible, macroporous cylinder for subsequent implantation, or better, to encapsulate each formulation in a bioerodible, macroporous membrane. In this latter approach, desired release profiles can be achieved by using appropriate mixtures of different capsules.

Actual pulsed release was simulated by placing two cells, each with a different polymer formulation, in parallel and in series. The placement in series simulates a situation where the materials are in close proximity and where degradation products from one domain can interact with the polymer in another domain. Comparison with the cells arranged in parallel allows a determination of whether this interaction is significant. Results of these studies are shown in Fig. 44 [50]. Clearly, hydrolysis products from the first polymer have no effect on the hydrolysis of the second polymer.

4.4 Poly (Ortho Ester) IV

4.4.1 Polymer Synthesis

The general synthetic procedure described in Sect. 4.3.1 can also be used in the preparation of solid polymers. To do so, it is only necessary to replace the

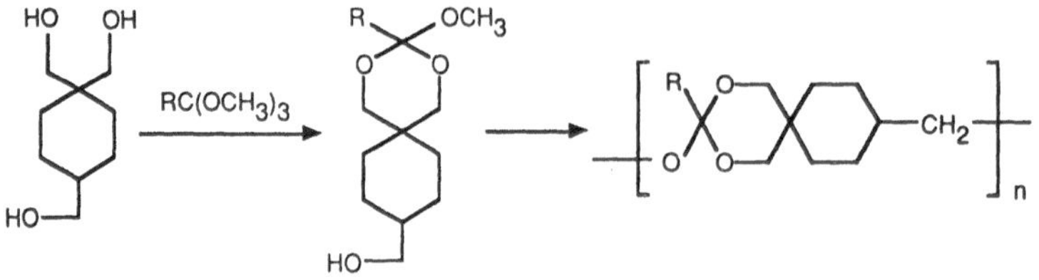

Scheme 19

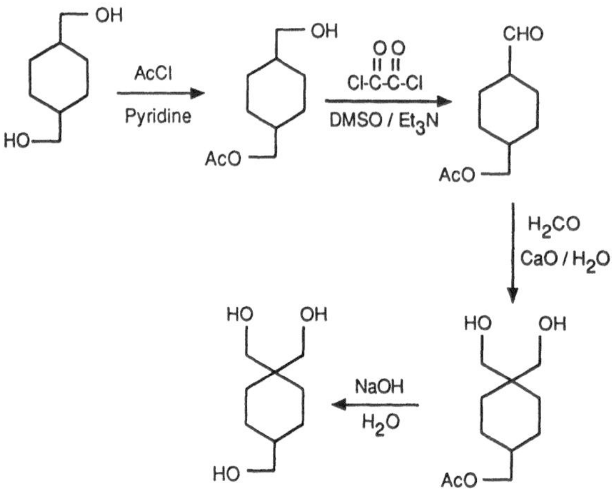

Scheme 20

flexible triol with a rigid one, such as 1,1,4-cyclohexanetrimethanol, as shown in Scheme 19 [58]. As before, the intermediate does not have to be isolated and continuing reaction produces a polymer [58].

The triol, 1,1,4-cyclohexanetrimethanol is readily prepared as shown in Scheme 20 [58].

4.4.2 Typical Experimental Procedure

Preparation of 1,1,4-Cyclohexanetrimethanol.

1,4-cyclohexanedimethanol monoacetate. A mixture of *cis* and *trans* 1,4-cyclo-hexane dimethanol, (930 g, 6,448 mol) was dissolved in 3 L tetrahydrofuran and 550 mL of pyridine, (7.69 mol) added. The solution was cooled in an ice bath and stirred under argon. An acetyl chloride solution, (506.4 g, 6.45 mol) in 500 mL of tetrahydrofuran was added dropwise over a 2 hour period. The ice bath was removed and the reaction mixture stirred for 2 hours at room temperature. It was then filtered to remove the pyridine HCl salt and evaporated to remove the tetrahydrofuran. The residue was dissolved in 2 L ethyl acetate and the solution extracted with diluted aqueous HCl (2×300 mL), warm water (2×300 mL) and aqueous $NaHCO_3$ solution (2×300 ml). The ethyl acetate solution was dried over anhydrous $MgSO_4$ and the ethyl acetate removed on a rotoevaporator. Vacuum distillation of the residue yielded 335 g of the product. GC analysis showed that the product contained 59% cyclohexanedimethanol monoacetate and 41% cyclohexanedimethanol diacetate. The overall yield of the monoacetate was 16.5%.

4-Acetoxymethyl-1-cyclohexanecarboxaldehyde. Under anhydrous conditions, oxalyl chloride (358 g, 2.82 mol) was dissolved in 2.5 L methylene chloride and the solution cooled to $-40\,°C$. Dimethylsulfoxide (407 g, 5.2 mol) dissolved in 200 mL methylene chloride was then added via a dropping funnel while the reaction mixture was vigorously stirred and the temperature maintained between $-40\,°C$ and $-20\,°C$. Next, a solution of cyclohexanedimethanol monoacetate (600 g, 59% pure, 1.9 mol) in 200 mL methylene chloride was added dropwise while the reaction temperature was kept below $-20\,°C$. After the addition of cyclohexanedimethanol monoacetate solution was completed, the reaction mixture was stirred for an additional 15 min. and triethylamine (658 g, 6.5 mol) added. The cooling bath was removed and the reaction mixture stirred for 2 h. It was then extracted successively with diluted aqueous HCl, aqueous $NaHCO_3$ and aqueous NaCl. After drying over anhydrous $MgSO_4$, the methylene chloride solution was distilled under argon to remove the solvent. The residue was distilled at $80\,°C$ at 0.4 mm to give the aldehyde (248 g, 70.8% yield).

4-Acetoxymethyl-1,1-cyclohexanedimethanol. A mixture of 4-acetoxymethyl-1-cyclohexanecarboxaldehyde, (248 g, 1.46 mol), a 37 wt% formaldehyde solution (700 mL, 8.6 mol) and tetrahydrofuran (200 mL) was cooled in an ice water bath. Calcium oxide was then added in small portions while the mixture was vigorously stirred with an overhead mechanical stirrer. After the addition of CaO was completed, the ice bath was removed and the mixture stirred for 2 h. It was then evaporated to dryness and the product extracted into acetone. Evaporation of the acetone solution produced a viscous oil.

1,1,4-Cyclohexanetrimethanol. The crude 4-acetoxymethyl-1,1-cyclohexanedimethanol was added to 1 L aqueous 2N NaOH solution and the mixture heated at $100\,°C$ for 2 h. After cooling to room temperature, the reaction mixture was neutralized with aqueous HCl and extracted with methylene chloride. The aqueous solution was evaporated to dryness and the residue extracted with acetone. After drying over anhydrous $MgSO_4$, the acetone solution was evaporated to dryness. Distillation of the crude product from the acetone solution at $175\,°C$ and a pressure of 13.33 Pa yielded a viscous liquid. Repeated trituration with methylene chloride produced a solid product (150 g, 59% yield, 98.8% purity by GC).

Preparation of Polymer. Under anhydrous conditions, 1,1,4-cyclohexanetrimethanol (3.524 g, 20 mmol), trimethyl orthoacetate (2.403 g, 20 mmol), p-toluenesulfonic acid (~ 3 mg) and distilled cyclohexane (80 mL) were added to a pre-dried flask. The flask was fitted with a spinning band column and heated to

100 °C under argon. Methanol was removed azeotropically at 56 °C at a fast rate and as the boiling point began to rise, the distillation rate was reduced to 4 drops/min. and heating continued for 15 h. The polymer thus prepared precipitated out of cyclohexane. The powdery polymer was crystalline with a melting point (DSC) of 212 °C. It was insoluble in the usual organic solvents such as methylene chloride, chloroform, ether, tetrahydrofuran, ethyl acetate, acetone, dimethylformamide, and dimethylsulfoxide.

In a similar manner, 1,1,4-cyclohexanetrimethanol (3.524 g, 20 mmol) was allowed to react with triethyl orthopropionate (3.634 g, 20 mmol). This reaction produced a polymer which remained in the cyclohexane solution. Precipitation into methanol yielded a polymer having a MW of 51 000 (GPC using polystyrene standards) and a T_g of 67.8 °C. The polymer was soluble in organic solvents with low or medium polarities such as methylene chloride, chloroform, ether, tetrahydrofuran and ethyl acetate.

4.4.3 Polymer Hydrolysis

When these poly (ortho esters) are placed in an aqueous environment they hydrolyze as shown in Scheme 21.

As with the other poly (ortho esters) discussed in the previous sections, the exact hydrolysis path depends on whether initial protonation and cleavage occurs at the exocyclic or endocyclic alkoxy groups. The two possible paths are shown in Schemes 22 and 23.

The relative importance of these two paths was again ascertained by careful hydrolysis of the polymer and analysis of the hydrolysis products by gas chromatography. Results are shown in Fig. 45 [58]. Figure 45a represents a chromatogram of products obtained by acetylating the triol and shows two sets of three peaks. These correspond to the three monoacetates and the three diacetates. Figure 45b represents the products obtained when the polymer is

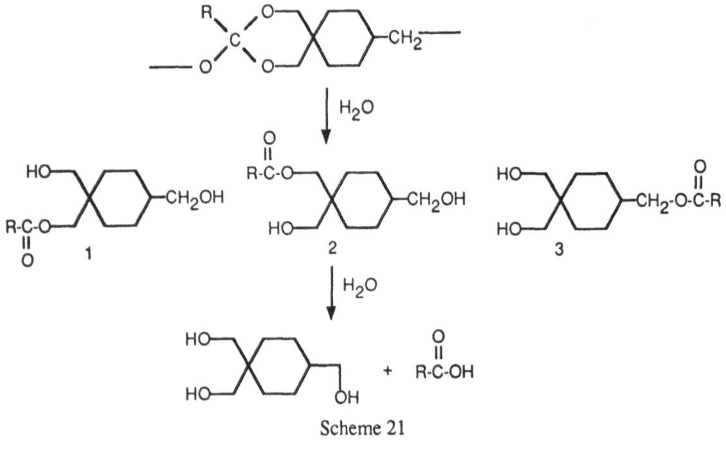

Scheme 21

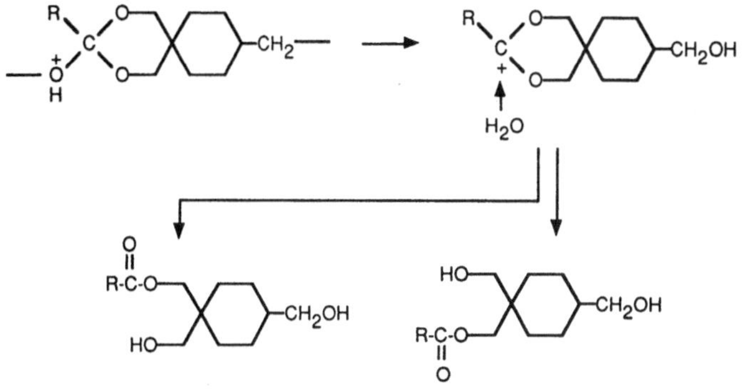

Scheme 22

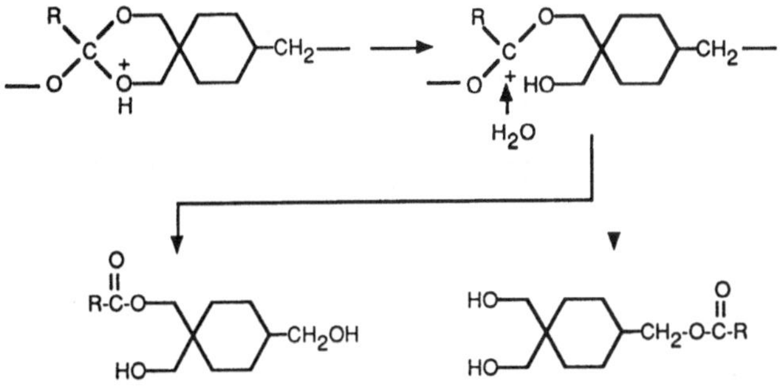

Scheme 23

carefully hydrolyzed in acid. Two major peaks are noted. Because the 3-isomer was available from the synthesis shown in Scheme 20, the two peaks in chromatogram shown in Fig. 45b have been identified as the 1-isomers and the 2-isomer. The three very minor peaks are the triol (10.016) and two diacetates (12.100 and 12.528). The complete absence of the 3-isomer indicates that this polymer, like the ointment-like polymer, also hydrolyzes exclusively by an exocyclic cleavage of the alkoxy groups.

4.4.4 Polymer Physical Properties

Very recent work has shown that when $R = CH_3$, the polymer is crystalline, but that the crystallinity disappears when $R = CH_3CH_2$. The effect of the alkyl group on crystallinity is illustrated in Fig. 46 which shows X-ray powder diffraction patterns for the two polymers.

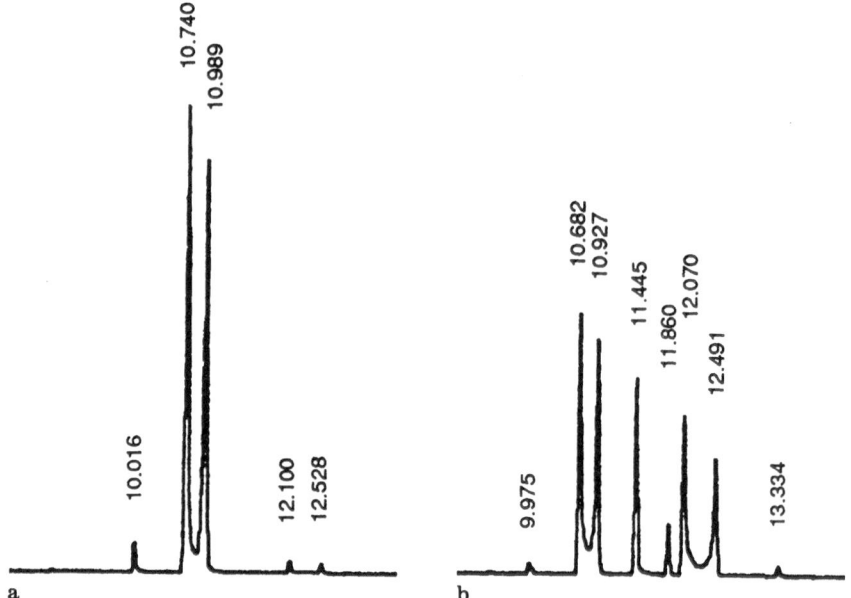

Fig. 45. Chromatographic trace of hydrolysis products of polymer prepared from 1,1,4-cyclohexan-etrimethanol and triethylorthoacetate. (**a**) polymer hydrolysis products, (**b**) products from esterific-ation of 1,1,4-cyclohexanetrimethanol with acetic acid

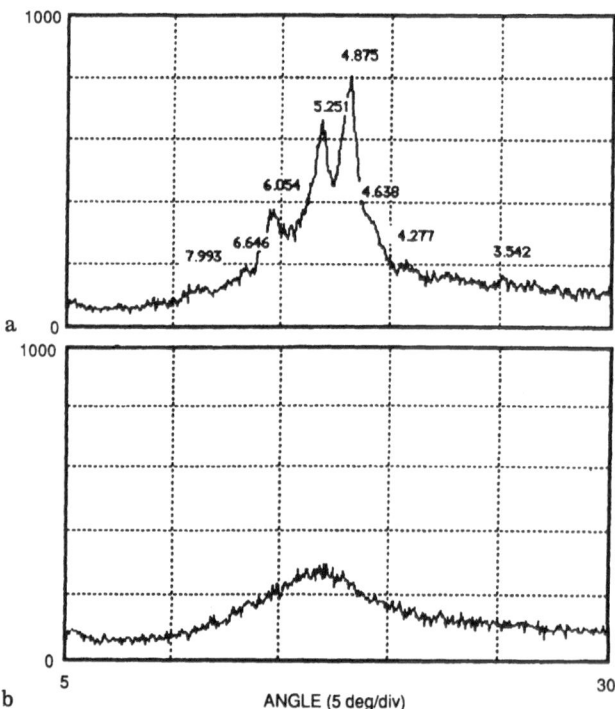

Fig. 46a, b. Powder X-ray diffraction patterns of polymers (**a**) $R = CH_3$ (**b**) $R = CH_2CH_3$

5 References

1. Heller J, Baker RW, Gale RM, Rodin JO (1978) J Appl Polymer Sci 22:1991
2. Chasin M, Domb A, Ron E, Mathiowitz E, Langer R, Leong K, Laurencin C, Brem H, Grossman S (1990) In: Chasin M, Langer R (eds) Biodegradable polymers as drug delivery systems, Marcel Dekker, New York, p 43
3. Heller J, Sparer RV, Zentner GM (1990) In: Chasin M, Langer R (eds) Biodegradable polymers as drug delivery systems, Marcel Dekker, New York, p 121
4. Lappas LC, McKeehan WJ (1962) J Pharm Sci 51:108
5. Lappas LC, McKeehan WJ (1965) J Pharm Sci 54:176
6. Lappas LC, McKeehan WJ (1965) J Pharm Sci 56:1257
7. Hopfenberg HB (1976) In: Paul DR, Harris FW (eds) Controlled release polymeric formulations, American Chemical Society, Washington DC, p 26
8. Cordes EH, Bull HG (1974) Chem Revs 74:581
9. Bronstead JN, Wynne-Jones WFK (1929) Trans Farad Soc 25:59
10. Choi NS, Heller J (1978) US Patent 4,079,038
11. Choi NS, Heller J (1978) US Patent 4,093,709
12. Choi NS, Heller J (1978) US Patent 4,131,648
13. Choi NS, Heller J (1979) US Patent 4,138,344
14. Choi NS, Heller J (1979) US Patent 4,180,646
15. DiVincenzo GD, Ziegler DA (1980) Toxicol Appl Pharmacol 52:10
16. Roth RH, Giarman NJ (1966) Biochem Pharmacol 15:1333
17. Capozza RC, Sendelbeck SL, Balkenhol WJ In: Kostelnik RJ (ed) Polymeric delivery systems, Gordon and Breach, New York, p 59
18. Gabelnick HL (1983) In: Mishell Jr DR (ed) Long-acting contraception, Raven, New York, p 149
19. 10th Annual World Health Organization Report (1981) p 62
20. 11th Annual World Health Organization Report (1982) p 61
21. Vistness LM, Schmitt EE, Ksander GA, Rose EH, Balkenhol WJ, Coleman CL (1976) Surgery, 79:690
22. Scheeren JW, Aben RW (1974) Tetrahedron Letters 12:1019
23. Yasnitskii BG, Sarkisyants SA, Ivanyua EG (1946) Zhurnal Obshchei Khimii 34:1940
24. Heller J, Penhale DWH, Helwing RF (1980) J Polymer Sci, Polymer Letters Ed, 18:82
25. Heller J, Fritzinger BK, Penhale DWH, Ng SY, Helwing RF (1985) J Controlled Release 1:225
26. Heller J, Penhale DWH, Fritzinger BK, Rose JE, Helwing RF (1983) Contracept Deliv Syst 4:43
27. Heller J, Fritzinger BK, Penhale DWH, Ng SY, Helwing RF (1985) J Controlled Release 1:233
28. Heller J, Ng SY, Penhale DWH, Fritizinger BK, Sanders LM, Burns RA, Gaynon MG, Bhosale SS (1987) J Controlled Release 6:217
29. Heller J (1985) J Controlled Release 2:167
30. Nguyen TH, Himmelstein KJ, Higuchi T (1985) Int J Pharmaceutics 25:1
31. Shih C, Himmelstein KJ (1984) Biomaterials 5:237
32. Sparer RV, Shih C, Ringeisen CD, Himmelstein KJ (1984) J Controlled Release 1:23
33. Thombre AG, Himmelstein KJ (1985) AIChE J 35:759
34. Joshi A, Himmelstein KJ (1991) J Controlled Release 15:95
35. Dott D, Roskos KV, Ng SY, Heller J, Duncan R (work in progress)
36. Heller J, Penhale DWH, Fritzinger BK, Ng SY (1987) J Controlled Release 5:173
37. Unger RH (1982) Diabetes 31:479
38. Heller J, Chang AC, Rodd G, Grodsky GM (1990) J Controlled Release 13:295
39. Franson NM, Heller J (unpublished results)
40. Soloway MS (1977) Cancer Res 37:2918
41. Gressel MG, Parrish RK, Folberg R (1984) Ophthalmology 91:378
42. Maa YF, Heller J (1990) J Controlled Release 13:11
43. Heller J, Maa YF, Wuthrich P, Ng SY, Duncan R (1991) J Controlled Release 16:3
44. Martin WR, Jasinski DR, Mansky PA (1973) Arch Gen Psychiatry 28:784
45. Chiang CN, Hollister LE, Kishimoto A, Barnett G (1984) Clin Pharmacol Ther 36:704
46. Maa YF, Heller J (1990) J Controlled Release 14:21

47. Heller J, Penhale DWH, Helwing RF, Fritzinger BK (1981) Polymer Eng and Sci 21 : 727
48. Heller J (1986) In: Chielini E, Giusti P, Migliaresi C, Nicolais L (eds) Polymers in medicine II,
    Plenum, New York, p 357
49. Heller J, Ng SY, Fritzinger BK, Roskos KV (1990) Biomaterials 11 : 235
50. Wuthrich P, Ng SY, Firitzinger BK, Roskos KV, Heller J (1992) J Controlled Release 21 : 191
51. Heller J, Maa YF, Wuthrich P, Ng SY, Duncan R (1991) J Controlled Release 16 : 3
52. Chiang Y, Kresge AJ, Salomaa P, Young CI (1974) J Am Chem Soc 96 : 4494
53. Shepp DH, Dandliker PS, deMiranda P, Burnette TC, Cederberg DM, Kirk LE, Meyers JD
    (1985) Ann Intern Med 103 : 368
54. Wuthrich P, Gaynon MG, Heller J (Unpublished results)
55. Wuthrich P, Heller J (Unpublished results)
56. Ng SY, Heller J (Unpublished results)
57. Shugar D (1950) Biochim Biophys Acta 8 : 302
58. Heller J, Ng SY, Fritzinger BK (1992) Macromolecules 25 : 3362
59. Bunton CA, DeWolfe RH (1965) J Org Chem 30 : 1371

Received April 22, 1992

Polyanhydrides: Synthesis and Characterization

Abraham J. Domb[1,2], Shimon Amselem[2], Jaymin Shah[3], and Manoj Maniar[2]
[1] Hebrew University of Jerusalem, School of Pharmacy, Faculty of Medicine, Jerusalem 91120, Israel
[2] Drug Delivery Laboratories, Nova Pharmaceutical Corporation, Baltimore, MD 21224, USA
[3] Medical University of South California, Department of Pharm. Sci. Charleston, SC 29425, USA

The delivery of drugs from biodegradable polymeric materials for human and animal use has attracted considerable attention of investigators throughout the scientific community. Various types of polymers have been synthesized and tested for this purpose which include: poly(α-esters), poly(aliphatic esters), polyorthoesters, polyphosphazenes, poly(phosphate esters), polymers based on amino acids, natural and synthetic peptides and proteins, polysaccharides and polyanhydrides. Comprehensive reviews on various biodegradable polymers and their advantages have been published [1–5]. This chapter concentrates on the polyanhydride class of polymers.

Polyanhydrides are useful bioabsorbable materials for controlled drug delivery. They are hydrolytically unstable and hydrolyze to diacid monomers in contact with body fluids. Since their introduction to the field of controlled drug delivery, about 10 years ago, extensive research has been conducted to study their chemistry as well as their toxicity and medical applications. Several review articles have been published on polyanhydrides and the focus has been on controlled drug delivery applications [1, 2].

A major part of this chapter will review recent developments in the chemistry and properties of polyanhydrides, which includes new synthetic methods, new polymeric structures, and in depth characterization of polyanhydrides. The degradation and drug release properties and applications that were not reviewed previously are included. A review article by the same authors concentrating on polyanhydride applications and toxicity is in preparation [6].

Abbreviations

ACDA	acetylenedicarboxylic acid
BTC	1,3,5-benzenetricarboxylic acid
Co	drug load in gm/cc
CPH	1,6-bis(p-carboxyphenoxy) hexane
CPP	1,3-bis(carboxyphenoxy) propane
CPV	carboxphenoxy valerate
DMF	N,N-dimethylformamide
DSC	Differential Scanning Calorimetry
Et3N	triethylamine
FA	fumaric acid
FAD	dimer fatty acid
Gelfoam	absorbable gelatin sponge
Gliadel	Polyanhydride brain tumor implant containing BCNU
GPC	gel permeation chromatography
4HC	4-hydroperoxycyclophosphamide
IPA	isophthalic acid
Ln	average length of sequence
MIT	Massachusetts Institute of Technology
Mw	weight average molecular weight
Mn	number average molecular weight
PA	poly(adipic acid)
PAA	poly(acrylic acid)
PAZ	poly(azelaic acid)
PLA	poly(lactic acid)
PCL	poly(caprolactone)
PDP	poly(phenylenedicarboxylic acid)
PHB	poly(hydroxybutyrate)
PSA	poly(sebacic acid)
PSU	poly(suberic acid)
SA	sebacic acid
SEM	Scanning Electron Microscope
Septicin	polyanhydride antibacterial bone implant
STDA	4,4′stilbenedicarboxylic acid
Surgicel	oxidized cellulose absorbable hemostat
TA	terephthalic acid
Tg	glass transition temperature
Tm	melting point
TMA-gly	trimellitimide-glycine
ToF-SIMS	time-of-flight secondary ion mass spectroscopy
VPO	vapor pressure osmometry
Vycryl	synthetic absorbable suture
Xc	degree of crystallinity
XPS	X-ray photoelectron spectroscopy

1 Introduction

Polyanhydrides were first reported in 1909 by Butcher and Slade [7] who discovered the formation of a high melting material when isophthalic or tereph-thalic acid were heated in acetic anhydride. About 20 years later, Hill and Carothers [8–10] in their course of developing new useful polymeric materials for textile applications, investigated polyanhydrides of simple aliphatic dicar-boxylic acids. They found that these polymers are hydrolytically unstable and degrade in room moisture. They discovered also that the polymers are thermally unstable and form cyclic dimers and polymeric rings when heated at high temperature. The research on polyanhydrides was renewed by Conix [11, 12] and Yoda [13–18] who synthesized more than a hundred new polymers based on aromatic, heterocyclic, and copolymers of aliphatic and aromatic diacid monomers that have been used in the synthesis of polyesters (Table 1). Their research was directed toward the synthesis of selected compositions designed to retain substantial hydrolytic and thermal stability and yet to have better plasticity than existing compounds as the condensation polymers like polyesters and polyamides. This goal was never reached, although there have been some progress in solving the problem. A few sporadic publications on polyanhydrides appeared during the late 1960s and 1980.

Table 1. Representative polyanhydrides synthesized during the years 1909–1980

Polymer structure	Melting point (°C)	Ref.
$\left[\text{-OC}-\bigcirc-\text{COO-}\right]$	400	13
$\left[\text{-OC}-\bigcirc_{\text{COO-}}\right]$	256	13
$\left[\text{OC-H}_2\text{C}-\bigcirc-\text{CH}_2\text{-COO}\right]$	90–91	13, 16
$\left[\overset{O}{\overset{\|}{C}}-\bigcirc-\overset{O}{\overset{\|}{C}}\text{-O-(CH}_2)_2-\text{O-}\overset{O}{\overset{\|}{C}}-\bigcirc-\overset{O}{\overset{\|}{C}}\text{-O-}\right]$	56	
$\left[-\overset{O}{\overset{\|}{C}}-\bigcirc-\overset{O}{\overset{\|}{C}}\text{-NH-(CH}_2)\text{-NH-}\overset{O}{\overset{\|}{C}}-\bigcirc-\overset{O}{\overset{\|}{C}}\text{-O-}\right]$	330	13
$\left[-\overset{O}{\overset{\|}{C}}-\bigcirc-(\text{CH}_2)_6-\bigcirc-\overset{O}{\overset{\|}{C}}\text{-O-}\right]$	151	11

Table 1. (continued)

Polymer structure	Melting point (°C)	Ref.
[-OC-CH₂-CH₂-S-⟨benzene⟩-S-CH₂-CH₂-COO-]	91	13, 17
[-OC-H₂C-O-⟨benzene⟩-O-CH₂-COO-]	160	70
[-OC-CH₂-CH₂-⟨furan, O⟩-CH₂-CH₂-COO-]	67	13, 14
[-OC-CH₂-CH₂-⟨thiophene, S⟩-CH₂-CH₂-COO-]	78	13, 14
[-OC-CH₂-CH₂-⟨pyrrole, N-CH₃⟩-CH₂-CH₂-COO-]	188	13, 14
[-OC-⟨benzene⟩-C(CH₃)₂-⟨benzene⟩-COO-]	230	13, 14
[-OC-⟨anthracene⟩-COO-]	> 300	71
[-OC-⟨benzene⟩-C(=O)-⟨benzene⟩-COO-]	338	72
[-OC-⟨benzene⟩-O-⟨benzene⟩-COO-]	295	11
[-OC-⟨benzene⟩-CH₂-⟨benzene⟩-COO-]	332	11
[-OC-⟨benzene⟩-O-(CH₂)₂-O-⟨benzene⟩-COO-]	237	11
[-OC-⟨benzene⟩-OCH₂-⟨benzene⟩-CH₂-O-⟨benzene⟩-COO-]	140	73

Table 1. (continued)

Polymer structure	Melting point (°C)	Ref.
$\left[-OC-\text{(naphthalene)}-COO- \right]$	450	72
$\left[-OC\text{-}(CH_2)_4CONH\text{-}C(\text{phenyl})_2\text{-}NH\text{-}CO\text{-}(CH_2)_4\text{-}COO- \right]$	> 300	74
$\left[-OC\text{-}(CH_2)_4CONH\text{-}CH_2\text{-}NH\text{-}CO\text{-}(CH_2)_4\text{-}COO- \right]$	285	74
[OC-(CH2)2-SO2-(CH2)2-SO2-(CH2)2-COO]	185	16
[OC-(CH2)2-S-(CH2)2-S-(CH2)2-COO]	81	13, 17
[OC-P-COO]		76
[OC-(CH2)x-COO] x=4-16	60–100	75
$\left[-\text{(phenyl)}-\overset{O}{\underset{}{C}}\text{-}O\text{-}\overset{O}{\underset{}{C}}-O\text{-}CH_2\text{-}CH_2\text{-}O\text{-}\overset{O}{\underset{}{C}}\text{-}O\text{-}\overset{O}{\underset{}{C}}- \right]$	155	37
$\left\{ \left[OC\text{-}(\text{phenyl})\text{-}COO\right]_x \left[OC-(CH_2)_8\text{-}COO-\right]_y \right\}_n$	220	15

In the 1970s when the field of controlled drug delivery started to gain attention, the need for absorbable materials for implantable controlled drug delivery was recognized. It was only in 1980 when Langer recognized the broad potential uses of this class of polymers as biodegradable materials for drug delivery and other medical applications [19, 20]. An extensive research program on the synthesis and applications of polyanhydrides was initiated at the Massachusetts Institute of Technology (MIT) which was significantly enhanced when NOVA Pharmaceutical Corp. engaged with MIT to develop these polymers as implantable drug carriers for human use. At present, two implantable devices for human use, the Gliadel implant for the treatment of brain tumors, and the

Septicin antibacterial implant for the treatment of chronic bone infections have been developed [21–24]. The multidisciplinary concept of polymeric implants has expanded to include research on the chemistry and characterization of polymers, experimental and theoretical polymer degradation and drug release, toxicology and metabolism, and research in specific fields of applications such as cancer, proteins and hormones delivery, infectious diseases, and brain disorders. This chapter concentrates on the chemistry and characterization of polyanhydrides with a brief description on recent applications of polyanhydrides.

2 Chemistry

2.1 Synthesis

Polyanhydrides have been synthesized by the following methods: a) bulk melt condensation of activated diacids, b) ring opening polymerization, c) reaction between dibasic acid and diacid chlorides, and d) interfacial polymerization. A detailed study of these polymerization methods and various polymerization conditions for a range of diacids were previously described [25–27].

The major drawback of all previous work on polyanhydrides was their low molecular weight, which made them impractical for many applications. A systematic study to determine the factors affecting the polymer molecular weight was conducted with the purpose of synthesizing high molecular weight polymers [28]. The highest molecular weight polymers were obtained using a melt condensation process, by operating under conditions which optimized the polymerization process while at the same time minimizing the depolymerization process. Pure prepolymers were individually prepared and purified by recrystallization. The reaction conditions, reaction temperature and time, the presence of coordinative catalysts, and the vacuum applied are very important for the production of high molecular weight polymers. In addition, these polymers are chemically pure and with a determined copolymer composition similar to the entry monomer composition. This is in contrast to the polymers obtained by polymerizing unisolated mixed diacid prepolymers which are characterized as non-pure, of low molecular weight, and with unpredictable copolymer composition that can vary in the range of ± 20% from the monomer entry composition [29].

2.1.1 Melt Polycondensation

The most widely used method is the melt condensation of dicarboxylic acids treated with acetic anhydride. This method was successfully used for the synthesis of aliphatic and aromatic polyanhydrides [25–28]:

$$HOOC-R-COOH + (CH_3-CO)_2O$$
$$\xrightarrow{\text{reflux}} CH_3-CO-(O-CO-R-CO-)_mO-CO-CH_3$$

(I)

$$(I) \xrightarrow{180\,°C/0.1\,mmHg} CH_3-CO-(O-CO-R-CO-)_nO-CO-CH_3$$

$$m = 1-20;\ n = 100-1000$$

The polycondensation takes place in two steps, in the first step the dicarboxylic acid monomers are reacted with excess acetic anhydride to form acetyl terminated anhydride prepolymers with a degree of polymerization (Dp) of 1 to 20, which are then polymerized at elevated temperature under vacuum to yield polymers with Dp of 100 to over 1000.

The Dp of prepolymers was affected by the nature of the monomer, the ratio between the monomer and acetic anhydride, and the reaction time. Short oligomers of 1–2 monomer units were prepared from the reaction of the diacid monomer with acetyl chloride in chlorinated hydrocarbon solution in the presence of an acid acceptor. Prepolymers were also prepared using propionic anhydride and butyric anhydride. However, these reagents boil at a higher temperature than acetic anhydride thus requiring vigorous conditions to remove the unreacted anhydride and acids by vacuum evaporation. Acetic acid mixed anhydride prepolymers were prepared also from the reaction of the diacid monomers with ketene [30].

The condensation reaction of diacetyl mixed anhydrides of aromatic or aliphatic diacids is carried out in the temperature range of 150 to 220 °C [28]. The optimum reaction temperature is 170 to 190 °C. A variety of catalysts have been used in the synthesis of a range of polyanhydrides. Some of these catalysts are listed in Table 2. Significantly higher molecular weights, in shorter reaction times, were achieved by utilizing cadmium acetate, earth metal oxides and $ZnEt_2-H_2O$ (Fig. 1). Except for calcium carbonate which is a natural material, the use of other catalysts for the production of medical grade polymers is limited

Table 2. Polymerization of poly(CPP-SA) 20:80 using coordination catalysts[a]

Catalyst	Polymerization time (min)	Viscosity [n] (dL/g)	Molecular weight Mw	Mn
no catalyst	90	0.92	116 800	18 200
cadmium acetate	31	1.25	245 000	29 420
calcium oxide	20	0.88	140 935	14 877
barium oxide	30	0.96	185 226	22 500
calcium carbonate	28	0.90	141 600	15 500
$ZnEt_2-H_2O$ (1:1)	60	1.18	199 060	25 312

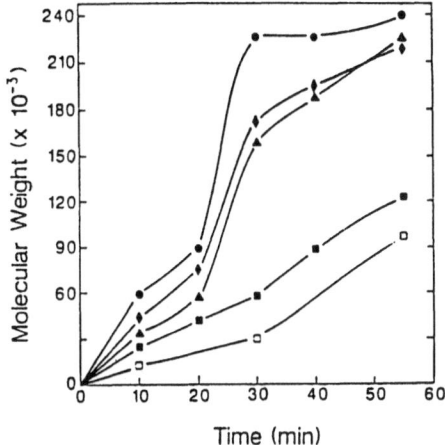

Fig. 1. Polymerization of poly(CPP-SA) 20:80 with various amounts of cadmium acetate. Polymerization at 180 °C (□) No catalyst, (■) 0.5% mole fraction, (△) 1%, (◇) 2%, (●) 3%. Molecular weight refers to weight average (from Reference [28])

because of their potential toxicity. Prior to the use of these catalysts in biopolymers, it would be necessary to ensure the removal of the catalyst.

Polyanhydrides can be synthesized by melt condensation of trimethylsilyl dicarboxylates and diacid chlorides to yield polymers with intrinsic viscosities up to 0.43 dl/g [31, 32] :

$$n\,Me_3Si-O-CO-R-CO-O-SiMe_3 + n\,Cl-CO-R'-CO-Cl$$

$$\rightarrow 2n\,Me_3SiCl + Me_3Si-(O-CO-R-CO-O-CO-R'-CO)_n-OSiMe_3$$

The polymerization is carried out at 100 °C under vacuum in the presence of benzyltriethylammonium chloride. This method possess several disadvantages over the acetic anhydride melt condensation method: it requires the pure chloride and the trimethylsilyl derivatives of the dicarboxylic acid monomers, and an equimolar ratio of monomers is required to affect the polymerization which makes it difficult to prepare copolymers of various ratios of comonomers. Also, the polymers obtained are of a lower molecular weight and contain trimethylsilyl or acid chloride as terminal groups.

Melt condensation of diacylium cations of tetrahaloterephthalic acids produced polymers with number average molecular weight of 10 400 in high yields [9]. Solid tetrabromoterephthalic diacilium bis-bisulfate salt (a), prepared from the reaction of tetrahaloterephthalic acid and a mixture of H_2SO_4 and SO_3 in a mass fraction of 85:15, was polymerized in SO_2 at 120 °C. The polymer can be synthesized directly from the reaction of a 1:1 molar ratio of tetrahaloterephthalic acid and SO_3. Copolymers of tetrachloro and tetrabromoterephthalic acid were prepared from the reaction of the bromodisulfate and the chloro acid derivatives to yield polymers of 5000 molecular weight in 50% yield.

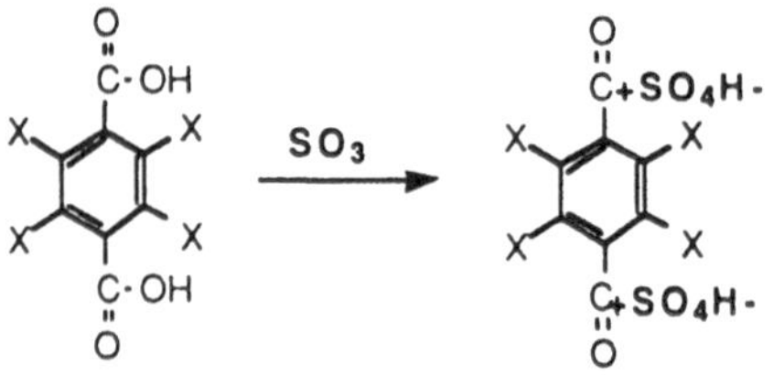

(a)

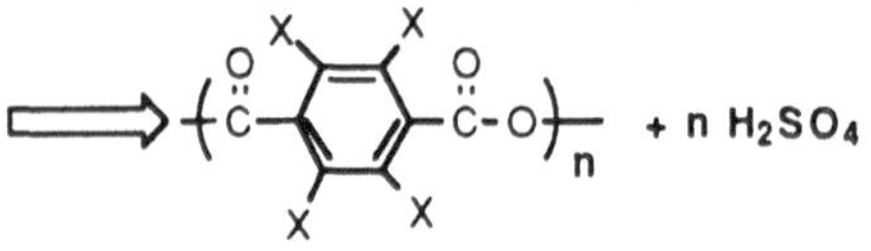

Direct polycondensation of sebacic acid and adipic acid at a high temperature under vacuum resulted in low molecular weight oligomers. These oligomers were used as hardeners for epoxy resins [34]:

$$n\,HOOC–(CH_2)_8–COOH \rightarrow (OOC–(CH_2)_8–CO)_n + n\,H_2O$$

$$n = 5 \text{ to } 10$$

2.1.2 Ring Opening Polymerization

In several studies on the behavior of aliphatic diacids of the structure:

$HOOC–(CH_2)_n–COOH$ where n is 4, 5, 6, 7, 8, 9, 10, 11, and 12

toward anhydride formation, Hill and Carothers reported [8–10] the formation of low molecular weight linear polymers which undergo transformations as follows:

$HOOC–(CH_2)_8–COOH \rightarrow \alpha$–anhydride(MW, 5000) $\rightarrow$

β–anhydride(cyclic dimers) + ω–anhydride (residue, MW, 20 000)

$\downarrow$ standing

δ–anhydride (cyclic MW 5000)

When these polymers were subjected to molecular distillation, cyclic monomers and dimers were distilled off and a high molecular weight polymer remained behind. The cyclic molecules were transformed to a polymer on standing, this polymer is thought to contain very large ring structures. In order to distinguish the cyclic monomer from the polymer and cyclic dimer, Hill and Carothers used the reaction with aniline. Monomeric anhydride can react with aniline to give only one product, the acid monoanilide, whereas the dimer or polymer may lead to three products, the dibasic acid, the acid monoanilide and the dianilide. The dimers are crystalline solids which polymerize instantly when heated above their melting point or upon standing. The monomers of 8, 10, and 12 atoms are very unstable and polymerize even below room temperature. The 7 atoms monomer of adipic acid is more stable but polymerizes upon heating at 100 °C for a few hours [9]. Polymerization of adipic acid diacetate prepolymers at 180 °C under vacuum yielded a polymer with a molecular weight of 14 000 that contains significant amounts of the cyclic adipic anhydride [28].

The preparation of adipic acid polyanhydride from cyclic adipic anhydride (oxepane- 2,7-dione) was investigated by Albertsson and Lundmark [35]. The monomer was prepared by the reaction of adipic acid and acetic anhydride followed by catalytic depolymerization under vacuum. The ring opening polymerization was affected by temperature, reaction time and concentration of catalyst (stannous 2-ethylhexanoate). H-NMR and IR studies indicated a non-ionic insertion polymerization mechanism at the beginning of the reaction, but after 2 hours at 80 °C, anhydride interchange appeared to be the dominating reaction which resulted in a low molecular weight polymer. The polymerization was carried out also in dichloromethane at room temperature in the presence of 1% $ZnCl_2$ and resulted in a polymer with Mn = 1700 in 75% yield [35]:

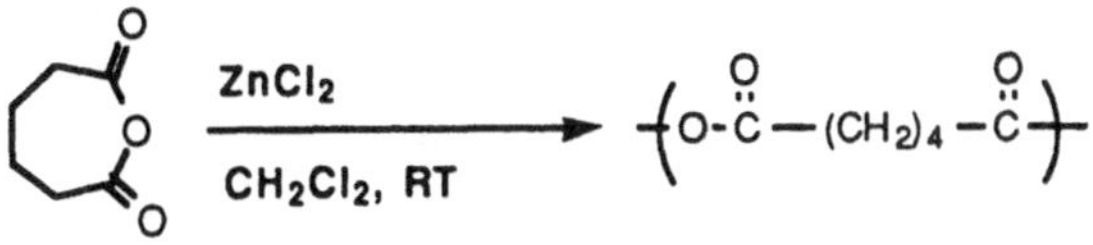

2.1.3 Solution Polymerization

A variety of solution polymerizations at ambient temperature have been reported (Scheme 1) [15, 25, 36]. Partial hydrolysis of terephthalic acid chloride in the presence of pyridine as an acid acceptor yielded a polymer of MW = 2100:

A. x HOOC–R–COOH + y Cl–CO–R′–CO–Cl

$$\xrightarrow{Et_3N/0\,°C} [(OC-R-COO)_x(OC-R-COO)_y]n$$

B. n Cl–CO–R′–CO–Cl + n/2 H_2O → [CO–R′–COO]n + n HCl

C. HOOC–R–COOH $\xrightarrow{dehydrat.\ agent}$ [CO–R′–COO]$_n$

D. HOOC–R–COOH + Cl–CO–Cl + Et_3N

$$\xrightarrow{0\,°C} [CO-R'-COO]_n + Et_3N \cdot HCl + CO_2$$

R = aliphatic, aromatic, and a heterocyclic organic residue

The use of dehydrating agents effected polyanhydride formation [25]. The use of N,N-bis(2-oxo-3-oxazolidinyl)phospharamido chloride, dicyclohexylcar- bodiimide, and chlorosulfonyl isocyanate as coupling agents produced impure and low molecular weight polymers. The isolation and purification of the polymer from the amine acid acceptor and the dehydrative agent by-products required extraction with water which evoked hydrolytic decomposition. The reaction between diacid chloride and dibasic acid or the dimethylester derivative in pyridine-toluene or ether in the presence of $ZnCl_2$ yielded white polymers [15, 38].

Two approaches for one step solution polymerization of polyanhydrides at ambient temperature were reported [39]. In the first approach, pure polymers (> 99.7%) were obtained by the use of sebacoyl chloride, phosgene, and diphosgene as coupling agents and poly(4-vinylpyridine) or K_2CO_3 as insoluble acid acceptors. The polymer is exclusively soluble in the reaction solution and the only by-product formed is the insoluble acid acceptor-hydrochloric acid salt. Polymerization of sebacic acid with phosgene either as a gas or in toluene solution, or diphosgene as a coupling agent with triethylamine or insoluble poly(4-vinylpyridine) as an acid acceptor yielded polymers with $MW = 16\,000$. The polymer was contaminated with a large amount of triethylamine-HCl, whereas the polymer prepared with poly(4-vinylpyridine) was pure. The triethylamine-HCl by-product can be removed by extraction with water which may hydrolize the polyanhydride. The second approach was the use of an appropriate solvent where the polymer is exclusively soluble but the corres- ponding by-product is insoluble or vice versa. Under this condition polymeriz- ation of sebacic acid gave the best results in N,N-dimethylformamide and in toluene.

2.1.4 Interfacial Polymerization

Homo and copolyanhydrides were synthesized in an aqueous and nonaqueous interfacial reaction. Various aromatic polymers were prepared from the reaction of equimolar amounts of the acid dissolved in aqueous base and the corresponding diacid chloride dissolved in an organic solvent [25, 38]. Because the reaction is between a dibasic acid in one phase and an acid chloride in the other phase, the copolymers may present a regularly alternating structure. In the reaction between sebacoyl chloride in chloroform and isophthalic acid sodium salt in water a copolymer that contain mostly sebacic acid units was obtained. These results can be explained by partial hydration of two acid chloride end groups to form an anhydride bond.

2.2 Polyanhydride Structures

Since the discovery of polyanhydrides in 1909, hundreds of polymers have been reported [26, 27]. A representative list of polymers developed up till the end of the 1970s is shown in Table 1. Polyanhydrides intended for use in medicine that have been developed since 1980 are discussed below.

2.2.1 Unsaturated Polymers

A series of unsaturated polyanhydrides were prepared either by melt or solution polymerization of fumaric acid (FA), acetylenedicarboxylic acid (ACDA), and 4,4'-stilbenedicarboxylic acid (STDA) [39, 40]. The properties of several of these polymers are listed in Table 3.

Weight average molecular weights of up to 30 000 were obtained in 30 minute of polymerization, longer polymerization time resulted in an insoluble material [40]. The double bonds remained intact throughout the polymerization process and were available for a secondary reaction to form a crosslinked matrix. The unsaturated homopolymers were crystalline and insoluble in common organic solvents. Copolymers of fumaric acid and acetylenedicarboxylic acid with aliphatic diacids were less crystalline and were soluble in chlorinated hydrocarbons. The ^{1}H NMR and IR data on these polymers and the isolation of the intact unsaturated monomers after polymer hydrolysis indicated that the monomers were not altered during the polymerization process. This is in contrast to early reports by Michael and Bucher [41] and Shopov [42] which indicated a change in the monomer structure of unsaturated diacids during polymerization reaction. Michael and Bucher [21] reacted acetylenedicarboxylic acid with acetic anhydride at reflux to yield acetoxymaleic anhydride. Shopov reported the formation of polyethynyleneketoanhydride of the formula:

$$\left\{ \underset{[-\overset{\overset{O}{\|}}{C}-CH=C-]_x}{} \underset{[-\overset{\overset{O}{\|}}{C}-CH=C-\overset{\overset{O}{\|}}{C}-O]_y}{} \right\}_n$$

from the reaction of acetylenedicarboxylic acid with phosgene and triehtylamine

Table 3. Unsaturated Polyanhydrides

P(FA)   P(ACDA)

Polymer	Tm [°C]	Mw	Mn
P(FA)	248–250		
P(FA-SA) 1:1	99–101	20 900	9 100
P(FA-SA) 1:4	70–72	29 350	13 250
P(ACDA-SA) 1:1	77–78	11 000	5 700
P(STDA-SA) 1:4	102–104	36 300	16 200

Data taken from Ref. [40].

in chloroform at room temperature. Both investigators studied the homopolymers which are insoluble and melt at high temperatures which made it difficult to analyze them properly.

The SA-FA copolymers displayed nearly constant degradation rates and drug release rates under physiological conditions. The time for complete degradation of poly(fumaric acid) and poly(sebacic acid) occured in 2 and 15 days, respectively, while their copolymers degraded within this range. Radical copolymerization of styrene and methylmethacrylate with the polyanhydrides resulted in crosslinked insoluble polymers [40].

Copolymers of fumaric acid or maleic anhydride and isophthalic acid were prepared by melt condensation at 250–300 °C for 8 hours using acetic anhydride or polyphosphoric acid as dehydrative agents [43].

2.2.2 Amino Acid Based Polymers

General methods for the synthesis of poly(amide-anhydrides) and poly(amide-esters) based on naturally occurring amino acids have been described [44]. The polymers were synthesized from dicarboxylic acids prepared by amidation of the amino group of an amino acid with a cyclic anhydride, or by the amide coupling of two amino acids with a diacid chloride. This approach was demonstrated by the synthesis of polymers based on alanine and proline. The polymers were of low molecular weight as a result of azlactone formation which terminates the polymerization [44]. Low molecular weight polymers from methylenebis(p-carboxybenzamide) were synthesized by melt condensation [45]. A series of amido containing polyanhydrides based on p-aminobenzoic acid were synthesized by melt condensation. The polymers melted at 58 to 177 °C and had a molecular weight of 2500 to 12 400 [46]. The properties of several of these polymers are listed in Table 4.

Eight imide-diacids synthesized from trimellitic anhydride, pyromellitic anhydride and aminoacids of the formula $HOOC-(CH_2)_n-NH_2$, with n = 1, 2, 3,

Table 4. Poly [methylenebis(p-carboxybenzamide)]

-R-	Tg	Tm	Mn
$(CH_2)_3$	38	58	2 300
$(CH_2)_2-CH(C_4H_9)-$	150		12 400
$(CH_2)_8$	102		11 600
$(CH_2)_2-O-(CH_2)_8-O-(CH_2)_2$	35	88	3 800
$(CH_2)_2-O-(CH_2)_6-O-(CH_2)_2$	39	114	7 100
$(CH_2)_2-O-(CH_2-CH_2-O)_2-(CH_2)_2$	28	50	4 800
$(CH_2)_2-O-CH(CH_3)-CH_2-O-(CH_2)_2$	37	106	6 900

Data taken from Ref. [46].

5 were polymerized by melt polycondensation [47, 48]. These co-polyimides have an aliphatic-aromatic structure and the relation between the general properties and the amount of the aliphatic part in the repeat unit was studied. These polymers were stable at temperatures above 300 °C and were soluble in *N,N*-dimethylformamide (DMF) and dioxan. The structure of these polymers and their properties are shown in Table 5.

The trimellitic-amino acid polymers and its copolymers were extensively studied for use as an erodible carrier for drugs [49–51]. The following amino acids were incorporated in a cyclic imide structure to form a diacid monomer: glycine, β-alanine, γ-aminobutyric acid, L-leucine, L-tyrosine, 11-aminoundecanoic acid and 12-aminododecanoic acid. These diacids were then converted into their corresponding polyanhydrides by melt condensation. The homopolymers of all *N*-trimellitylimido acids containing amino acids were rigid and brittle with MW below 10 000 [49]. Higher molecular weight polymers were obtained by incorporation of flexible segments, i.e. copolymers with aliphatic diacids, in the polymer backbone (Tables 6, 7). Maximal molecular weight was generally obtained at 180 °C after 1 to 2 hours. Addition of earth metal oxides and metal salts, known catalysts for anhydride synthesis [28], resulted in higher molecular weight polymers in shorter reaction times.

Table 5. Poly(imide-anhydrides)

TRIMELLITIC-AMINO ACID

PYROMELLITIC-AMINO ACID

Polymer (x=)	Degradation[a] temp. (°C)	$[\eta]^{b}$ (dl/g)	DMF	Solubility CHCl$_3$	Dioxan
Trimellitic					
1	330	0.16	+ +	−	±
2	320	0.16	+ +	±	±
3	360	0.19	+	±	±
5	330	0.23	+ +	±	+ +
Pyromellitic					
1	365	−	±	−	±
2	312	−	±	−	±
3	347	0.17	+	−	±
5	377	0.19	+ +	±	+ +

Data taken from Refs. [47, 48] + soluble; + + very soluble; ± slightly soluble; − insoluble.
[a] 10% weight loss in a thermogravimetric test.
[b] determined in DMF at 25 °C in an Ubbelohde viscometer.

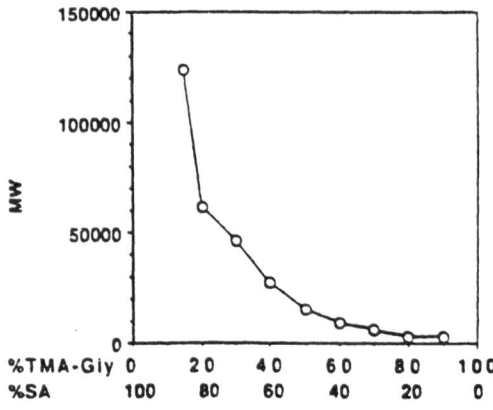

Fig. 2. Poly(TMA-gly:SA) of various ratios, melt polymerized for 2 h at 180°C (from Ref. [49])

Copolymers of N-trimellitylimido-glycine or aminodecanoic acid with either sebacic acid (SA) or 1,6-bis(p-carboxyphenoxy)hexane (CPH) were prepared in defined ratios. High molecular weight copolymers (> 100000) were generally obtained with an increasing content of the SA or CPH comonomer (Fig. 2). Similar phenomena were found for copolymers of aromatic and unsaturated diacids with sebacic acid. Increasing content of SA or CPH also resulted in low melting polymers.

Table 6. Poly(imide-anhydride) based on trimellitic-imide diacids

X	Tg	Mn	crystallinity (%)
1	98	< 5000	66.0
2	102.7	< 5000	
3	98.8	6500	
4	81.5	6700	
5	63.2	14800	
10	21.4	11450	66.0
11	25.9	19250	

$R = -CH_2-CH-(CH_3)_2$
$R = -CH_2-C_6H_5-OH$
$R = -CH_2-S-CH_2-C_6H_5$

Table 7. Poly(imide-anhydride) copolymers with sebacic acid

Polymer (X=)	%SA	Mw	Tg [°C]	Elongation at break [%]	Tensile strength [kg/cm²]
1	50	< 5000	29.3	44	205
2	50	21 390	26	298	1872
3	50	18 060	29.9	339	3531
4	0	14 600	81	54	1202
5	0	12 350	63	21	1890
10	47	17 760	9.3	37	1069

Data taken from Refs. [48–51].

2.2.3 Aliphatic-Aromatic Homopolymers

Polyanhydrides of diacid monomers containing aliphatic and aromatic moieties, poly[p-carboxyphenoxyalkanoic anhydride], were synthesized by either melt or solution polymerization with molecular weights of up to 44 600 (Table 8) [52].

The aliphatic-aromatic diacid monomers were prepared from the reaction of bromoalkanoic acid methyl ester and p-hydroxy benzoic acid methyl ester. The polymers of carboxyphenoxy alkanoic acid of n = 3, 5, and 7 methylenes were soluble in chlorinated hydrocarbons and melted at temperatures below 100 °C. Copolymers of these monomers melted at lower temperatures than the respective homopolymers. These polymers displayed zero-order hydrolytic degradation profile ranging from 2 to 10 weeks. Increasing the length of the alkanoic chain, decreased the degradation rate of the polymer (Fig. 3).

Table 8. Aliphatic-aromatic homopolyanhydrides

X	Tm	Mw	Mn	[n]
1	204–5			
5	50–1	44 600	18 950	0.58
7	53–4	33 300	15 300	0.46
1 + 5	62–5	21 800	10 100	0.32
1 + 7	37–40	13 630	5 640	0.30

Data taken from Ref. [52].

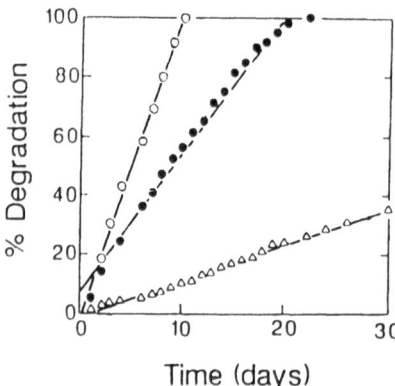

Fig. 3. Degradation of poly(ω-(p-carboxy-phenoxy)alkanoic anhydrides). (o) poly(CPA), ($\bullet$) poly(CPV); ($\triangle$) poly(CPO) (from Ref [52])

2.2.4 Soluble Aromatic Polymers

Aromatic homopolyanhydrides are insoluble in common organic solvents and melt at high temperatures (Table 1). These properties limit the use of purely aromatic polyanhydrides, since they can not be fabricated into films or microspheres using solvent or melt techniques. Synthesis of soluble and low melting copolymers of common aromatic diacids have been recently reported [53, 54].

Copolymers of isophthalic acid (IPA), terephthalic acid (TA), 1,3- bis(carboxyphenoxy)-propane (CPP), or hexane (CPH), and fumaric acid (FA) were synthesized. Copolymers containing a mass fraction of 20% of a second aromatic monomer became soluble and melted at temperatures below 120 °C. The polymers had a molecular weight of 35 000. The copolymer properties were dependent on the nature and ratio of the comonomers in the polymer as seen in Table 9. Polymers of isophthalic acid containing a mole fraction of 10 to 90% fumaric acid or a mole fraction of 10 to 60% CPP were soluble, non-crystalline, and melted below 120 °C. The copolymers of terephthalic acid (TA), however, were soluble only between the mole fraction range of 15 to 40% TA content as shown in Table 10.

Table 9. Soluble and low melting aromatic copolymer compositions[a]

Copolymer of:	Compositions, % monomer	Copolymer of:	Compositions, % monomer
TA-CPP	20–30 TA	TA-SA	0–30 TA
TA-FA	10–35 TA	CPP-SA	0–65 CPP
TA-IPA	10–40 TA	IPA-SA	0–70 IPA
CPP-IPA	10–60 CPP	FA-SA	0–70 FA
CPP-FA	15–50 CPP		
IPA-FA	10–90 IPA		

[a] Polymers in this range have a solubility of > 1% in dichloromethane and melting points below 150 °C. TA- terephthalic acid; IPA- isophthalic acid; CPP- bis(p-carboxyphenoxy propane); FA- fumaric acid; SA- sebacic acid. The estimated range is in mole fractions with an error of $\pm$ 5%.

Table 10. Copolyanhydrides of terephthalic acid and isophthalic acid

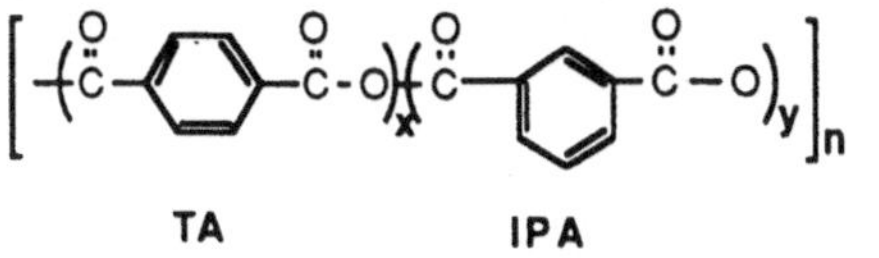

polymer of:	Melt. point (°C)	Crystallinity (%)	Solubility (% w/v)
TA	400	60	< 0.1
IPA	256	61	< 0.1
TA-IPA 1:4	110	< 5	20
TA-IPA 3:7	120	< 5	15
TA-IPA 1:1	230	> 30	< 1

Data taken from Ref. [53].

In a recent US patent, Ziegast described a series of polyanhydrides (of the formula shown below) which are useful for controlled drug delivery of drugs [55]. These polymers have a similar structure to the polymers synthesized by Conix (Table 1) [11, 12], but with different chain structures between the two aromatic rings. The polymers were soluble in chlorinated hydrocarbons with a Tg ranging from 6 to about 90 °C. The in vitro degradation in phosphate buffer of these polymers is about 100 days with a lag time of up to 30 days for some polymers [55]. If these homopolymers are implanted in animals, they will produce large diacid monomers which have to be eliminated from the body.

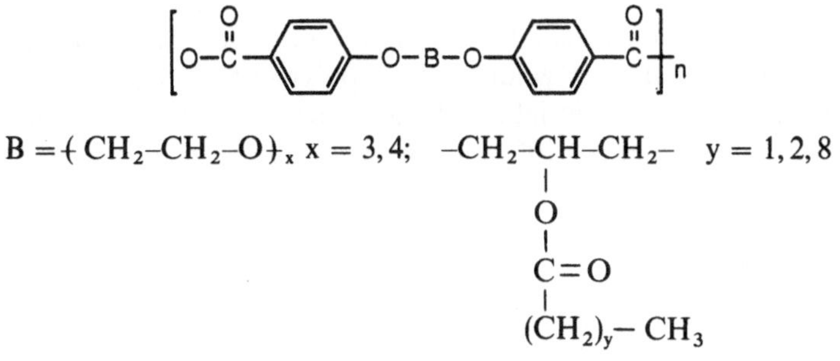

$$B = (CH_2-CH_2-O)_x \quad x = 3,4; \quad -CH_2-CH-CH_2- \quad y = 1,2,8$$

$$\begin{array}{c} | \\ O \\ | \\ C=O \\ | \\ (CH_2)_y - CH_3 \end{array}$$

2.2.5 Poly(ester-anhydrides)

4,4'-Alkaline- and oxaalkanedioxydibenzoic acids were used for the synthesis of polyanhydrides [56]. The polymers melted at a temperature range of 98 to 176 °C and had a molecular weight of up to 12 900 as listed in Table 11.

2.2.6 Fatty Acid Based Polyanhydrides

Polyanhydrides were synthesized from dimer and trimer of unsaturated fatty acids. The dimers of oleic acid and erucic acid, prepared by radical coupling via

Table 11. Ester containing polyanhydrides

$$\left[\!\begin{array}{c} \text{C} \end{array}\!\!-\!\!\bigcirc\!\!-\!\text{O}-\underset{\text{O}}{\overset{\text{O}}{\text{C}}}\text{-R}-\underset{\text{O}}{\overset{\text{O}}{\text{C}}}\text{-O}-\!\bigcirc\!\!-\underset{\text{O}}{\overset{\text{O}}{\text{C}}}\text{-O}\!\right]_n$$

-R-	Tg	Tm	Mn
$(CH_2)_2$	45	176	8720
$(CH_2)_4$	33	159	6080
$(CH_2)_8$	33	135	12900
$(CH_2)_2-O-(CH_2)_6-O-(CH_2)_2$		125	8900
$(CH_2)_2-O-(CH_2)_2-O-(CH_2)_2$		105	9300
$(CH_2)_2-O-(CH_2-CH_2-O)_2-(CH_2)_2$	50	110	5900
$(CH_2)_2-O-CH_2-CH(C_2H_5)-O-(CH_2)_2$		98	8550

Data taken from Ref. [56].

the double bond, are liquid oils containing two carboxylic acids available for anhydride polymerization. Table 12 summarizes the molecular weights and melting points of these polymers [57]. The homopolymers are viscous liquids, copolymerization with increasing amounts of sebacic acid forms solid polymers with increasing melting points as a function of SA content. The polymers are soluble in chlorinated hydrocarbons, tetrahydrofuran, 2-butanone, and acetone. The degradation and drug release from these polymers is discussed in Sect. 5

The properties of a polyanhydride were modified by the incorporation of long chain fatty acids, such as stearic acid, in the polymer composition which alters its hydrophobicity and decreases its degradation rate [58]. Since natural fatty acids are monofuctional they would act as polymerization chain terminators and control the molecular weight. A detailed analysis of the polymerization reaction show that up to about 10% mole fraction content of stearic acid, the final product is essentially a stearic acid terminated polymer. Whereas, at

Table 12. Typical molecular weights and melting points of Poly(FAD-SA)[a]

Polymer	Peak MP	Mw	Mn	$[\eta]$
P(FAD)	liquid	34 400	8 600	0.25
P(FAD-SA) 80:20	30	24 400	9 100	0.26
P(FAD-SA) 60:40	49	36 500	14 000	0.32
P(FAD-SA) 50:50	68	280 000	28 500	0.96
P(FAD-SA) 40:60	72	88 000	20 800	0.67
P(FAD-SA) 30:70	74	103 700	22 600	0.77
P(FAD-SA) 20:80	76	113 000	20 900	0.74
P(FAD-SA) 10:90	78	120 400	22 600	0.84
P(SA)	82	130 500	23 300	0.88

[a] molecular weight was determined by GPC, the melting point was determined by DSC.

higher amounts of acetyl stearate in the reaction mixture resulted in the formation of increasing amounts of stearic anhydride by-product with minimal effect on the polymer molecular weight which remained less than 5000. The rate of drug release from these polymers decreased as the stearic acid content of the polymer was increased. Mixtures of polyanhydrides with triglycerides and fatty acids or alcohols did not form uniform blends.

2.2.7 Modified Polyanhydrides

The physical and mechanical properties of polyanhydrides can be altered by modification of the polymer structure with a minor change in the polymer composition. Several such modifications include the formation of polymer blends, branched and crosslinked polymers, partial hydrogenation and reaction with epoxides.

Biodegradable polymer blends of polyanhydrides and polyesters have been used as drug carriers [59]. Poly(lactic acid) (PLA), poly(hydroxybutyrate) (PHB), and poly(caprolactone) (PCL), of 2000 and 50 000 molecular weights were mixed with poly(sebacic anhydride) (PSA), and the properties of these mixtures were studied. Mixtures of PHB and low molecular weight PLA or PCL formed uniform blends with various amounts of PSA. These blends possess different physical and mechanical properties compared to the parent polymers. The release rate of drugs from these polymeric blends increases with the increase in the content of the rapidly degrading component, PSA.

Branched and crosslinked polyanhydrides were synthesized by reacting the prepolymers of diacid monomers with tri- or polycarboxylic acid branching monomers [60]. Sebacic acid was polymerized with 1,3,5 benzenetricarboxylic acid (BTC) and poly(acrylic acid) (PAA) to yield random and graft-type branched polyanhydrides (Fig. 4). The polymerization was followed until the gel point, and the resulting polymers were evaluated for their physico-chemical properties and degradation behaviour. The molecular weights of the branched polymers were significantly higher (mol. wt. 250 000) than the molecular weight of the respective linear polymer (mol. wt. 80 000) (Fig. 5). In the case of poly(acrylic acid) branched polymers, the molecular weight increased linearly with increasing concentration of poly(acrylic acid). The specific viscosities of the branched polymers were lower than linear polyanhydrides with similar molecular weights. Except for the difference in molecular weights, there were no noticeable changes in the physico-chemical or thermal properties of the branched polymers and the linear polymer [60]. Release of morphine was much higher from the poly(acrylic acid) branched polymers compared to the 1,3,5 benzenetricarboxylic acid branched polymers and increased with increasing concentrations of the branching agent (Fig. 6).

Polyanhydrides can be transformed into poly(anhydride-esters) without degradation of the polymer. The reaction between propylene oxide and a polyanhydride yielded a poly(anhydride-ester) [43]. The reaction was carried out at

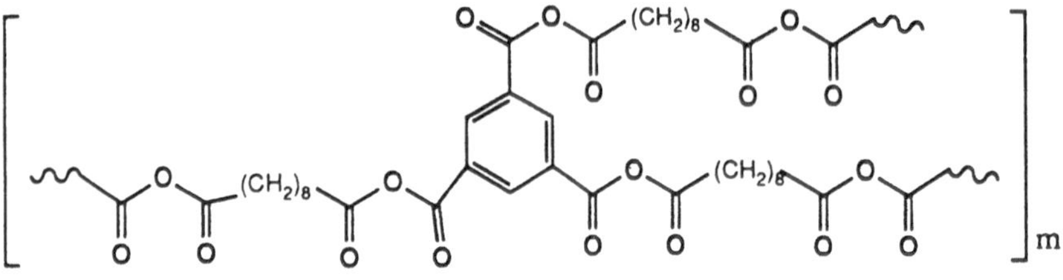

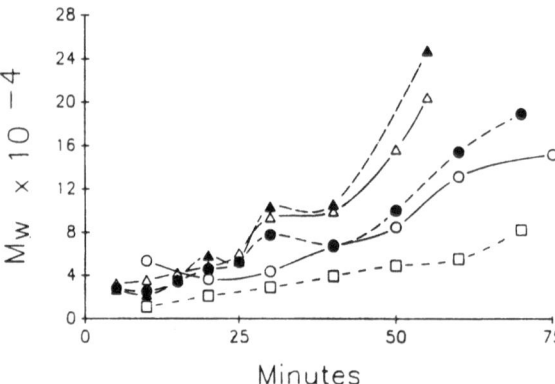

Poly(sebacic anhydride) branched with 1,3,5 benzenetricarboxylic acid

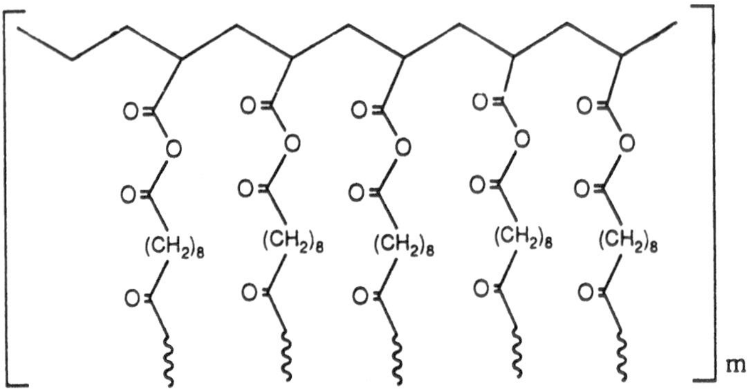

Poly(sebacic anhydride) branched with poly(acrylic acid)

* m is the number of branching molecules
  ∿ is a polymeric chain

Fig. 4. Structures of branched polymers (from Ref [60])

Fig. 5. Molecular weight of branched polyanhydrides; effect of the concentration of poly(acrylic acid) (PAA) branching agent on gel time, and molecular weight. (□) 0%; (o) 0.5% (●) 1.0%; (△) 1.5%;, (▲) 2.0%. (from Ref [60])

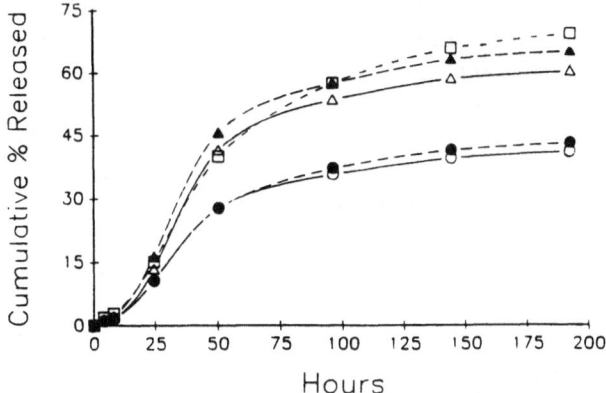

Fig. 6. Release of morphine from randomly branched polyanhydrides; effect of the concentration of the branching agent, benzenetricarboxylic acid (BTC) on the cumulative release of morphine. (□) 0%; (o) 0.5%; (●) 1.0%; (△) 1.5%; and (▲) 2.0% (from Ref [60])

60 °C/69 kPa in an autoclave with propylene oxide and triethylamine/propylene glycol as catalyst as follows:

The polymer of molecular weight range from 1500 to 3000 was polymerized with styrene via the double bonds to form a crosslinked composite.

3.1 Characterization

In the past decade, extensive research has been carried out on the characterization of polyanhydrides. This section will describe the methods used for the characterization of polyanhydrides and data obtained about their chemical composition and structure, crystallinity and thermal properties, mechanical properties, and thermodynamic and hydrolytic stability.

3.1.1 Composition by $^1H\,NMR$

The composition of polyanhydrides has an important role in enabling different erosion rates of these polymers. The composition and structure of polyanhydrides has been determined by NMR, Raman, and IR spectra analysis [61–64], and by masspecrometry, SIMS, and XPS [65, 66].

The polymer composition and the sequence of the comonomers can be determined by ^{1}H NMR as demonstrated in the analysis of the aromatic-aliphatic homopolymers [52]. The aliphatic-aromatic diacids can be connected in a polymer in three ways, aliphatic moiety to aliphatic moiety, aliphatic to aromatic, and aromatic to aromatic moieties. The ^{1}H NMR spectra of the carboxyphenoxy valerate (CPV) diacetate prepolymer and polymer are shown in Fig. 7. The methylenic protons of the aliphatic residue conjugated to the anhydride bond in the polymer appeared as two triplets (e) at 2.54 ppm and 2.72 ppm chemical shifts. The aromatic protons *ortho* to the anhydride bond appeared in two chemical shifts (a), a doublet at 7.90 ppm and at 8.10 ppm. The peaks at 2.72 and 8.10 were not observed in the spectrum of the prepolymer (Fig. 7). These peaks were explained by a chemical shift effect across the anhydride bond, affecting the absorbancies of the α-protons to the anhydride bond. These peaks were attributed to the three types of anhydrides present in the polymer: the 2.54 ppm signal corresponds to the aliphatic-aliphatic anhydride bond, 2.72 ppm to aliphatic-aromatic, and 8.10 ppm to aromatic-aromatic anhydride bonds. Examination of the integration of these peaks revealed a ratio of 1:2:1, aliphatic-aliphatic, aliphatic-aromatic, and aromatic-aromatic moieties. This ratio implies an equal statistical distribution of alternating aromatic-aromatic and aliphatic-aliphatic units throughout the polymer backbone [52].

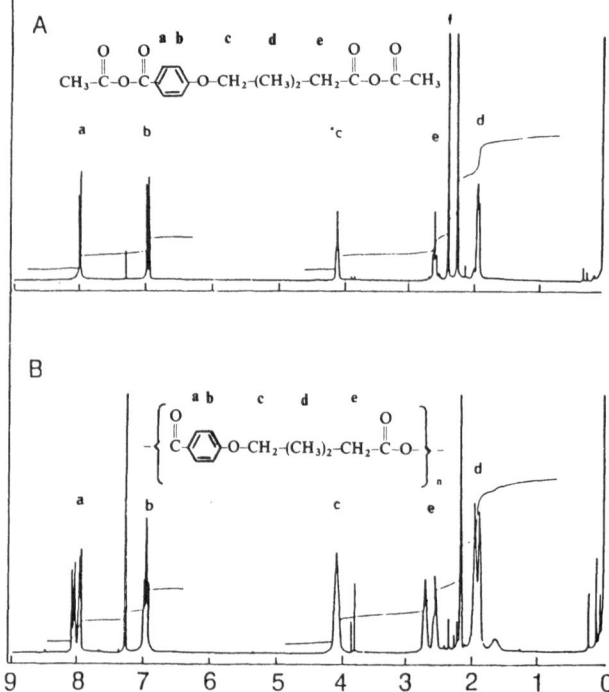

Fig. 7. H NMR spectra (in ppm) of (A) CPV prepolymer, (B) poly(CPV) (from Ref [52])

A detailed study on the use of NMR to identify statistical segments in a polyanhydride chain was reported by Ron et al. [61]. The following copolymer characteristics were studied: 1. the degree of randomness that suggests whether the polyanhydride is either a copolymer or a mixture of homopolymers, 2. average length of sequence (Ln), and 3. frequency of occurrence of specific comonomer sequences. Copolymers of carboxyphenoxy propane (CPP) and sebacic acid (SA) were used as model polymers (Fig. 8). As observed in the aliphatic-aromatic polymers described above, the protons close to the electronegative groups, as the aromatic comonomers, experience a lower density of shielding electrons and absorb at lower frequency. On the other hand, the protons next to aliphatic comonomers, absorb at higher frequency. Accordingly, the CPP-CPP and CPP-SA diads were represented by peaks at 8.1 and 8.0 respectively, and the triplets at 2.6 and 2.4 represent the SA-CPP and SA-SA diads, respectively. By integration of the ^{1}H NMR spectra of poly(CPP-SA) of various compositions the degree of randomness, average block length, and the probability of finding the diad SA-SA or SA-CPP were calculated as shown in Table 13 and Fig. 9.

Similar data analysis was applied also to copolymers of sebacic acid with fumaric acid [40], CPH [61], and trimellitimide derivatives [51]. Based on the ^{1}H NMR analysis, a FA-FA or SA-SA dimer alternating structure (-SA-SA-FA-FA-SA-SA-FA-FA) was suggested for P(FA-SA) 1:1. In the case of copolymers of the asymmetric trimellitimide comonomer (represented by the letters B1 or B2) with the symmetric comonomers sebacic acid or CPH (represented by the letter A), 6 diad sequences were expected to be present in the polymer backbone (A-A, A-B1, A-B2, B1-B1, B1-B2, B2-B2) and they were identified by ^{1}H NMR spectroscopy. Due to partially overlapping peaks in some of the copolymers, some assumptions were made. A typical data from this analysis is shown in Table 14.

3.1.2 Molecular Weight

The molecular weight of polyanhydrides was determined by viscosity measurements and gel permeation chromatography (GPC) [61]. Attempts to determine the molecular weight using vapor preasure osmometry (VPO) were not successful as discussed below. The weight average molecular weight (Mw) of polyanhydrides range from 5000 to 300 000 with a polidispersity of 2 to 15 which increases with the increase in Mw molecular weight. The intrinsic viscosity [η] increases with the increase in Mw.

The Mark-Houwink relationship for poly(CPP-SA) was calculated from the viscosity data and the Mw values, as determined by universal calibration of the GPC data using polystyrene standards:

$$[\eta]_{CHCl_3}^{23\,°C} = 3.88 \times 10^{-7} Mw^{0.658}$$

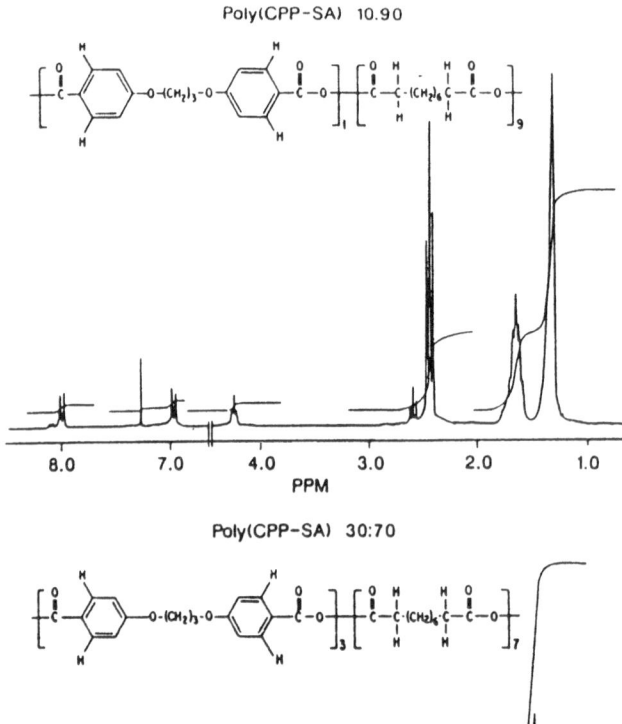

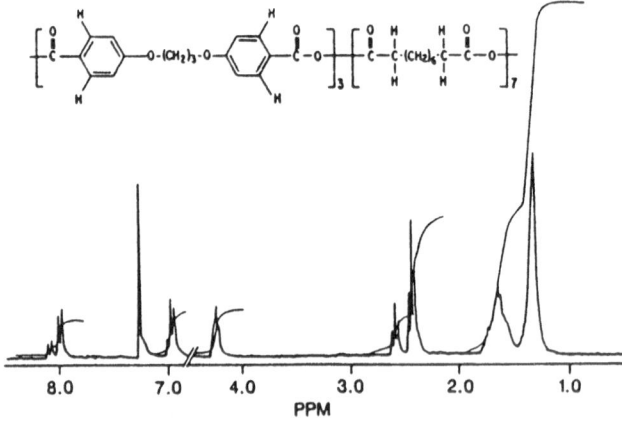

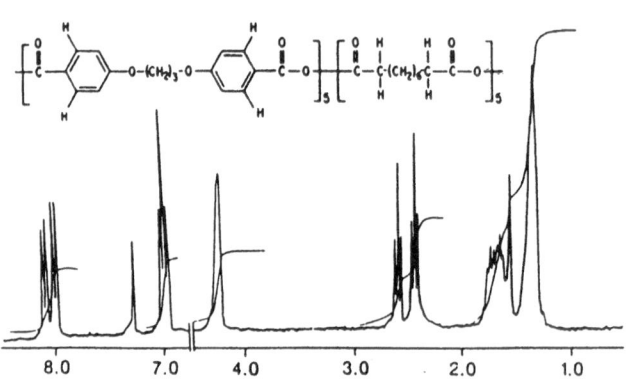

Fig. 8. H NMR spectra (in ppm) of CPP-SA copolymers (from Ref [61])

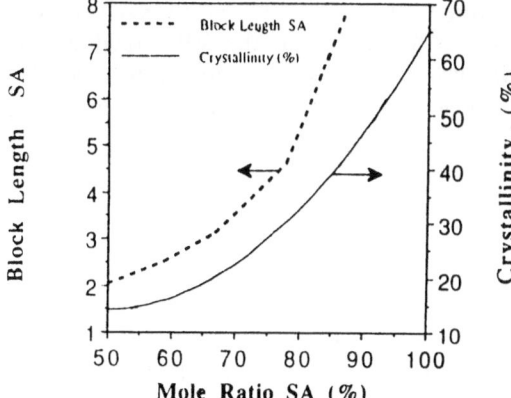

Fig. 9. Crystallinity, the average block length of SA vs mole fraction of SA for poly(CPP-SA). (from Ref [61])

Table 13. Comonomer sequence distribution of the poly(CPP-SA) series

mole ratio of SA-CPP in the polymer, p(SA)	probability of finding the diad SA-SA, p(SA-SA)	probability of finding the diad SA-CPP, p(SA-CPP)	average block length L(SA)	degree of randomness
0.96	0.86	0.14	12.3	0.3
0.87	0.76	0.22	7.8	0.4
0.82	0.67	0.30	5.5	0.6
0.63	0.45	0.36	3.5	0.7
0.59	0.36	0.47	2.5	0.9
0.49	0.24	0.49	2.0	1.0

Data taken from Ref. [61].

Table 14. Observed diad probabilities in copoly(trimellitimide-glycine: sebacic acid) (TMA-gly:SA) of various ratios

%TMA-gly	%SA	PSA-SA	PSA-B1	PSA-B2
0.00	1.00	1.00	0.00	0.00
0.11	0.89	0.906	0.051	0.043
0.30	0.70	0.800	0.100	0.100
0.47	0.53	0.579	0.211	0.211
0.60	0.40	0.115	0.471	0.414

Data taken from Ref. [51].

The universal calibration concept for GPC was confirmed for poly(CPP-SA). The acetic acid end group determination for molecular weight estimation was not used because the polymer may contain cyclic macromolecules with no acetate end groups [28].

Attempts to determine the Mn using VPO resulted in a decrease in the polymer molecular weight during measurements. At a given concentration and temperature, the polymer sample being measured is expected to reach equilibrium in approximately 1–3 minutes. An increase in the VPO readings, as exhibited by the aliphatic polyanhydrides, indicated a decrease in molecular weight during measurements as revealed by GPC analysis of samples taken during VPO determination.

3.1.3 Crystallinity

Since crystallinity is an important factor in controlling polymer erosion, analysis of the effect of polymer composition on crystallinity was studied [61, 62]. Polymers based on sebacic acid (SA), (p-carboxyphenoxy)propane (CPP), (p-carboxyphenoxy)hexane (CPH), and fumaric acid (FA) were investigated. The crystallinity was determined by: 1. X-ray diffraction, 2. a combination of X-ray and DSC, and 3. data generated from ^{1}H NMR spectroscopy and Flory's equilibrium theory. Homopolyanhydrides of aromatic and aliphatic diacids were crystalline (> 50% crystallinity). Copolymers possess high degree of crystallinity at high mole ratios of either aliphatic or aromatic diacids. A typical X-ray powder diffraction of CPP-SA copolymer series is shown in Fig. 10. The heat of fusion and crystallinity of poly(CPP-SA) is shown in Table 15. The glass transition, Tg, the melting point, Tm, and the heat of fusion were determined by DSC. The crystallinity, Xc, was calculated from the DSC and X-ray powder diffraction. Heat of fusion values for the polymers demonstrated a sharp decrease as CPP is added to SA or viceversa. The trend of decreasing crystallinity, as one monomer is added, appeared using the X-ray or DSC methods. The decrease in crystallinity is a direct result of the random presence of other units in the polymer chain. A detailed analysis of the copolymers of sebacic acid with the aromatic and unsaturated monomers, CPP, CPH, FA, and trimellitic-amino acid derivative was reported [51]. Copolymers with high ratios of SA and CPP, TMA-gly, or CPH were crystalline while copolymers of equal ratios of SA and CPP or CPH were amorphous. The poly(FA-SA) series displayed high crystallinity regardless of comonomer ratio.

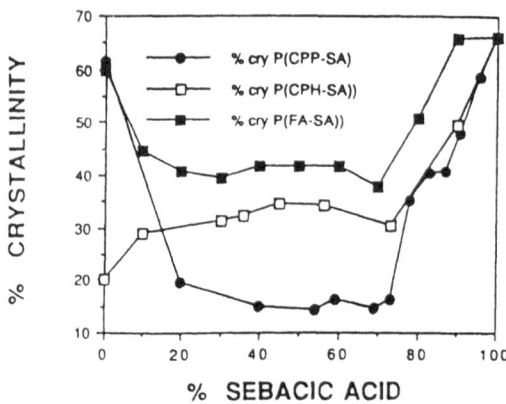

Fig. 10. Percent crystallinity of P(CPP-SA), P(CPH-SA), and P(FA-SA) polymers. (from Ref [61])

Table 15. Heat of fusion and crystallinity of poly(CPP-SA)

Polymer	Tm °C	Tg °C	Heat of fusion cal/g	Crystallinity Xc %	Wc %
Poly(SA), 100%	86.0	60.1	36.6		66.0
poly(CPP-SA) 4:96	76.0	41.7	24.9	46.5	58.7
poly(CPP-SA) 9:91	78.0		25.7	48.5	58.7
poly(CPP-SA) 13:87	75.0	47.0	20.7	39.5	40.5
poly(CPP-SA) 17:83	72.0	47.0	19.3	37.0	40.2
poly(CPP-SA) 22:78	66.0	47.0	15.3	30.0	35.0
poly(CPP-SA) 27:73	66.0	44.0	10.2	20.0	16.2
poly(CPP-SA) 31:69	66.0	40.0	5.1	10.6	14.5
poly(CPP-SA) 41:59	178.0	4.2	2.0	4.0	16.2
poly(CPP-SA) 46:54	185.0	1.8	3.1	6.1	14.2
poly(CPP-SA) 60:40	200.0	0.2	6.0	13.9	15.0
poly(CPP-SA) 80:20	205.0	15.0	8.2	17.6	19.5
poly(CPP), 100%	240.0	96.0	26.5		61.4

Data taken from Ref. [61]. Tm, Tg and heat of fusion were determined by DSC. The crystallinity was determined from the X-ray diffraction and the heat of fusion.

3.1.4 Infra Red and Raman Analysis

Anhydrides present characteristic peaks in the IR and Raman spectra. In general, aliphatic polymers absorb at 1740 and 1810 cm^{-1} and aromatic polymers at 1720 and 1780 cm^{-1}. A typical IR spectra of aliphatic and aromatic polymers that contain aliphatic and aromatic anhydride bonds may present 3 distinct peaks, where the aromatic peak is shown at 1780 cm^{-1} and the aliphatic peaks at 1720–1740 cm^{-1} in general overlap. The presence of carboxylic acid groups in the polymer can be determined from the presence of a peak at 1700 cm^{-1}. The degradation of polyanhydrides can be followed by IR from the ratio between the anhydride peak at 1810 and 1700 cm^{-1}. The significance of this analysis is that it measures the hydrolysis of the anhydride bonds and not the dissolution of the degradation products which is dependent on the solubility of the degradation products.

The Raman spectra for a variety of polyanhydrides was studied by Davies et al. [63, 64]. The Raman spectra of polyanhydrides were similar to the IR spectra for the same compounds. Polyanhydrides show two distinctive carbonyl Raman bands corresponding to the symmetric and asymmetric vibrations of the carbonyl groups, the separation of the pair being generally 50–70 cm^{-1}.

Fourier-Transform Raman Spectroscopy (FTR) was used to characterize a homologous series of aliphatic poly(anhydrides), poly(carboxyphenoxy)alkanes, and copolymers of carboxyphenoxy propane (CPP) and sebacic acid. All anhydrides show two diagnostic carbonyl bands, the aliphatic polymers has the carbonyl pairing at 1803/1739 cm^{-1}, and the aromatic polymers have the band pair at 1764 and 1712 cm^{-1}. All the homo- and copolymers showed methylene bands due to deformation, stretching, rocking and twisting; the spectra for the aromatic poly(anhydrides) such as P(CPP) also showed diagnostic benzene para-substitution bands. It was possible to differentiate between aromatic and aliphatic anhydrides bonding and in conjunction with other diagnostic bands to monitor the change in individual monomer composition within a copolymer mixture.

Fourier transform Raman (FTR) was used to study the hydrolytic degradation of polyanhydrides [63]. PSA rods exposed to water for 15 days were analyzed daily by FTR (Fig. 11). The carbonyl anhydride band pair (1803/ 1739 cm^{-1}) diminished in intensity from day zero to 15, with the emergence of the complimentary acid carbonyl band (1640 cm^{-1}) which increased in intensity over the same period. Similarly, the increase in the intensity of the C–C deformation at 907 cm^{-1} with hydrolysis reflects the increased freedom of the methylene chain in the low molecular weight oligomers.

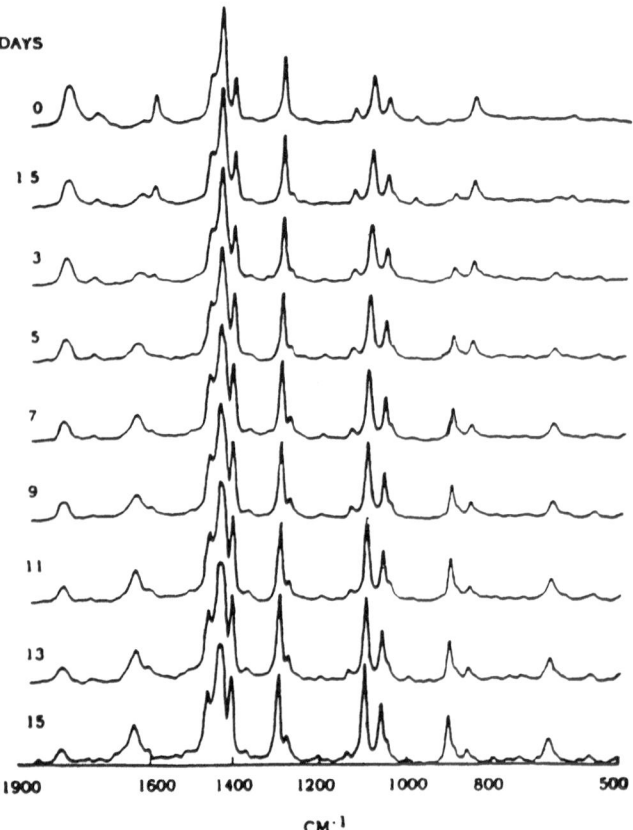

Fig. 11. Raman spectra of PSA at various time intervals during a degradation study. Note change in peak sizes at (in cm^{-1}) 1803/1739 (anhydride); 1640 (acid); 907 (monomer); 850 (polymer) (from Ref [63])

3.1.5 Surface and Bulk Analysis

The morphology of polyanhydride was studied by Scanning Electron Microscope (SEM) to elucidate the mechanism of polymer degradation and drug release from polyanhydrides [67]. Microspheres prepared by three different

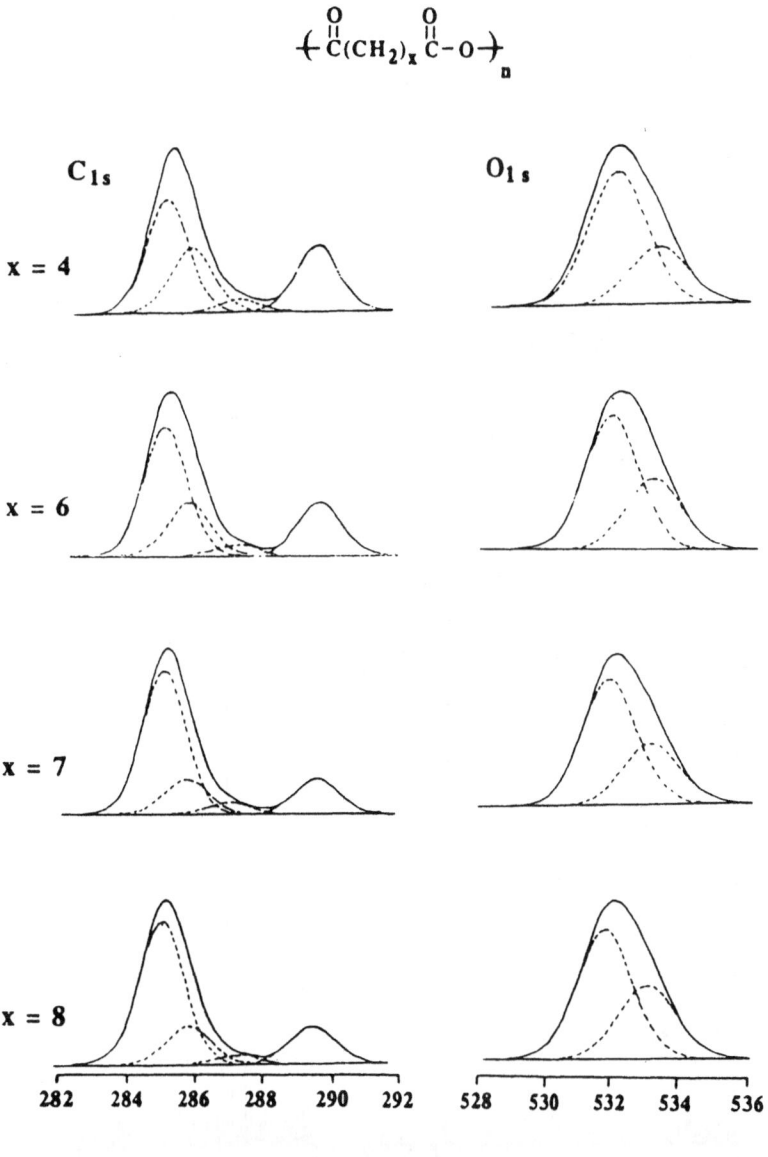

Binding Energy / eV

Fig. 12. The C1s, O1s core level envelopes for polyanhydrides of adipic (x = 4), suberic (x = 6), azelaic (x = 7), and sebacic acid (x = 8) (from Ref [66])

techniques, solvent removal, solvent evaporation and melt encapsulation, were analyzed by SEM. The degradation process in buffer was followed by SEM. Microspheres showed distinctive morphological characteristics induced by the fabrication method. SEM could be used in characterizing the drug release profiles and polymer degradation [67].

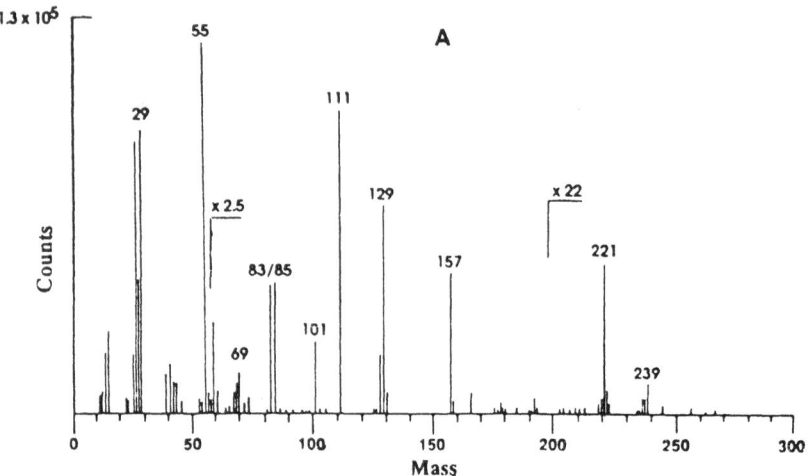

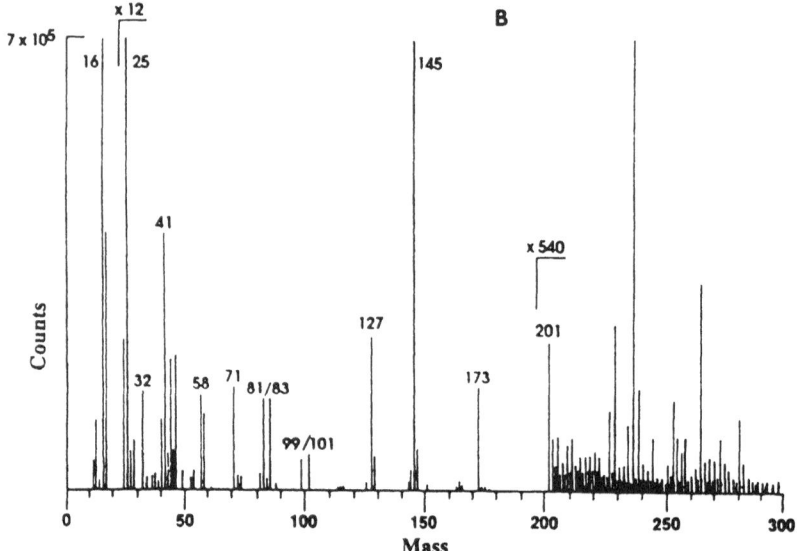

Fig. 13. Positive (A), negative (B) ion spectra of poly(adipic anhydride) (from Ref [66])

Over the last decade, SIMS and XPS have been shown to be powerful complementary techniques for determining the interfacial chemistries of polymers [68]. The surface chemical structure of aliphatic polyanhydride films has been examined using time-of-flight secondary ion mass spectroscopy (ToF-SIMS) and X-ray photoelectron spectroscopy (XPS) [65, 66]. The C1s and O1s core level spectra are displayed for the homologous series of aliphatic polyanhydrides in Fig. 12. The main peak at 285 eV corresponds to the C–H. The peak at 289.5 eV arises from $O–C=O$. The XPS data confirmed the purity of the surface, and the experimental surface elemental ratios were in good general agreement with the known stoichiometry of polyanhydrides.

The ToF-SIMS spectra of polyanhydrides are shown to reflect the polymer structure. The SIMS data confirms a systematic fragmentation, in both negative- and positive-ion SIMS spectra, occuring throughout the entire series of the polyanhydrides examined (Fig. 13). Radical cations were observed in the positive-ion spectra. The lower mass ranges of the negative-ion SIMS spectra contain ions at m/z 12, 13, 16, 17, 24, 25, 41, 43, and 45 that may be assigned to $C–, CH–, O–, OH–, C_2– C_2H–, C_2HO–, C_2H_3O–$, and $CHO_2–$. The ion at m/z 71 arise from the fragmentation of the anhydride unit, $CH2=CHCOO–$ and it was seen for all polyanhydrides. At higher mass a general fragmentation pattern was observed, and the major ions are noted in Table 16.

The combined use of ToF-SIMS and XPS have shown to provide a detailed insight into the interfacial chemical structure of polyanhydrides.

Table 16. Major ions in polyanhydride SIMS spectra

Ion	PA	PSU	PAZ	PSA
positive ions				
M + H +	129	157	171	185
M +	128	156	170	184
M–OH + /MH–H$_2$O +	111	139	153	167
MH + CO +	157	185	199	213
M–CO$_2$ ±	83/85	111/113	125/127	139/141
negative ions				
M–H–	127	155	169	83
M + OH–	145	173	187	201
MH–CO$_2$–	173	201	215	229
M–CO$_2$ ± H–	83/85	–/113	–/127	–/141
M–Co ± H–	99/101	127/129	141/143	155/157
MH + CO$_2$ + CO–	201	–	243	–

Data taken from Ref. [66]. PA-poly(adipic acid); PSU-poly(suberic acid); PAZ-poly(azelaic acid); and PSA-poly(sebacic acid).

4 Stability

The stability of polyanhydrides in solid state and dry chloroform solution was studied [69]. Aromatic polymers such as poly(CPP) and poly(CPM) maintained their original molecular weight for at least one year in solid state [69]. In contrast, aliphatic polyanhydrides such as poly(SA) and poly(phenylenedipropionic acid) (PDP) decreased in molecular weight over time (Figs. 14, 15). The decrease in molecular weight shows first-order kinetics, with activation energies of 7.5 Kcal/mole- K. The decrease in molecular weight was explained by an internal anhydride interchange mechanism, as revealed from elemental and spectral analysis (IR and ^{13}C, and ^{1}H NMR) (Fig. 16). This mechanism was supported by the fact that the decrease in molecular weight was reversible and heating of the depolymerized polymer at 180 °C for 20 min yielded the original high molecular weight polymers. Under similar conditions, an hydrolyzed polymer did not increase in molecular weight. To confirm a depolymerization process rather than hydrolysis, polymer solutions were mixed with tritiated water for 24 h, and if hydrolysis occurs, the formation of a radioactive polymer would be expected. However, no radioactive polymer was obtained although a significant decrease in molecular weight was observed [69]. The depolymerization in solution can be catalyzed by metals. Among several metals tested, copper and zinc were the most effective. It was found that the stability of polymers in the solid state or in organic solutions does not, in many cases, correlate with their hydrolytic stability. The aliphatic-aromatic homopolymers as well as the imide containing polymers also decreased in molecular weight with time which was explained by a depolymerization process [50, 52].

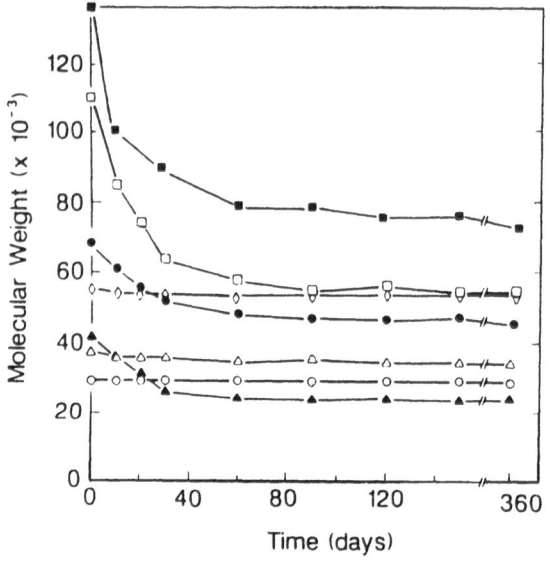

Fig. 14. Solid state depolymerization of poly(anhydrides). Weight average molecular weight of bulk poly(anhydrides) stored under vacuum at 21 °C as determined by GPC: (■) poly(SA); (□) poly(CPP-SA) 20:80; (●) poly(CPP-SA) 35:65, (◇) poly(CPP-SA) 50:50; (△) poly(CPH); (●) poly(PDP); (△) poly(CPM) (from Ref [69])

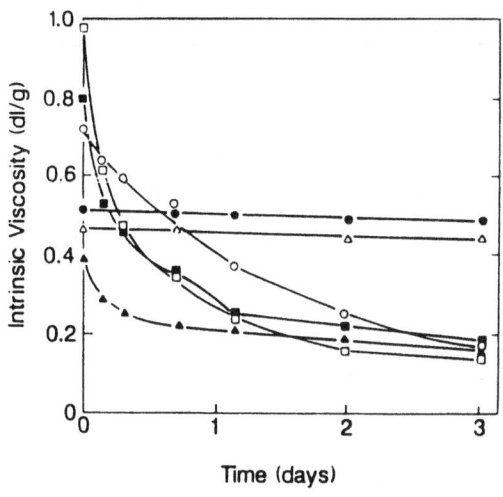

Fig. 15. Solution depolymerization of poly(anhydrides) in chloroform (10 mg/mL) stored under nitrogen at 37 °C: (□) poly(SA); (■) poly(CPP-SA) 20:80; (o) poly(CPP-SA) 50:50; (o) poly(CPH; (▲) poly(PDP); (△) poly(CPM). Viscosity was measured at 23 °C (from Ref [69] Fig. 3)

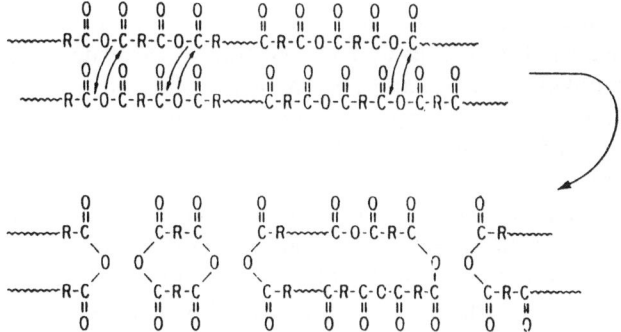

Fig. 16. Mechanism of the depolymerization of poly(anhydrides) (from Ref [69] Scheme I)

The effect of γ-irradiation on polyanhydrides for sterilization purposes has been studied. Several polymers were γ-irradiated with 2.5 Mrad dose and the properties of the polymer before and after radiation were monitored. All polymers did not change in color or pliability as well as the ^{1}H-NMR and IR spectra remain the same. The polymers did not change in molecular weight after irradiation as shown in Table 17.

Table 17. γ-Irradiation of polyanhydrides[a]

Polymer	control		γ-irradiated	
	Mw	Mn	Mw	Mn
PSA	94 300	6 500	78 900	6 000
P(FAD-SA) 1:4	40 100	8 700	42 500	10 100
P(FAD-SA) 1:1	38 000	10 100	40 100	10 000
P(FA-SA) 1:4	5 100	1 900	6 200	2 600
P(ISO-SA) 25:75	34 700	6 100	36 100	5 100

[a] Molecular weight was determined by GPC

5 In Vitro Hydrolysis and Drug Release

A drug is released from biodegradable devices by one or a combination of three processes: diffusion through the polymer matrix, dissolution of the dispersed solid drug directly into the release medium, or release of the dispersed drug by polymer surface erosion, bringing the drug along with it into solution [78, 79]. In the bioerodible systems, erosion is the release rate controlling process [79, 80]. Erosion due to polymer degradation has been classified into two types, homogeneous or bulk degradation, and heterogeneous or surface erosion [79, 80].

In homogeneous or bulk erosion, the release medium penetrates the entire matrix and degradation occurs at the same rate on the surface and in the bulk. In this scenario, drug release is initially delayed and quite slow; however, as the bulk matrix starts eroding, the release rate increases significantly. Therefore, in bulk erosion, the release is not zero order, and it is essentially independent of the device geometry. The poly(lactic acid) and poly(lactic-glycolic acid) polymers belong to the class of bulk eroding polymers, and the release from these devices exhibits the profile expected for bulk erosion [79].

In contrast, the surface eroding polymers are hydrophobic, resist penetration of water into the bulk matrix, and hence the release medium degrades and erodes only the surface of the matrix [79]. Therefore, the release rate is affected by the surface-to-volume ratio, the geometry of the device, and drug loading. If the surface area of the eroding surface remains constant, drug release would be zero order. In addition, release rate is directly proportional to drug loading, and the lifetime of the device is directly proportional to device thickness [79].

Poly(ortho ester)s were the first polymers used to prepare surface eroding devices, but additional excipients such as acid anhydrides or basic excipients in the interior of the matrix, had to be used to prevent bulk erosion [79, 81].

The degradation of polyanhydrides, in general, varies with a number of factors, such as:
1. The chemical nature of the monomer used to produce the polymer
2. In a series of copolymers, the relative amounts of the two monomers used to produce the polymer

3. The hydrophobicity and loading of the drug in the polymeric matrix
4. The pH of the surrounding medium (the higher the pH, the more rapidly the materials degrade)
5. The shape and geometry of the implant (the degradation is dependent on the surface area)
6. The accessibility of the implant to water (porous materials will degrade more rapidly than non-porous)
7. The method of manufacture for the material (e.g. compression-molded materials will degrade more rapidly than injection-molded)

The degradation rates for a number of polyanhydrides are available in the literature [1, 82, 83]. However, recently, a new class of polyanhydrides poly(FAD-SA) have been synthesized from non-linear hydrophobic dimer of oleic acid or erucic acid and relatively hydrophilic sebacic acid. This copolymer can be prepared in various ratios of the monomers to achieve the desired degree of hydrophobicity; increasing the percentage of FAD, a more hydrophobic copolymer, is obtained. Another advantage of this copolymer is its ability to be formulated as films, microspheres, and beads [84].

To study the release characteristics and the mechanism of drug release from P(FAD-SA) devices, a study was conducted employing bupivacaine as a model drug. Bupivacaine, a hydrophobic weak base (pKa = 8.2) is used for local anesthesia [85]. In several clinical situations, a prolonged local anesthetic effect is desired. Hence, a controlled release formulation of bupivacaine would be useful. Five copolymers, P(FAD-SA) 10:90, 20:80, 30:70, 50:50 and 80:20, with varying degree of hydrophobicity and erosion rates, (10:90 being more hydrophilic than 30:70) were used to fabricate the devices. Bupivacaine hydrochloride was dispersed in the melted copolymers at 10% W/W load. The devices, 200 mg in weight, in the shape of rectangular blocks, with dimensions of $1.3 \times 0.7 \times 0.5$ cm were prepared by the melt casting technique. The drug loading per unit volume was therefore, 0.044 gm/cm^3 of the device. The drug release profiles were determined at pH 7.4, and analyzed by the surface erosion based model [78, 80] as described below.

Various mathematical models have been developed to predict drug release behavior from degradation controlled monolithic systems [78–80]. A model for predicting drug release from erodible devices of various geometries developed by Hopfenberg was used for data analysis of this study [78]. This model assumes an agent uniformly dispersed in the matrix, with the drug release primarily occurring due to degradation/erosion of the matrix [78, 80].

The equation for cumulative drug release from an erodible cylinder shaped device with radius r and length h is given by Eq. (1):

$$M_t = 2\pi C_o hBt[r - (Bt/2)] \tag{1}$$

where,

$\quad$ Co = Drug load in gm/cm^3
$\quad$ B = Erosion rate of the polymer device in cm/day
$\quad$ t = Time in days

The equation for the fractional agent release versus time is given by the following expression:

$$M_t/M_\infty = 2(t/t_\infty) - (t/t_\infty)^2 \tag{2}$$

The devices prepared in our study were almost cylindrical in shape because the two smaller dimensions were very similar (0.5 and 0.7 cm). The devices were equivalent to a cylinder both in terms of surface area and volume, with a radius of 0.33 cm and a length of 1.3 cm, and they could thus be classified as erodible cylindrical devices with dispersed drug. Therefore, Eqs. (1) and (2) are applicable and should describe the release profile. To determine if drug release from the devices was controlled by surface erosion, the release profiles were fitted to Eq. (1). Since all the parameters in Eq. (1) except B (the erosion rate) are known, nonlinear regression was used on the release profiles to obtain the optimized values of B.

The release profiles were also independently evaluated for the type of release kinetics observed; i.e. zero order, SQRT of time and first order release. It appeared that for all the copolymers studied, the release was best described by first order release kinetics. The release profiles were thus fitted to the following equation to obtain the first order release rate constant, Kr:

$$\%\text{Cum. Drug Rel.} = M_t = 100[1 - \exp(-k_r t)] \tag{3}$$

The Kr observed was then correlated with the erosion rate constant (B) to determine if drug release is controlled by polymer erosion.

The cumulative release profile of bupivacaine from the various copolymer devices are shown in Figs. 17a and 17b. The drug release occurred very slowly over a period of 15–25 days depending on the polymer used. The fractional agent release (M_t/M_∞) was very well described by Eq. (2) and the coefficients of the various terms in Eq. (2) were determined by polynomial regression for all the copolymers studied. The equation best describing the release profiles is as shown below:

$$M_t/M_\infty = (1.99 \pm 0.11)t/t_\infty - (1.19 \pm 0.53)(t/t_\infty)^2 \tag{4}$$

n = 8 and $R^2 > 0.998$ for all copolymers.

Equation (4) is very similar to the theoretically derived Eq. (2) in the form and the coefficients, suggesting that the release is surface erosion controlled.

The release profiles were also fitted to Eq. (1) using a nonlinear regression program to obtain estimates of B, the erosion rate. R^2 in all cases was greater than 0.99 and hence the good fit of the release profile to Eq. (1). The optimized values of B obtained by nonlinear regression are listed in Table 18. The erosion rate ranged from 0.0037 cm/day for the most hydrophobic copolymer. P(FAD-SA) 80:20 to 0.028 cm/day for the most hydrophilic copolymer, P(FAD-SA) 10:90. However, the erosion rate was not linearly dependent on the hydrophilicity of the polymer, expressed as % of sebacic acid in the copolymer (Fig. 18).

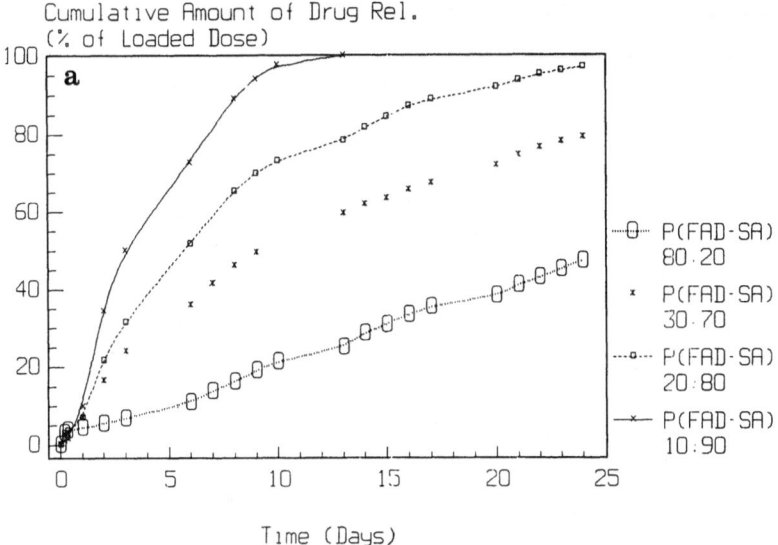

Fig. 17a. Release profiles of bupivacaine hydrochloride from polyanhydride copolymer devices at pH 7.4

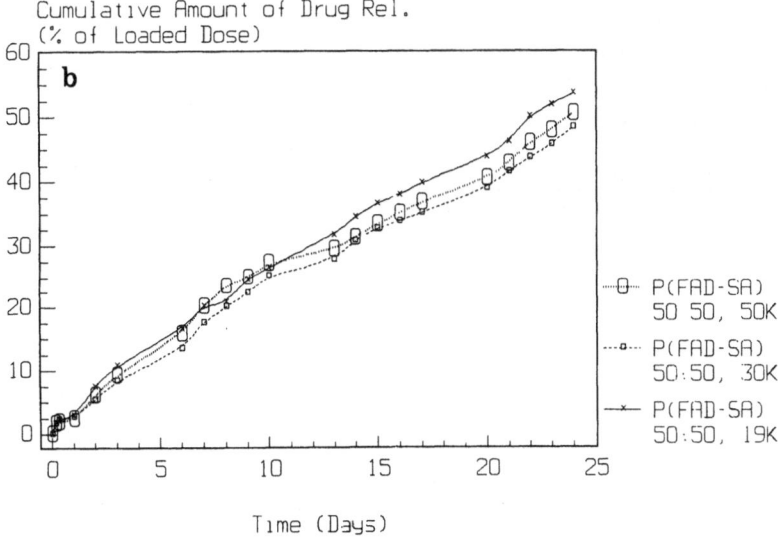

Fig. 17b. Release profiles of bupivacaine hydrochloride from P(FAD-SA) 50:50 devices with different initial molecular weights of the copolymer, at pH 7.4

The dependence was hyperbolic with very little effect of the sebacic acid content on erosion rate between 20–50% sebacic acid. However above 50% sebacic acid, the erosion rate increased very rapidly with a small increase in the hydrophilicity of the copolymer. This results also demonstrate that by varying the ratio of

Table 18. Parameters for release of bupivacaine hydrochloride from polyanhydride devices at pH 7.4

Copolymer used to prepare the device	Erosion rate cm/day	Drug release rate due to erosion, B * Co gm/cm²/day	First order release rate constant, Kr, Day⁻¹
P(FAD-SA) 80:20 MW 72000	0.0037	0.00016	0.025
P(FAD-SA) 50:50 MW 19000	0.0044	0.00019	0.031
P(FAD-SA) 50:50 MW 30000	0.0039	0.00017	0.0027
P(FAD-SA) 50:50 MW 50000	0.0041	0.00018	0.029
P(FAD-SA) 30:70 MW 55000	0.009	0.00037	0.07
P(FAD-SA) 20:80 MW 107600	0.015	0.00068	0.13
P(FAD-SA) 10:90 MW 225000	0.028	0.0012	0.24

monomers in the copolymer, a wide range of erosion rates can be produced to get the desired drug release rates [87].

The release profiles could not be adequately described by either zero or SQRT of time order kinetics, but the release profiles were well described by first

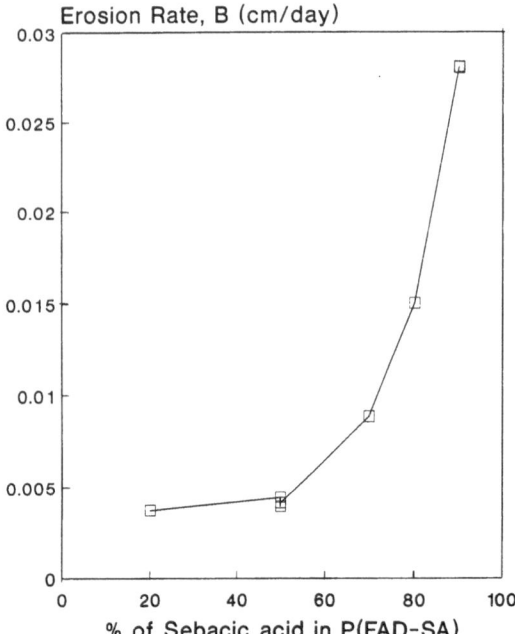

Fig. 18. Dependence of erosion rate estimated from Eq. (1) on the hydrophilicity of the P(FAD-SA) copolymer expressed as % of the sebacic acid content

order kinetics as seen from Fig. 19. The R^2 values for all the release profiles were greater than 0.99, and the optimized values of K_r, the first order release rate constant obtained are also listed in Table 18. K_r appeared to depend on the hydrophilicity of the copolymer in the same fashion as erosion rate. There was excellent linear correlation between Kr and the erosion rate as shown in Fig. 19. The linear relationship between erosion rate and Kr could be described by the following equation:

$$K_r = 8.73B - 0.007, \quad R^2 = 0.9996 \tag{5}$$

This indicates that Kr is linearly dependent on erosion rate of the polymer suggesting that erosion rate solely controls and determines the rate of drug release from the device.

The polymer degradation data expressed as a decrease in the molecular weight and weight loss of the device was available for only one of the copolymers studied, P(FAD-SA) 10:90. The release of drug was correlated with polymer degradation by plotting the % cumulative drug released and the % decrease in molecular weight as shown in Fig. 20. As can be observed from this figure, while the molecular weight of the copolymer decreases by 80% in the first two days, only 20% drug is released. Subsequently, the copolymer degradation slows down, yet drug release continues at a fairly slow first order rate for 24 days. This

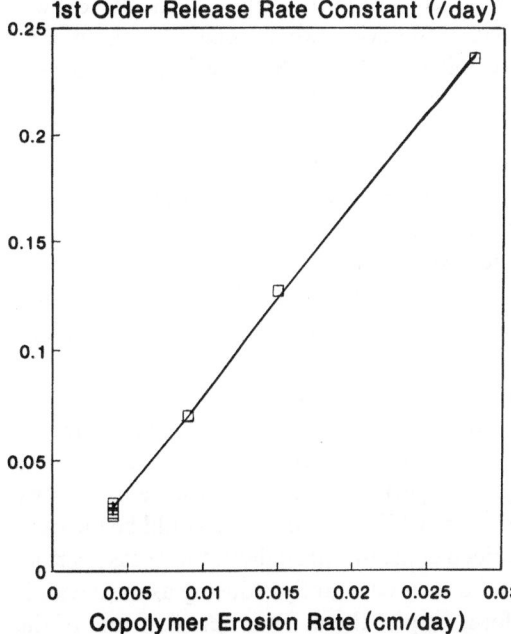

Fig. 19. Correlation of first order release rate constant with erosion rate of the P(FAD-SA) device

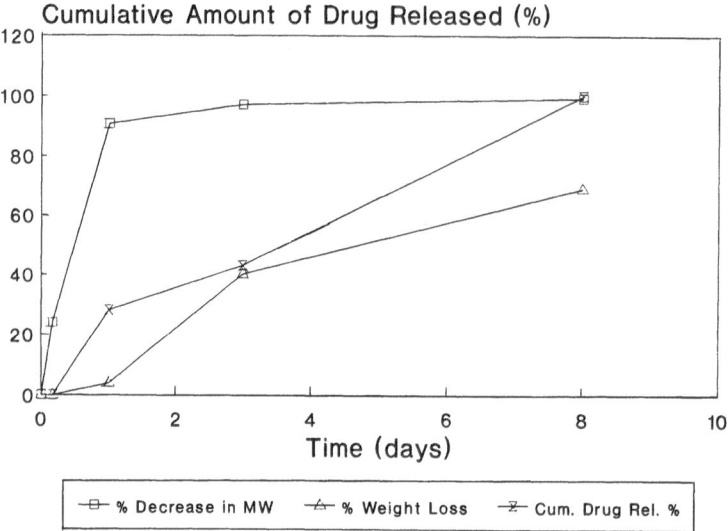

Fig. 20. Correlation of drug release with copolymer degradation expressed as % decrease in molecular weight of the copolymer P(FAD-SA) 10:90 and the weight loss of device

was not a direct correlation between the % decrease in molecular weight and % drug released. However, much better correlation between drug release and polymer degradation (expressed as % decrease in the weight of the device) was observed (Fig. 20). The P(FAD-SA) 50:50 devices were prepared with different initial molecular weights of the copolymer (Table 18). The initial molecular weight of the copolymer appeared to have no effect on drug release (Fig. 17b), and also on erosion rate (Fig. 18, Table 18).

The erosion in polyanhydride matrices has been reported to be purely surface of the heterogeneous type, and the results of this study support the past observation [82]. No lag time for drug release was found, as observed also for poly-lactic acid and poly-lactic acid-glycolic acid, which have been known to undergo bulk erosion [79]. In addition, fit of the data to Eqs. (1) and (2) (which have been derived for a surface eroding polymer), confirm the surface eroding nature of the polyanhydrides.

The erosion rates estimated are as expected and drug release rates calculated from the erosion rates are in the same order of magnitude as those reported by Leong et al. [83]. The expected drug release rates based on erosion are in units of gm/cm^2/day, and thus one would expect zero order release rates for any incorporated drug from a unit surface area of the device. This would be the case, if the surface area of the device remained constant throughout the release period. In the case of the cylinder, the surface area is constantly reducing as the device is eroding at a constant rate. Therefore, despite the constant erosion rate of the polymer, release rate of drug decreases steadily. This results in a first order rate of drug release. If the devices were designed in the shape of a slab or a disk, the surface area exposed to the release medium would be constant, as well as the release rate, i.e. zero order release.

The strong correlation observed between K_r and erosion rate indicates that erosion is the major mechanism for drug release. In fact, a priori knowledge of the erosion rate of a polymer could be used to predict the release profile of the drug from a similar device or from a device of a known geometry. The erosion rate appears to be dependent on the hydrophilicity of the copolymer but not linearly. Increasing content of sebacic acid increases the hydrophilicity of the polymer [82], which results in a higher erosion rate and hence higher release rate. This could be explained by the fact that the anhydride linkages in the polymer are hydrolyzed subsequent to penetration of water into the polymer. The penetration of water or water uptake depends on the hydrophobicity of the polymer and therefore, the hydrophobic polymers which prevent water uptake, have slower erosion rates and lower drug release rates [88, 89]. This is valuable information since one can alter the hydrophobicity of the polymer by altering the structure and/or the content of the copolymer, thereby being able to alter the erosion rate. Since in the P(FAD-SA) series of copolymers, a 10 fold increase in erosion rate was achieved by alteration of the ratio of the monomers, P(FAD-SA) can be used to deliver drugs over a wide range of release rate.

There was no correlation between drug release and polymer degradation expressed as % decrease in the molecular weight, which may appear to be self-contradictory. However, on closer examination, it appears that drug dispersed in the polymer matrix is released when the polymer matrix erodes bringing the drug with it into solution. Thus the release rate would depend on the rate of erosion expressed as volume of the matrix dissolved per unit time, times the drug load. The implication being that weight loss should correlate with drug release and it is a more appropriate indicator of erosion rate than the % decrease in molecular weight. Another fact is that in surface erosion, the molecular weight of the polymer at the surface may decrease but the interior of the device may still have the same molecular weight. Secondly, the lower molecular weight fragments may not diffuse out or dissolve into the release medium. Therefore, it is not the decrease in molecular weight but the weight loss subsequent to the decrease in molecular weight, and the diffusing out of the low molecular weight fragments, which correlates with drug release. This also explains why drug release from P(FAD-SA) 50:50 devices, was independent of the initial molecular weight of the copolymer (Fig. 17b and Table 18).

6 Biocompatibility and Toxicology

The eventual clinical use of any synthetic polymeric material requires adequate testing for safety, toxicity, and biocompatibility of the specific polymer matrix, as a direct contact between the polymer and biological tissues is evident, and potential undesired tissue-implant interactions might occur. In the case of biodegradable matrices, not only the possible toxicity of the polymer has to be evaluated, but also the potential toxicity of breakdown products. The last section of this chapter reviews existing data about the biocompatibility, safety, and toxicity of polyanhydride polymers actually available for biomedical applications.

The polyanhydrides constitute so far the only class of surface-eroding polymers approved for clinical trials by the Food and Drug Administration. In one of the first detailed reports on biocompatibility and toxicology of polyanhydrides published in 1986, several accepted criteria and tests to evaluate new biomedical materials were used to assess the safety of polyanhydrides [90]. In this study, poly [bis(p-carboxy-phenoxy) propane anhydride [P(CPP)], poly (terephtalic acid anhydride) (PTA), and their copolymers were tested. Neither mutagenicity nor cytotoxicity was associated with the polymers or their degradation products, as evaluated by mutation assays [90]. The products also gave a negative response in an in vitro teratogenicity test [91]. The tissue response of these polyanhydrides was studied by subcutaneous implantation in rats and in the cornea of rabbits. The polymers did not provoke inflammatory responses in the corneas over a six week implantation period [90]. The authors also reported no evidence of inflammatory cells after the subcutaneous implantation in rats over a six month period, and only slight tissue encapsulation by layers of fibroblastic cells was observed [90]. Growth of two types of mammalian cells in tissue culture was also not affected by the polyanhydride polymers; both the cellular doubling time and cellular morphology were unchanged when either bovine aorta endothelial cells or smooth muscle cells were grown directly on the polymeric substrate.

Additional evidence of polyanhydride biocompatibility was provided from 8 weeks subcutaneous implantation in rats of high doses of the 20:80 copolymer of CPP and SA. Prenecropsy examination of all rats revealed no clinical evidence of induced changes in physical appearance or activity due to implantation of the polymer [92]. Histological evaluation indicated relatively minimal tissue irritation with no evidence of local or systemic toxicity [92]. Systemic response to the polymer was evaluated by monitoring of blood chemistry and hematologic values, and by comprehensive examination of organ tissues. Both methods revealed no significant response to the polymer [92].

Since the PCPP-SA polyanhydride copolymer was designed to be used clinically to deliver an anticancer agent directly into the brain for the treatment of brain neoplasms, in vivo safety evaluations and brain biocompatibility were assessed with rats [93], rabbits [94], and monkeys [95]. In the rat brain study, the tissue reaction of the polymer (PCPP-SA 20:80) was compared to the reaction observed with two standard materials used in surgery, which have been extensively studied. These materials are Gelfoam (absorbable gelatin sponge), and Surgicel (oxidized cellulose absorbable hemostat commonly used in brain surgery). Histological evaluation of the tissue demonstrated a small rim of necrosis around the implant, and a mild to marked cellular inflammatory reaction limited to the area immediately adjacent to the implantation site, slightly more pronounced than Surgicel at the earlier time points, but noticeably less marked than Surgicel at the later times [93]. The reaction to Gelfoam was essentially equivalent to that observed in control rats.

In the rabbit brain safety study using P(CPP-SA) 50:50 copolymer, even less of an inflammatory reaction was observed, and the polymer was essentially equivalent to Gelfoam [94]. In a similar brain biocompatibility study conducted in monkeys, no abnormalities were noted in the computer tomography scans and magnetic resonance images, nor in the blood chemistry or hematology evaluations [95]. No systemic effects of the implants were observed on histological examinations of any of the tissues tested [96]. No unexpected or untoward reactions to the treatments were observed.

Recently, new classes of polyanhydrides have been synthesized and are undergoing extensive preclinical testing, including a wide range of toxicity and biocompatibility studies. Examples of these new polyanhydrides materials are polymers of sebacic acid (SA), and 1:1 copolymers of SA with fatty acid dimer [P(FAD:SA)], fumaric acid [P(FA:SA)], and 20:80 copolymer of SA with isophthalic acid [P(IPA:SA)] [97]. These new polyanhydrides were implanted intramuscularly, subcutaneously and in the cornea of rabbits, and ocular and muscle irritation studies were performed. The results were compared to those obtained with the previously mentioned standard materials Gelfoam and Surgicel, and with Vicryl, a synthetic absorbable suture [97]. No significant clinical signs or abnormalities of the incision sites were observed during the study period (4 weeks). No meaningful differences could be seen in reaction between the various polymer implants tested and the control materials [97]. In the cornea study, no evidence of inflammatory response was observed with any of the implants at any time. On an average, the bulk of the polymers disappeared completely between 7 and 14 days after implantation [97].

The new FAD-SA anhydride copolymers were used to deliver water soluble anticancer agents such as, carboplatin, 4-hydroperoxycyclophosphamide (4HC), and methotrexate to the rat brain [98]. In vitro release kinetics studies conducted on polymer matrices containing these drugs have shown sustained release of the drugs over a period of 3 to 5 weeks. This is particularly important for the delivery to the brain in that the blood-brain barrier effectively blocks most water soluble drugs from entering the brain. The safety and biocompatibility of these polymers in the rat brain were compared with that of Surgicel and Gelfoam. The localized inflammatory response generated by the p(FAD-SA) polymer was comparable to that of Surgicel, but more pronounced than the reaction evoked by Gelfoam [99].

With these preclinical toxicology and biocompatibility studies carried out in animals having demonstrated both the efficacy and safety of the polyanhydrides, studies involving these materials moved toward human clinical use. In 1987, the Food and Drug Administration approved experimental use of these polyanhydrides in humans, under an Investigational New Drug clinical trial application. A Phase I/II clinical trial of 21 patients in five U.S. hospitals was carried out

showing the safety and no toxicity of these polymers either clinically or patho-
logically, and extending patient life time beyond conventional drug treatment
[23]. In these clinical studies, a polyanhydride dosage form (Gliadel) consisting
of wafer polymer implants of 20:80 poly(CPP-SA) and containing the
chemoherapeutic agent Carmustine (BCNU) were used for the treatment of
glioblastoma multiforme, a universally fatal form of brain cancer. In these
studies, up to eight of these wafer implants were placed to line the surgical cavity
created during the surgical removal of the bulk of the brain tumor in patients
undergoing a second operation for surgical debulking of either a Grade III or IV
anaplastic astrocytoma. Following surgery the BCNU is then released directly
into adjoining tissues that may contain cancer cells not removed during surgery.
The safety of this polyanhydride copolymer implanted into these patients has
been demonstrated. No central or systemic side effects of doses of BCNU which
would produce marked effects on the hemopoietic system when injected intra-
venously were observed. No adverse reactions to the BCNU wafer treatment
itself were found. Based on these results, further studies designed to measure the
efficacy of this approach to the treatment of brain cancer is currently underway
in a Phase III study in 32 U.S. and Canadian hospitals [95, 96].

7 Applications

Applications of Polyanhydrides have been reviewed [1, 22]. A comprehensive
review on polyanhydride applications is in preparation [6].

Anticancer agents were incorporated in polyanhydride wafers and used for
site-specific chemotherapy for the treatment of brain tumors [22, 91–102].
BCNU has been the primary drug in this application. In the past 3 years
investigations have expanded to new polymers and other drugs such as 4HC,
cisplatin, carboplatin, and several alkaloid based drugs to develop a better
system for the treatment of brain tumors [100]. Carboplatin incorporated in
poly(FAD-SA), prepared by mixing the drug in the melted polymer, showed
promising result in treating brain tumors in laboratory animals [102]. The same
polymer has been used for the delivery of gentamicin sulfate for the treatment of
osteomyelitis [21]. Gentamicin was released for more than two weeks both in
vivo and in vitro. This device in a form of linked beads is now considered for
human clinical trials. The effect of long term glutamic acid stimulation of
trigeminals motoneurons, using poly(FAD-SA) microspheres was studied. The
study was undertaken to determine the role of glutamate in possible growth
disorders of the craniofacial skeleton. Rats receiving glutamate showed pronounced skeletal changes in the snout region, showing that sustained release of
glutamic acid in vivo can effect the developing skeleton in growing rats [103].
Poly(FA-SA) microspheres containing an antiinflammatory drug dispersed in
a biocompatible organic vehicle have been used as eye drops for extended drug

action. The microspheres were not irritating to the rabbit eye and lasted in the eye for 6 to 10 hours. The polymer was completely degraded and dissolved within 24 hours both in vitro and in vivo.

8 Conclusion and Future Directions

The chemistry of polyanhydrides has been considerably developed and a large selection of polymers are available for the applications at hand. We now have a better understanding on how these polymers degrade and release the drug under physiological conditions. The drug release from these polymers is controlled by polymer hydrolysis and by diffusion, where the ratio between the two processes depends largely on the chemistry of the polymer, the fabrication method, the geometry of the device and the drug properties. Extensive toxicology information on a range of polyanhydrides is available, which indicate that most polyanhydrides prepared from pure common diacids are biocompatible. Various fabrication and preparation methods are available, and some polymers and devices have been produced in large quantities in a cost effective process. Future studies should concentrate on finding new medical applications using polyanhydrides. These polymers are particularly useful in local delivery of drugs for periods of several weeks in a form of an implant or film.

Acknowledgments. This work was supported in part by The Eliahu and Tatiana Leszczynski Research Foundation and The Robert szold institute for Applied Science of the P.E.F Israel Endowment Funds Inc. Dr. Domb is affiliated with the Blum Center for Pharmacy, The Hebrew University of Jerusalem.

9 References

1. Chasin M, Langer R (eds) (1990) Biodegradable polymers as drug delivery systems. Marcel Dekker, New York
2. Tarcha PJ (1991) Polymers for controlled drug delivery. CRC Press, Boca Raton, Florida
3. Domb A, Amselem S, Maniar M (1992) In: Dumitriu S (ed) Polymeric biomaterials. Marcel Dekker, NY (in press)
4. Leong KW, Domb AJ, Langer R (in press) Polyanhydrides, Encycl Biotech
5. Domb A, Langer R, Maniar M (1992) Site specific delivery of drugs using absorbable Polymers. Polym Adv Tech special issue (in press)
6. Domb A, Amselem S, Langer R, Maniar M (1992) In: Shalaby S (ed) Designed to degrade biomedical polymers. Carl Hanser, New York (in preparation)
7. Bucher JE, Slade WC (1909) J Am Chem Soc 31:1319
8. Hill JW (1930) J Am Chem Soc 52:4110
9. Hill JW, Carothers HW (1932) J Am Chem Soc 54:1569
10. Hill JW, Carothers HW (1933) J Am Chem Soc 55:5023
11. Conix A (1958) J Polym Sci 29:343
12. Conix A (1957) Makromol Chem 24:76
13. Yoda N (1963) J Polym Sci Part A 1:1323

14. Yoda N (1962) Makromol Chem 55:174
15. Yoda N Miyake A (1959) Bull Chem Soc Japan 32:1120
16. Yoda N (1962) Kobunshi Kagaku (Chemistry of High Polymer) 19:495 (Chem Abst 61:13472h, 1964)
17. Yoda N (1962) Makromol Chem 56:36
18. Yoda N (1959) Makromol Chem 32:1
19. Rosen H, Langer R (1983) Biom. 4:131
20. Langer R et al. (1989) US Pat 4,886,870
21. Fait C, Rock M, Maniar M, Domb A (October 1991) ACCP meeting in Atlanta
22. Chasin M, Lewis D, Langer R (February 1988) Biopharm Manufacturing
23. Brem H, Mahaley MS, Vick N, Black K, Schold C, Burger PC, Friedman AH, Ciric IS, Eller TW, Cozzens JW, Kenealy JN (1991) J Neurosurg 74:441
24. Nelson CL, Hickmon SG, Skinner RA (Feb 1992) 38th annual meeting, Orthop Res Soc
25. Leong KW, Simonte V, Langer R (1987) Macromolecules 20:705
26. Cottler RJ, Matzner M (1967) Chemisch Weekblad 11:133
27. Polyanhydrides in Encyc. Polym Sci Tech John Wiley, NY, NY, (1969) 10:630
28. Domb AJ, Langer R (1987) J Polym Sci Polym Chem 25:3373
29. Domb AJ (1988) US Pat 4,789,724
30. Albertsson A, Lundmark S (1988) J Macromol Sci Chem A25:247
31. Kricheldorf HR, Lubbers D (1990) Macromol Chem, Rapid commun 11:83
32. Gupta B US patent 4,868,265
33. Knobloch JO, Ramirez F (1975) J Org Chem 40:1101
34. Rohdes MS (1977) Insulation/Circuits December:39
35. Albertsson A, Lundmark S (1990) J Macromol Sci Chem A27:397
36. Subramanyam R, Pinkus AG (1985) J Macromol Sci Chem A22:23
37. Domb AJ, Ron E, Langer R (1988) Macromols 21:1926
38. Hideo S, Awashi Y (1964) Kogo Kagaku Zasshi 67:1444, 1449
39. Domb A, Ron E, Giannos S, Flores C, Kim R, Dow L, Langer R (1987) 14th International Symposium on the Controlled Release of Bioactive Materials, Toronto
40. Domb A, Mathiowitz E, Ron E, Giannos S, Langer R (1991) J Polym Sci Part A: Polymer Chemistry 29:571
41. Michael A, Bucher JE (1895) Ber 28:2511
42. Shopov I (1969) Vysokomol Soyed 9:2012
43. Thompson RM (1973) US patent 3,766,145
44. Domb A (1990) Biomaterials 11:680
45. Naoya Ida (1964) Japanese Pat 19,244 (Chem Abst 62:9260g)
46. Hartmann M, Schultz V (1989) Makromol Chem 190:2133
47. Gonzalez JI, de Abajo J, Gonzalez-Babe S, Fontan J (1976) Angew Makromol Chem 55:85
48. de Abajo J, Gonzalez-Babe S, Fontan J (1972) Angew Makromol Chem 19:1259
49. Staubli A, Mathiowitz E, Lucarelli M, Langer R (1991) Macromols 24:2283
50. Staubli A, Ron E, Langer R (1990) J Am Chem Soc 112:4419
51. Staubli A, Mathiowitz E, Langer R (1991) Macromols 24:2291
52. Domb AJ, Gallardo CF, Langer R (1989) Macromols 22:3200
53. Domb AJ (1992) Macromols 25:12
54. Domb AJ (1991) US Pat 4,997,904
55. Ziegast G (1988) US Pat 4,792,598
56. McIntyre JE (1964) British Pat 978,660
57. Domb AJ, Maniar M (1992) J Polym Sci, Part A: Polymer Chemistry (in press)
58. Xie X, Adam M, Maniar M, Domb AJ (1991) Pharm Res 8:193
59. Domb AJ, Haffer A, Xie X, Maniar M (1991) ACS, MARM 25th Conf., Montchanic, DE
60. Maniar M, Xie X, Domb AJ (1990) Biomaterials 11:690
61. Ron E, Mathiowitz E, Mathiowitz G, Domb AJ, Langer R (1991) Macromolecules 24:2278
62. Laurencin CT, Domb AJ, Morris C, Harris M, Lopez L, Langer R (1990) Proceed. Intern. Symp. Control Rel Bioact Mater 17:466
63. Davies MC, Tudor AM, Hendra PJ, Domb AJ, Langer R (1990) Proceed Intern Symp Control Rel Bioact Mater 17:236
64. Tudor AM, Charch S, Domb AJ, Hendra PJ, Langer R, Melia CD, Davies MC (1992) J Polym Sci (in press)
65. Davies MC, Domb A, Lynn RAP, Khan MA, Paul A, Langer R (May 1991) ISPAC, 4th International Symposium Baltimore MD

66. Davies MC, Khan MA, Domb A, Langer R, Watts JF, Paul A (1991) J Apl Polym Sci 42:1597
67. Mathiowitz E, Kline D, Langer R (1990) Scanning Microscopy 4:329
68. Davies MC et al. (1989) Surf Interface Anal 14:115
69. Domb A, Langer R (1989) Macromols 22:2117
70. Sawada H, Yasue A (1964) Kogyo Kagaku Zasshi 67:1444
71. Hinterhofer O (1980) Macromol Chem 181:83–88
72. Goodman I, McIntyre JE (1964) Brit Pat 978,715
73. Matsuda Y, Nakahama Y, Yasue T, Swada H, Suzuki S Japan Patent 10,944.
74. Ida N (1964) Japan Pat 19,244
75. Carothers WH, Hill HW (1932) J Am Chem Soc 54:1579
76. Loucheux MH, Banderet A (1960) J Polym Sci 48:405
77. Windholz TB US pat 3,200,097
78. Hopfenberg HB (1976) In: Paul DR, Harris FW (eds) Controlled release polymeric formulations. Am Chem Soc, Washington, D.C. chap 3
79. Heller J (1984) Crit Rev Ther Drug Carrier Syst 1:39
80. Baker R (1987) In: Controlled release of biologically active agents, John Wiley, New York, p 84
81. Heller J, Sparer RV, Zentner GM (1990) In: Chasin M, Langer R (eds) Biodegradable Polymers as Drug Delivery Systems, Marcel Dekker, New York, p 121
82. Chasin M, Domb A, Ron E, Mathiowitz E, Langer R, Leong KW, Laurencin C, Brem H, Grossman S (1990) In: Chasin M, Langer R (eds) Biodegradable polymers as drug delivery systems, Marcel Dekker, Inc, New York, p 43
83. Leong KW, Brott BC, Langer R (1985) J Biomed Mater Res 19:941
84. Maniar M, Xiadong X, Domb A (1990) Pharm Res 7:179
85. Wilson TD (1990) In: Analytical profiles of drug substances. Academic, New York, 19:59
86. Higuchi T (1963) J Pharm Sci 52:1145
87. Maniar M, Shah J, Domb A (1991) Pharm Res 8:197
88. Shih C, Higuchi T, Himmelstein KJ (1984) Biomaterials 5:237
89. Nguyen TH, Himmelstein KJ, Higuchi T (1985) Int J Pharm 25:1
90. Leong KW, D'Amore P, Marletta M, Langer R (1986) J Biomed Mat Res 20:51
91. Braun AG, Buckner CA, Emerson DJ, Nichinson BB (1982) Proc Natl Acad Sci USA 79:2056
92. Laurencin C, Domb A, Morris C, Brown V, Chasin M, McConell R, Lange N, Langer R (1990) J Biomed Mat Res 24:1463
93. Tamargo RJ, Epstein JI, Reinhard CS, Chasin M, Brem H (1989) J Biomed Mater Res 23:253
94. Brem H, Kader A, Epstein JI, Tamargo RJ, Domb AJ, Langer R, Leong KW (1989) Selective Ther 5:55
95. Brem H, Ahn H, Tamargo RJ, Pinn M, Chasin M (1988) Am Assoc Neurol Surg 349
96. Brem H, Tamargo RJ, Pinn M, Chasin M (1988) Am Assoc Neurol Surg 381
97. Rock M, Green M, Fait C, Geil R, Myer J, Maniar M, Domb A (1991) Polym Preprints 32:221
98. Domb A, Bogdansky S, Olivi, A., Judy, K., Dureza C, Lenartz D, Pinn ML, Colvin M, Brem H (1991) Polymer Preprints 32
99. Brem H, Domb A, Lenartz D, Dureza C, Olivi A, Epstein JA (1992) J Control Rel 19:325
100. Domb A, Maniar M, Bogdansky S, Chasin M (1991) Critic Rev Therap Drug Carrier Syst 8:1
101. Langer R (1991) J Control Rel 16:53
102. Domb A, Olivi A, Judy K, Pinn LM, Ewend MG, Goodman JH, Brem H (1991) ACS meeting, Philadelphia, PA
103. Hamilton-Byrd EL, Sokoloff AJ, Domb AJ, Terr L, Byrd KE (1992) Polym Adv Tech (in press)

Received 12 August 1992

Microencapsulation of Live Animal Cells Using Polyacrylates

Michael V. Sefton[1], William T. K. Stevenson[2]
[1] Department of Chemical Engineering and Applied Chemistry,
University of Toronto, Toronto, Ontario, M5S 1A4, Canada
[2] Department of Chemistry, Wichita State University,
Wichita, Kansas, 67208, USA

Organ and tissue transplantation, traditionally supported by immunosuppressant therapy, has entered a new era through the use of semipermeable microcapsules to circumvent rejection pathways by selective isolation of the implant from the hosts immune system. The bewildering structural and behavioral latitude and proven biocompatibility of synthetic polymer based systems make them materials of choice for this application. Herein are described a number of thrusts towards the development of permselective microcapsule based systems for the eventual treatment of Type I diabetes, which are based on the structurally diverse and biocompatible methacrylate family of polymers. Thoroughly explored concepts based on the use of uncharged and water insoluble hydroxyalkyl methacrylates are reported, along with studies of polyelectrolyte complex based systems. The preliminary development of systems based on the surface modification of alginate by precipitation of a cationic, emulsion or through formation of a covalently crosslinked network are detailed along with others based on the formation of a cohesive precipitate of alginate stabilized polymethacrylate emulsion. Encapsulated cells include erythrocytes, fibroblasts, lymphoma and CHO cells, and islets of Langerhans. Future work will be geared towards an understanding of interactions between the host and the encapsulated cells and the long term maintenance of normoglycemia in diabetic mammals.

1 Why Microencapsulation

By making use of the natural control mechanism, live animal cells can be used as drug delivery systems to provide biologically active agents at variable rates in direct response to natural physiological stimuli. Insulin delivery by pancreatic islets in response to glucose (for diabetes) or dopamine release from adrenal cells in response to potassium (for Parkinson's disease) are but two examples of this concept, which has more in common with transplantation technology than with conventional drug delivery systems. Other agents or situations that can be approached in this way include blood coagulation factors (hemophilia), growth factors (wound healing), interleukins (cancer and immune disorders), and an artificial liver.

The primary limitation of cell transplantation or drug delivery from cells is the immune mediated rejection phenomenon. Our approach to this problem is to microencapsulate the cells in a semipermeable membrane which prevents the large (> 150 kDa) antibodies from coming into contact with the cells ("immunoisolation"). The membrane is permeable however to the low molecular weight nutrients, metabolites and more importantly, the active agents (e.g., dopamine) or intermediate molecular weight (< 50 kDa) protein products (e.g., insulin, growth factors, etc.). These encapsulated cells can then be transplanted into the appropriate target site (peritoneal cavity, brain, etc.) to release drug as needed using the natural physiological stimuli. Hence the capsule wall must also be biocompatible and this is the advantage of using polyacrylates.

Based on the pioneering work of Chang (1) numerous biologically active species (e.g., enzymes and whole microbial or plant cells) have been encapsulated or immobilized in polymeric matrices. Mammalian cells are more difficult to encapsulate than these other species because they have an active metabolism and a fragile cell membrane which must be preserved during encapsulation.

Although other water soluble polymers have also been used, viable mammalian cells have been encapsulated most frequently in stabilized calcium alginate. Sun pioneered this process to microencapsulate pancreatic islets (2). The thin calcium alginate/polylysine capsule wall was able to protect the transplanted islets from the immune system of the host rats (chemically induced diabetes) for 2–3 weeks initially and with further refinement for as much as two years (3, 4). The feasibility of this approach has been shown in various animals (5–8). The long term success in rats has been questioned, however, because of the problem of pancreas regeneration in chemically induced diabetic animals (9).

1.1 Applications

Pancreas transplantation, with immunosuppression therapy, has become clinically acceptable (10) at least for those who also receive kidney transplants (despite the nephrotoxicity of cyclosporin). Islet transplants are simpler surgically, reduce the possible complications due to exocrine tissue, may be immunologically manipulated (11, 12) and open up the potential for xenografts. Islet transplants are beginning to be successful clinically (13, 14); in animals even reversal of the degenerative complications have been observed (15, 16). The major difficulty with studies in larger animals (e.g., dogs) or humans has been the lack of islet tissue. Methods for isolating large numbers of porcine (17) or other islets have been devised (although reproducibility is a problem) as have cryopreservation protocols (18), the latter allowing for long term storage of islets.

As an alternative to immune suppression or immune alteration, immunoisolation units have been used to separate islets or β-cells from the host's immune system. Such "hybrid artificial organs" have taken a variety of forms but the most popular has been the use of hollow fibers (19–23) with (or without, 24, 25) direct blood contact with the fiber. While glucose and insulin freely diffuse across the membrane, the membrane prevents the antibodies from rejecting the donor tissue. Microencapsulation, with a high surface to volume ratio and by avoiding contact with blood, provides a means of circumventing the problems of coagulation and vascular anastomoses. Microencapsulated cells could be injected or implanted intraperitoneally (ip), for example, although other sites may be feasible or preferred. The issues of retrievability and the number of islets needed (as many as $3 \times 10^5 - 10^6$ [26]) and capsule size, will influence the choice of implant site.

Immunoisolation may also have an advantage in preventing the autoimmune destruction of the transplanted islets (27) by separating them from the host (T-cells or autoantibodies) by the permselective membrane. Although the detailed mechanism of islet cell damage is unknown, evidence from spontaneously diabetic BB rats indicates that the rapid failure of islet transplants, even in twins (28), may be due to a recurrence of autoimmune disease rather than the conventional transplant rejection phenomenon (29).

Microencapsulated cells are potentially useful in other situations such as the treatment of Parkinson's disease (dopamine delivery), liver failure (hepatocytes), wound healing (growth factor delivery) and immune modulation/tumor therapy (interleukin 2 delivery). Other neurological applications under consideration include chronic pain relief (30) and Huntington's and Alzheimer's diseases (31). Such technology may also play a similar role in gene therapy to allow the use of genetically corrected/modified cells.

Alginate-polylysine has been used to encapsulate hepatocytes (32–34), parathyroid cells (35) and growth hormone transfected fibroblasts (36). Poly (acrylonitrile/vinyl chloride) (PAN/PVC) "macrocapsules" have been used with PC12 (37, 38), embryonic mesencephalon tissue (39), thymic epithelial cells (40), adrenal chromaffin cells (41) and islets (25) using preformed hollow fibers or more recently coextrusion techniques (41) similar to those we have developed; microcapsules cannot be made since DMSO is used as the solvent. All these studies have concluded from the maintenance of viability of the islets or cells that immunoprotection provided by the capsule membrane was compatible with normal metabolic diffusion processes. A practical concern with the use of PC12 cells (and other cell lines) is the ultimate production of a tumor should the immunoisolation barrier be breached. There are other regulatory issues related to the use of primary tissue.

1.2   Intracapsule Microenvironment: Transport Limitations

Minimization of the diffusion limitations associated with microencapsulation is critical to the success of encapsulation. The capsule wall must exclude immuno-globulins. On the other hand, by restricting the diffusion of lower molecular weight solutes the capsule wall can adversely affect cell behavior in a number of ways. Because of depletion of a particular nutrient (e.g., oxygen) cells may grow at a lower, diffusion limited rate or the centre of a cell cluster (e.g., islet) may become necrotic. The presence of a thick, stagnant liquid layer (largely water) between the islet and the capsule wall would be expected to exacerbate this problem so that viability should be enhanced in smaller capsules. Alternatively the capsule wall can have a diffusion limiting effect on the secretion of a product (e.g., insulin) or an increased time lag in the response to a change in external stimulus (e.g., glucose concentration). The effect of alginate-polylysine capsule size on insulin secretion was measured (42) but the results are now considered to be compromised by differences in capsule structure. The behavior under diffusion limiting conditions may be more complex still. Oxygen may turn out to have little effect on central necrosis, but a low pO_2 at the centre of a cell cluster is likely to have an adverse effect on secretion by the hypoxic portion of the cluster. A direct relationship between hypoxia and low insulin secretion rates has been quantified recently (43). Hence small and large islets may behave differently upon transplantation (44) with the capsule, through the alteration of both local pO_2 and intracapsule insulin concentration, modulating the overall behavior. The low peritoneal pO_2 may also exacerbate the problem for ip implanted capsules.

For growing cells (e.g., PC12, HepG2) diffusion constraints appear to result in the cells growing in a shell (~ 100 μm thick) immediately below the capsule wall, with a necrotic central core. This is similar to what occurs with tumor spheroids (45). The differences among the cells is presumed to be related to the gradients of nutrients and growth factors into the spheroid and metabolites and toxic products (from dead cells) out of the spheroid. The details of these gradients are not known but this model may be a useful means of predicting microencapsulated cell behavior.

1.3   Biocompatibility

Biocompatibility is defined as "the ability of a material to induce the appropri-ate host response in a particular application" (46). Here the appropriate host response implies the absence or at least minimization of tissue reaction and fibrous capsule formation (negligible diffusion resistance), macrophage activation (negligible IL1, cytokine or cytotoxic agent release) (47) and capsule biodegradation. Although no fibrous capsule may ultimately be desirable, this may not be necessary if the (presumably thin) fibrous capsule is a negligible additional resistance to diffusion (relative to the capsule itself) or if it or the surrounding tissue is vascularized. Biocompatibility is not a "yes" or "no" property and care must be taken in characterizing a material/implant as biocompatible. Although the presence of cells may be thought to influence the tissue reaction to the capsule wall through the shedding of antigens or the products of dying cells, assessments of PAN/PVC macrocapsules at explantation show no difference in tissue reaction between cell-containing and empty capsules (38). Although there are some *in vitro* assays for specific aspects related to soft tissue biocompatibility (48), normally biocompatibility is assessed qualitatively by histological sections after explantation.

It appears axiomatic that a prerequisite for biocompatibility is the stability of the capsule *in vivo*. Hence the capsule wall must first of all be water insoluble. Slowly dissolving, or even degrading polymers, will elicit a continuing inflammatory response which will act as a transport barrier to drug delivery. Capsule dissolution will ultimately expose the transplanted cells (dead or alive) to the host and presumably cause the immune response which was to be avoided in the first place. For this reason, the reported "biocompatibility" of the alginate-polylysine capsule is controversial. Both alginate and polylysine are water soluble polymers which individually would be expected to elicit severe inflammatory or foreign body responses. However, the polyelectrolyte complex formed during encapsulation is clearly more stable (less soluble) than the individual components and, based on implant experiments, is more biocompatible than the individual components. Presumably, the reactivity of the individual components has been altered by the formation of the capsule wall complex. Whether these reactivities have been fully neutralized or are sensitive to some, as yet unknown, aspect of the encapsulation process is not known. Evidence (48–50) indicating the failure of alginate-polylysine encapsulated islets transplanted in BB rats and NOD mice, in contrast with the results of Sun (7) and others (5, 51) support these reservations regarding the biocompatibility of alginate-polylysine. Differences in biocompatibility were noted between empty and cell containing capsules in NOD mice (48, 49) with one group (50) concluding that the reaction was directed to the encapsulated islets and not the capsules.

However, these differences may reflect the inherent difficulty of producing reproducible cell-containing capsules. Published reports have not been very clear about interanimal or batch-to-batch variability in tissue response or even variability within a single animal. Free floating capsules may behave or be reacted to differently than those imbedded in fat tissue. Alginate purity, source and chemical heterogeneity may also play a role in the disparate reports. High guluronic acid containing alginates are purported to have a better biocompatibility than other alginates (52). Others suggest a lack of biostability with some alginate-polylysine capsules disintegrating after 3 months; this would lead to significant, yet undetected foreign body reactions when only intact capsules are recovered at explantation. Unfortunately the fraction of recovered capsules relative to those implanted is rarely reported. Sun's "new" capsules are smaller and stronger (53) and this may account for their better performance in more recent reports. The limited biocompatibility of alginate-polylysine has led another group to work on polyethylene glycol modified alginate-polylysine capsules (54). These discrepancies clearly warrant further study of the alginate-polylysine system.

Nevertheless we have chosen an alternative approach: microencapsulation using synthetic polyacrylates which may be stronger and more biocompatible (or at least have a more reproducible biocompatibility). Most of the effort has been with a particular water insoluble synthetic copolymer (HEMA-MMA), although other polyacrylates (water soluble or insoluble) have been used. In general, two approaches have been used: complex coacervation with water soluble polycations and polyanions to generate an insoluble complex and simple coacervation with a water insoluble polymer precipitated from an organic solvent.

2 Chemistry of Polymethacrylates

2.1 Synthesis

Hydrogel chemistry is dominated by the methacrylate family of monomers for good reason; they are cheap, readily available, and are susceptible to simple free radical polymerizations. Several reviews are available (55, 56). Some common methacrylate based monomers are included in Fig. 1. The methacrylate building block is preferred over the corresponding acrylate because of its better hydrolytic stability (57). Polymer properties may be tailored through appropriate choice of the pendant side chain which may be acidic (in the case of methacrylic acid: MAA), basic (dimethylaminoethyl methacrylate: DMAEMA), uncharged and hydrophobic (methylmethacrylate: MMA), or uncharged and hydrophilic (2 - hydroxyethyl methacrylate: HEMA). All these have been used to make capsules. Other monomers that may prove to be of interest in the future include the di- or tri- ethylene glycol methacrylate or the methoxy polyethylene glycol monomethacrylate series (58). Polymer properties may be continuously varied over a wide spectrum by co-polymerizing these and other monomers. Polymerizations may be performed in polar organic solvent such as ethanol in which instance azobisisobutyronitrile (AIBN) is often the preferred initiator. Polymerizations may also be performed in aqueous media using a persulfate based redox system. Azobismethylisobutyrate (AMIB) can be used to avoid nitrile groups in the final polymer. Hydrophobic impurities may be removed by solution precipitations

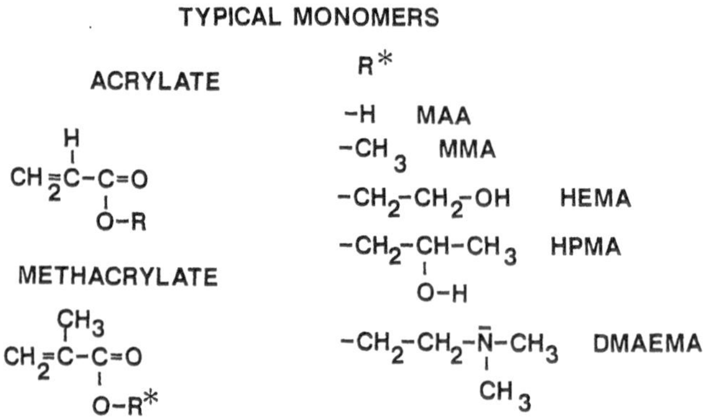

TYPICAL MONOMERS

ACRYLATE   R*

$$CH_2{=}C{-}C{=}O$$
with H above and O–R below

-H   MAA

-CH_3   MMA

-CH_2-CH_2-OH   HEMA

-CH_2-CH-CH_3   HPMA
     |
    O-H

METHACRYLATE

$$CH_2{=}C{-}C{=}O$$
with CH_3 above and O-R* below

-CH_2-CH_2-N̄-CH_3   DMAEMA
            |
           CH_3

Fig. 1. Readily available methacrylate based monomers. MAA: methacrylic acid, MMA: methyl methacrylate, HEMA: 2-hydroxyethyl methacrylate, HPMA: one isomer in commercial 2-hydroxypropyl methacrylate, DMAEMA: dimethylaminoethyl methacrylate

into, or extraction by a nonpolar solvent such as hexane. Hydrophilic impurities may be removed by simple dialysis against distilled water or buffer if the polymer is water soluble, or, if insoluble, by precipitation into or extraction into aqueous solution through a dialysis bag. Water soluble polyelectrolyte derivatives may then be protonated or deprotonated by acid-base titration or by passage through an appropriately conditioned ion exchange column (59). Dissolved or water swollen polymer may be recovered as a friable solid by freeze drying the solution or gel. Such purification steps are an essential part of the synthesis process, for though methacrylate based polymers are exceptionally well tolerated by the mammalian host, residual monomer impurities invariably elicit an inflammatory response (60).

Polymer synthesis is complicated if the diol or higher adduct used to prepare the hydroxyalkyl methacrylate, condenses with more than one molecule of methacrylic acid to form a di or higher methacrylate crosslinking agent impurity. For example, technical grade HEMA is contaminated with a few percent of ethylenglycol dimethacrylate (EGDMA), and hydroxypropyl methacrylate (HPMA) by propylenglycol dimethacrylate (PGDMA). Such impurities may be quantified by gas liquid chromatography (61). The volatility difference between HPMA and PGDMA is such that purification may be achieved by careful vacuum distillation. In contrast, EGDMA appears to azeotrope with HEMA and so may be reduced but not eliminated by simple distillation (62).

EGDMA is a preferred crosslinking agent for the production of crosslinked hydrogel, so that its presence is rarely a problem. In contrast, residual crosslinker has enormous influence on the molecular weight and branch content of thermoplastic polymer (63). A partial removal of EGDMA may be achieved through a combination of distillation and extraction with hexane (64). A complete removal of EGDMA may be achieved through the use of column chromatography (62). We have shown that reduction of EGDMA to acceptable levels in large batches of HEMA may be routinely achieved through a combination of distillations and extractions with hexane and glycerol triolate (61). In this instance, edible grade corn oil has been shown to be more acceptable than industrial grade glycerol triolate. Experiments with MMA and added corn oil have determined that the latter is neither a chain transfer reagent or a coreactant in free radical polymerizations (61).

Simple test tube bulk polymerizations of HEMA, HPMA, and mixtures containing either are sufficient to determine that these very viscous monomers (compare MMA: 0.51 cSt, HPMA: 5.23 cSt, both at 35 °C) autoaccelerate at very low conversions. Viscosity is most conveniently reduced by dilute solution polymerization in a polar or hydrogen bonding solvent such as ethanol (61). The polymer may be conveniently recovered by coagulation in low boiling point petroleum ether. In most instances, methacrylate based monomer pairs appear to form random co-polymers, leading to a minimal compositional drift even in high conversion polymerizations. Of course, the quantitative drop in monomer concentration during such a reaction will considerably broaden the molecular weight distribution, but such could be minimized through the use of continuous stirred tank reactors or continuous monomer feed systems (65).

These generalizations may be illustrated by example (61). Consider the experiments outlined in Table 1. PHEMA, PMMA and copolymers of the two were prepared by dilute solution polymerization. Copolymer compositions (Table 1A [parentheses]), obtained by application of proton Nuclear Magnetic Resonance spectroscopy (as detailed later in the discussion), are almost identical to the monomer feed ratios, despite high conversions. The latter increase through series 3 to 1, due to the square root relationship between initiator concentration and instantaneous rate of polymerization (63). Close to random co-polymerizations have been observed also in monomer mixtures containing

Table 1. Preparation and characterization of some thermoplastic methacrylate based hydrogel polymers
(A) Monomer charges and polymer composition by NMR (in parentheses)

series	mole fraction of HEMA in %				
	a	b	c	d	e
1	0(0)	25(24)	50(51)	75(73)	100(100)
2	0(0)	25(26)	50(49)	75(79)	100(100)
3	0(0)	25(27)	50(51)	75(79)	100(100)

(B) Polymer yields (% of monomer charge) and intrinsic viscosities (L/kg, in parentheses)

series	a	b	c	d	e
1	71.4(12.7)	70.2(18.6)	57.1(22.7)	69.0(23.5)	80.4(31.9)
2	29.0(22.6)	30.1(32.7)	35.0(65.4)	33.8(71.5)	52.6(124.5)
3	12.0(72.6)	15.1(107.3)	16.4(162.6)	16.2(439)	20.3(1124)

Reaction conditions: 10% monomer HEMA and/or MMA (w/v) in ethanol; 70 ± 0.05 C; 4 h mole fraction in % AIBN initiator: Series 1: 1%; Series 2: 0.1%; Series 3: 0.01%.

HEMA, MMA, and MAA or DMAEMA (66, 67). Polymer molecular weight control was attempted by adjustment of initiator concentrator through the presumed inverse square root relationship between initiator level and kinetic chain length, (63) and this was reflected in measured intrinsic viscosities. Comparisons across a series are made difficult by changing Mark-Houwink-Sakurada coefficients. However direct comparisons may be made between polymers of similar composition, for example 1c, 2c, and 3c, to show an expected increase in hydrodynamic volume from series 1 through series 3.

Compositional drift becomes a deciding factor if methacrylates are copolymerized with other classes of monomer. Even such similar systems as methyl methacrylate (M1) and methyl acrylate (M2) copolymerize with reactivity ratios r1 and r2 close to 2.2 and 0.4 respectively. An even more extreme case was encountered during attempts to prepare a water soluble polyepoxide by copolymerizing glycidyl methacrylate (GMA) with hydrophilic monomers by high temperature free radical polymerization (68). Preliminary experiments had shown that acidic comonomers such as MAA formed a cross linked product, that basic comonomers led to a thermoplastic which crosslinked on storage, and that protic solvents quantitatively ring-opened the epoxide. A comonomer had to be found which was sufficiently hydrophilic to solubilize the hydrophobic GMA, inert to epoxy functionality, and soluble after polymerization in aprotic solvent. N-vinyl pyrrolidinone (NVP) satisfied these requirements, but was shown by simple Q,e type calculations to encourage a massive compositional drift when copolymerized with methacrylate based monomers (69). Reactivity ratios: rGMA about 4.4, rNVP about 0.014 in a range of solvents and solvent mixtures; were obtained by relating copolymer compositions in very low yield copolymerizations to average monomer feed ratios, (70) emphasizing that monomer combinations which encourage extreme compositional drift are susceptible to careful analysis. Compositional drift was minimized in the laboratory through the use of very low yield polymerizations.

Special mention must be made of problems associated with the preparation of polyampholyte copolymers containing acidic and basic monomers. It has been reported that copolymers from mixtures of acidic and basic monomers often possess alternating character over a wide range of feed mixture compositions through facile polymerization of the acid-base pair (71). We have shown that soluble polyampholytes display unusual solution properties reminiscent of proteins, and that polyampholyte gels possess unusual but explainable pH sensitive swelling characteristics (72). In view of these unusual and potentially useful attributes, it would appear that the development of methodologies for the rational modification of monomer sequence distributions in methacrylate based polyampholyte polymers and hydrogels is a goal with merit.

New developments in group transfer polymerization have made possible the living polymerization of acrylate and methacrylate monomers using silyl ketene acetal initiators with a nucleophilic or Lewis acid catalyst (73). By this method we may circumvent the side reactions which accompany conventional anionic polymerizations of acrylates and methacrylates and prepare almost monodisperse homopolymer, and acrylate and methacrylate based block copolymers. Monomers with active hydrogens such as AA, MAA, and HEMA may not be polymerized by this method. However, (for example) it should be possible to polymerize methyl acrylate then hydrolyse the acrylate residues to acrylic acid under mildly basic conditions (58). In similar fashion, it should be possible to esterify HEMA, (74) polymerize the derivative, and recover the HEMA residue by hydrolysis under mildly basic conditions.

2.2   Chemical Characterization

Proton Nuclear Magnetic Resonance (NMR) spectroscopy is usually the method of choice if a quick reliable estimate of copolymer composition is required (75). We have found dimethyl formamide (DMF) to be the best "universal" solvent for these hydrophilic polymers, however d7 DMF is prohibitively expensive for use in routine measurement. We have achieved best overall success, i.e. we routinely achieve narrow signal widths and reproducible signal positions, through the use of d4 methanol as a solvent. Deuterated dimethylsulfoxide (DMSO) will dissolve all but the most hydrophilic (e.g. PHEMA) polymers, but will often complex with functional groups in the polymer to broaden and shift the corresponding NMR signals. Deuterated chloroform will often dissolve DMAEMA containing polymers to yield high quality spectra.

The tactic placement of monomer residues in hydrophilic methacrylate based polymers can influence solution or swelling properties. For example, isotactic PHEMA has been shown to display markedly different swelling and solution behavior than its syndiotactic counterpart (74, 76, 77). Polymer tacticity can be measured also by NMR spectroscopy (67, 78) since the in chain methyls of methacrylate based polymer are very sensitive to tactic placement. Consider the signals clustered around 1 ppm in the proton NMR spectrum of Fig. 2. In homopolymer spectra, the tactic content may be calculated by simple comparison of peak areas. As is often the case with copolymer spectra, signal broadening in this instance produces some overlap between peaks from isotactic (i), heterotactic (h), and syndiotactic (s) triads. Luckily, relaxation times for i, h, and s carbons are sufficiently close that peak areas in ^{13}C spectra obtained by a standard 2 pulse sequence may be directly compared. For example ^{13}C signals associated with in chain methyl carbons (insert Fig. 2) show a significant splitting and indicate that this polymer contains about 3% i, 38% h, and 59% s material. This distribution is typical of methacrylates prepared by free radical polymerization around 70 °C. A higher syndiotactic content may be realized by photoinitiated free radical polylmerization at sub zero temperatures (74). A higher isotactic content may be achieved by living anionic polymerization at low temperatures (74, 77).

NMR has also been used to monitor chemical changes over time in the 75% HEMA/25% MMA copolymer, because of concerns regarding transesterifi-

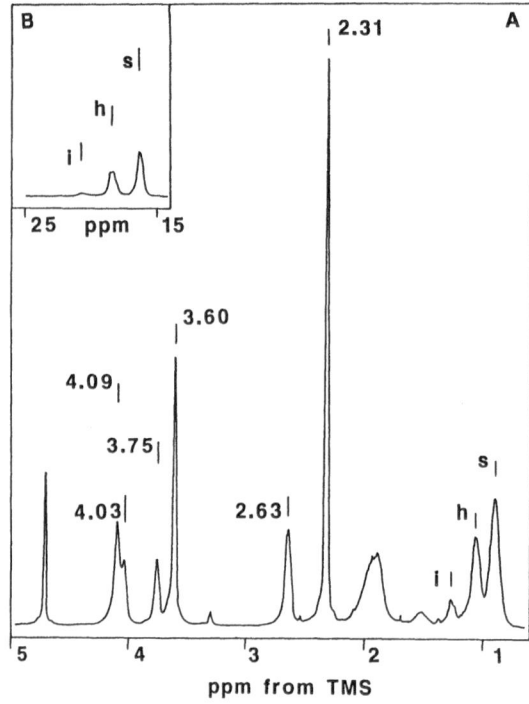

Fig. 2. (A) Proton Nuclear Magnetic Resonance (NMR) spectrum of a soluble terpolymer containing 38% mole fractions of DMAEMA, 22% HEMA, and 40% MMA. (B) ^{13}C-NMR spectrum of the same polymer. *i*: Isotactic, *h*: Heterotactic, *s*: Syndiotactic triads

cation and corresponding loss of HEMA functionality. No changes in copolymer composition were noted with storage (as PBS swollen capsules) at 37 °C for 4 weeks. There was also no change in polymer molecular weight as measured by intrinsic viscosity.

Infrared (IR) spectroscopy is perhaps the most convenient complementary technique for use with NMR. For example, we show in Fig. 3(a) (61) an IR spectrum of a soluble PHEMA. The polymer contains hydroxyls (3400 cm^{-1}), saturated hydrocarbon functionality (circa 3900 cm^{-1} and 1500–1300 cm^{-1}), and ester functionality at 1725 cm^{-1}. Deuterium exchange brought about by exposure to d4 methanol vapor may be used to show that the in chain C-C skeletal vibration of PMMA at 1070 cm^{-1} which has been associated with atactic polymer, (79) has an analogue in PHEMA at 1080 cm^{-1} (Fig. 3b). Spectral subtraction after deuteration reveals also the primary alcohol C-O stretch of PHEMA at 1025 cm^{-1}.

IR analysis after treatment with "signal enhancing" reagents can also be useful. For example, the presence of carboxylic acid impurities in PHEMA and PHPMA was confirmed by IR analysis after exposure to sulfur tetrafluoride gas, (a useful technique if the 1600–1700 cm^{-1} region is "cluttered" with extraneous absorbances) converting carboxylic acids to the corresponding fluoride which absorbs at about 1825 cm^{-1} (80, 81). The acidic impurity was confirmed in both HEMA and HPMA as MAA by GC-MS. The original content of MAA in

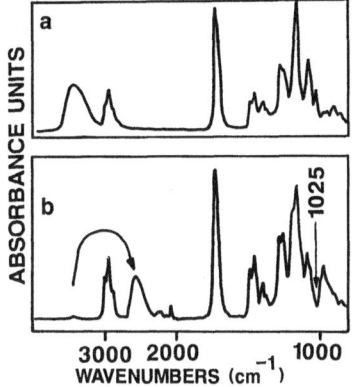

Fig. 3. (a) Infrared spectrum of poly 2-hydroxyethyl methacrylate (PHEMA) on a salt plate glued into a gas phase IR cell. (b) Spectrum after exposure to d4 methanol vapor. *Arrows* indicate major peaks moved by deuteration

HEMA was reduced by distillation to around 1% and by repeated distillations and extractions to around 0.7%. MAA polymerizes randomly with HEMA and MMA (67).

The equilibrium water uptakes of series 2 polymers in distilled water are illustrated in Fig. 4 along with that of soluble PHPMA, (f) which was prepared by a similar method. As expected, water content of all three series increased monatonically with the HEMA content in the polymer. We were encouraged that the EWC of PHEMA appeared very similar to that of crosslinked PHEMA hydrogel. Unexpectedly, the EWC of these polymers changed dramatically in phosphate buffered saline (PBS). Some of the more hydrophilic polymers in series 1 dissolved in PBS, the EWC of hydrophilic series 2 polymers increased dramatically and that of series 3 polymers to a lesser extent. Differential Scanning Calorimetry (DSC) of hydrated polymer indicated that the content of "freezable" water (82, 83) in the more hydrophilic polymers increased dramatically also in PBS. We attribute these changes to ionization of trace quantities of acidic impurity in the PBS buffer.

The massive effect of this relatively small (about 0.7%) concentration of MAA on polymer properties (swelling, viscosity, etc.) indicates that monomer purity should be a driving concern in the synthesis of soluble hydrogel polymers, and that EWC measurements in unbuffered solution should be treated with caution. On the other hand, the incorporation of trace quantities of acidic and basic monomers into this type of polymer may prove a useful method of behavioral modification.

2.3 Interfacial Properties

If we neglect the influence of cytotoxic leachable impurities, the success or failure of an implant (its "biocompatibility") resides primarily with the interactions between its surface, or in many instances, between the protein modified surface,

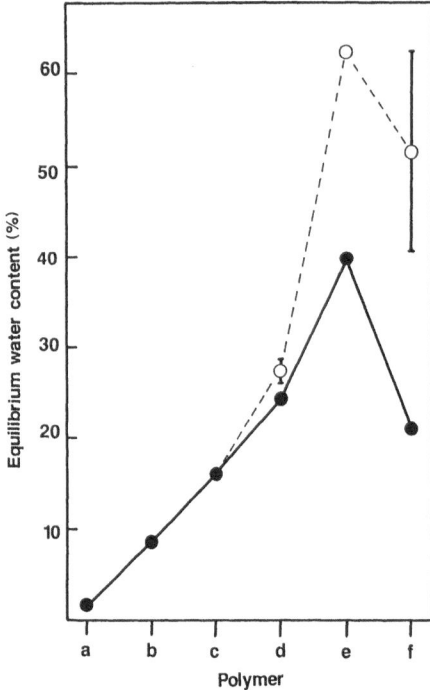

Fig. 4. Equilibrium water content of solution cast films of high molecular weight polymers in distilled water (*filled circles*) and in phosphate buffered saline (*open circles*). (*a*) poly MMA, (*b*) 25/75 HEMA/MMA, (*c*) 50/50 HEMA/MMA, (*d*) 75/25 HEMA/MMA, (*e*) poly-HEMA, (*f*) poly-HPMA

and the inflammatory system: i.e., the nature and degree of the foreign body reaction. Although protein adsorption and cell attachment can be studied as a function of material surface chemistry (e.g., 84, 85), prediction of the tissue reaction on the basis of surface chemical and mechanical properties is largely impossible at the current state of the art. One generally relies on crude extrapolation from the relatively few materials that have been used without obvious difficulty as implants. For many applications surface "wettability" appears to be of prime importance, usually quantified by measurement of surface and interfacial free energy using the contact angle method (86). The chemistry of the surface is also important and this is routinely measured by X Ray Photoelectron Spectroscopy (XPS or ESCA) (87). The primary limitations of this technique include uncertainties introduced during sample preparation (does the surface reorganize on removal from the biological medium, etc?) which is exacerbated for hydrogels by the need to dry the sample prior to analysis.

In brief, biological interactions with hydrophilic methacrylate based polymers appear to be a complex function of surface water content and charge balance. The more hydrophilic polymers such as PHEMA do not encourage cell attachment, primarily, it is thought, because they resist deposition of appropriate "conditioning" proteins and fail to provide sufficient mechanical support to the cells (88). In contrast, it is often desirable to prepare an implant suitable for direct attachment to tissue and these conditions may be met through use of less hydrophilic, usually surface modified, methacrylate based polymers (89).

An interesting observation with 75% HEMA/25% MMA capsules was the propensity to harden when incubated in tissue culture phosphate buffered saline (PBS). Unlike the biochemist's PBS, tissue culture PBS (Dulbecco's PBS) contains 0.9 mM Ca^{+2} (and 0.5 mM Mg^{+2}). This is a relatively unstable solution and incubation of HEMA-MMA capsules in this medium for even a few days lead to the formation of a calcium phosphate crust on the outside of the capsule. Presumably the polymer acted as a nucleation site for calcification. Such calcium deposits are absent when capsules are incubated in tissue culture medium or calcium free PBS.

2.4 Solution Properties

Equilibrium water uptakes of methacrylate based polymers may be varied in monatonic fashion from effectively zero in the case of PMMA to around 40% for PHEMA (80, 90). The crossover to a water soluble polymer may be made through use of more hydrophilic monomers such as glycerol monomethacrylate, (GMMA) which is conveniently prepared by acid hydrolysis of glycidyl methacrylate (62). Hydrophilic but insoluble methacrylate based hydrogels have been studied extensively and need not be mentioned here (90, 91). It is expected that the solubility of PGMMA and other water soluble methacrylate based polymers will vary somewhat with added salt concentration due to changes in the solubility parameter and hydrogen bonding index of the solvent. In general they are not expected to be overly pH sensitive.

Polymethacrylic acid (PMAA), on the other hand, is very pH sensitive undergoing a compact-extended coil transformation at about 25% ionization (92). Manipulation of raw data from the titration of PMAA from high to low pH allows us to construct plots of "apparent" pKa or pKb as a function of degree of ionization or protonation (66, 67). Taking the reduced specific viscosity at alpha = 1 as an upper limit, we may construct also a plot of reduced specific viscosity relative to that for the fully ionized polymer as a function of pH (72). These constructs, along with a derivative of the normalized reduced specific viscosity, for a polymer containing a mole fraction of 36% HEMA and 64% MAA are reproduced in Fig. 5, to show that (as expected) the polymer coil expands with ionization of the MAA residues, that the plateau in the pKa curve of PMAA (93) is reproduced in soluble copolymers containing MAA and HEMA, and that, as with PMAA, the plateau in the pKa curve corresponds to a region in which polymer chain dimensions are a strong function of alpha. We have shown that the magnitude of the compact-extended coil transformation diminishes with reduced MAA content in the polymer and vanishes at mole fractions of MAA of around 10% (66). In contrast, homo and co-polymers containing DMAEMA were found to expand and contract in monotonic fashion with the level of amine protonation, with no apparent evidence for any pronounced conformational transition (67). Despite extensive investigations of equilibrium and kinetic aspects of the pH mediated swelling properties of weakly acidic crosslinked

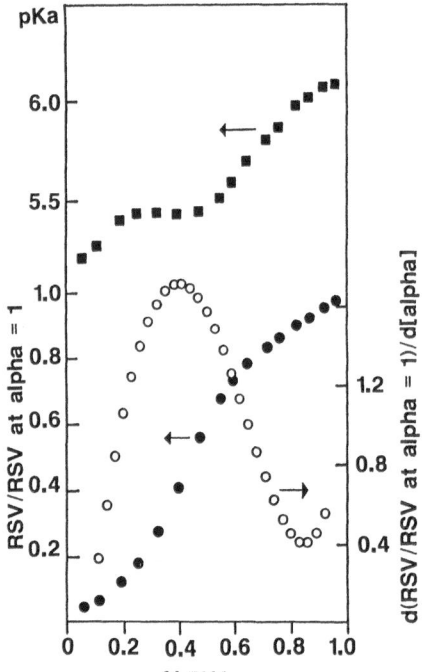

Fig. 5. Reduced specific viscosity of a copolymer containing mole fractions of 36% HEMA and 64% MAA, normalized to value at alpha = 1 at high pH, plotted as a function of alpha. Also as a function of alpha: derivative of normalized viscosity and apparent pKa of the polymer

hydrogels containing MAA, most notably by Peppas (94), it remains to be seen if this abrupt conformational transition can be incorporated into a crosslinked membrane so that a very small change in pH would result in a pronounced expansion or contraction of the polymer (as in NIPAM copolymers). Recent work based on PMAA-g-PEO hydrogels has served notice that such may be possible (95).

In Fig. 6 we compare and contrast polymer solution dimensions as a function of pH for PMAA, a polymeric weak acid, PDMAEMA, a polymeric weak base, and a soluble polyampholyte (B), prepared from a feed mixture containing mole fractions of about 30% MAA and DMAEMA and 60% HEMA. At high pH, PMAA is fully ionized and expanded by coulombic repulsions between the charges. At succeedingly lower pH the PMAA coil contracts due to protonation of the carboxyls. Soluble PDMAEMA, in contrast, is fully protonated and chain expanded at lower pH and shrinks at higher pH due to deprotonation of the tertiary amines. The polyampholyte, on the other hand adopts a compact conformation from about pH = 5 to pH = 8 and expands to an extended conformation above and below that pH range. B was shown to contain mole fractions of about 26% MAA, 30% DMAEMA, and 44% HEMA. Given the tendency of acid-base pairs towards alternating polymerization (96), the DMAEMA and MAA are probably sequenced in pairs along the chain although more evidence of this is needed. It would appear that

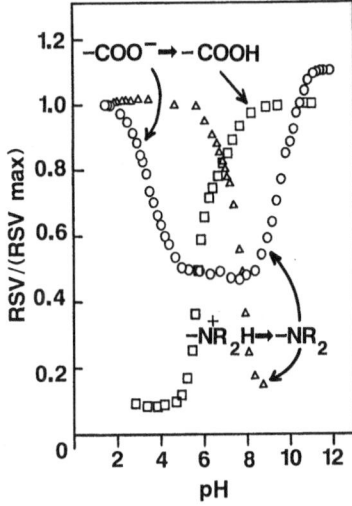

Fig. 6. Normalized reduced specific viscosities as a function of pH: Polymethacrylic acid (PMAA: *open squares*), polydimethylaminoethyl methacrylate (PDMAEMA: *open triangles*) and a polyampholye containing mole fractions of 26% MAA, 30% DMAEMA, and 44% HEMA (*open circles*)

polyampholytes offer the possibility of opening up new pH "windows" for the swelling process through use of conventional biocompatible monomers and without making use of potentially bioreactive polymers containing such as aromatic amines which would protonate at lower pH than PDMAEMA.

The pH mediated swelling of soluble polyampholyte polymers is reflected also in crosslinked hydrogels of similar composition. For example, shown in Fig. 7 are equilibrium water uptakes as a function of pH for three different crosslinked polyampholyte hydrogels. Although absolute water contents are shown to increase with the MAA and DMAEMA content, the swelling process appears to be more pH sensitive at lower levels of MAA and DMAEMA. Our rationalization of this process is presented in Fig. 8. Formation of the acid base pair (A) stabilizes both the protonated form of the base and the deprotonated form of the acid and acts through secondary bonding to limit coil expansion. At low pH (B) the acid protonates, allowing the free protonated base to expand the polymer coil. At high pH, (C) the base is deprotonated, allowing the acid anion to expand the polymer coil. These salt bridges were shown to be relatively insensitive to added saline (data not shown) but to be very sensitive to the presence of multivalent ions such as Ca^{2+} (D) which promotes chain expansions by selectively binding with acid anions and disrupting the ion pair complex.

Polymer molecular weights may be measured by vapor phase or solution osmometry, or more usually by size exclusion chromatography (SEC or GPC). If the polymer is hydrophilic but neutral, measurements will usually be made in a polar organic solvent such as 1,4 dioxane or dimethyl formamide. If insoluble in organic solvents, measurements may be performed in water, a less than ideal solvent for this type of measurement. The situation is further complicated if dealing with a polyelectrolyte polymer (especially containing "weak" pH sensitive functionality). SEC measurements become a function of nominal charge

Michael V. Sefton and William T. K. Stevenson

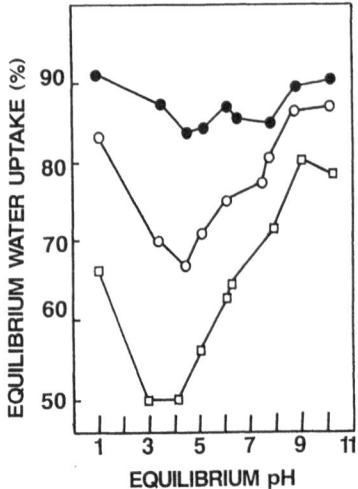

Fig. 7. Equilibrium water uptakes of chemically cross-linked polyampholyte hydrogel as a function of pH, compositions in mole fractions MAA/DMAEMA/ HEMA/EGDMA. Formulation A: 10/10/79.5/0.5 (*open squares*). Formulation B: 30/30/39.5/0.5 (*open circles*). Formulation C: 49.75/49.75/0/0.5 (*filled circles*)

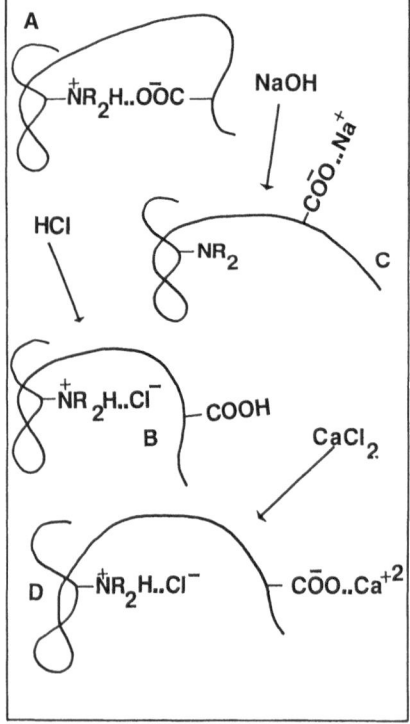

Fig. 8. (*A*) Schematic of polyampholyte showing salt bridge. (*B*) Broken by lowering the pH and protonating the carboxylic acid. (*C*) Broken by raising the pH and deprotonating the tertiary amine. (*D*) Broken by preferential complexation of the acid with Ca^{2+}

content through its influence on effective hydrodynamic volume. Aqueous phase osmometry becomes complicated by virtue of changes in the number of counterions and their association with the polymer chain (i.e., in the effective mole fraction of solute) as a function of pH and dilution. Even so, meaningful results have been obtained with these techniques. For example, dialysis dilution techniques have allowed for the estimation of the true molecular weights of polyelectrolyte polymers (97). In similar fashion, meaningful universal calibration plots for aqueous polyelectrolytes have been obtained through control of solution pH and ionic strength (98).

2.5 Derivatization Reactions

The application of methacrylate polymers may be extended by introducing new functionality to a preformed polymer or by protecting existing functionality on a monomer to allow for its polymerization by an unconventional method. For example, as a route to highly isotactic PHEMA, Gregonis derivatized HEMA to benzoxyethyl methacrylate and successfully polymerized the derivative using *n*-butyl lithium-copper complex, an anionic initiator which has been shown to be relatively unreactive towards carbonyl functionality (74, 99). PHEMA was then recovered by preferential aqueous base hydrolysis of the benzoyl ester (100). It would appear that this method should be effective also in protecting hydroxyl functionality prior to group transfer polymerization. In similar fashion, trimethyl silyl methacrylate should prove of use in the preparation of poly-methacrylic acid by group transfer polymerization of the derivative, followed by a very mild hydrolysis to the free acid (101). The hydroxyl group may also serve as a functionality susceptible to derivatization, for example, through esterification with acid chlorides or phosphorylation with nitrophenyl phosphates (102).

Epoxidized methacrylates such as glycidyl methacrylate (GMA) make available a number of syntheses involving the versatile oxirane ring. For example, the epoxy may be ring opened by acid hydrolysis prior to polymerization to prepare glycerol monomethacrylate which may be polymerized to a water soluble polymer or a high water content crosslinked gel (62). In contrast, the GMA may be polymerized as is, then ring opened to the diol by acid hydrolysis. This latter has been applied to the derivatization of cross linked polymer (103). It remains to be seen if soluble polymer may be transformed to a soluble derivative by this latter method. GMA has also been block copolymerized with MMA and subsequently derivatized to the hydroxysulfonate by reaction with sodium sulfite in the presence of a phase transfer catalyst (104). In addition to these less conventional reactions, GMA containing polymers may be derivatized or crosslinked by conventional epoxy resin chemistry, i.e. by reaction with amines, mercaptans, hydroxyls, etc. (105).

The quaternization of amines to the quaternary ammonium salt has proved a very useful method of converting the "soft" amine base to the pH insensitive quaternary ammonium salt. The amine is reacted with an appropriate measure of an alkyliodide as first described by Menshutkin (106). This reaction was systematically examined by Arcus and Hall in the 1960s as a means of quaternizing preformed polymers (107), and has emerged as a method of choice. For example, thwarted in our efforts to solubilize a high molecular weight copolymer containing mole fractions of 90% HEMA and 10% DMAEMA in isotonic phosphate buffered saline due to a only partial ionization of the base under those conditions, we quaternized the polymer with methyl iodide to the very soluble quaternary methyl derivative (108).

2.6 Final Comments

To end, we must emphasize that methacrylates in medicine are made use of in diverse areas including bone and dental cements (109), and as surface modifiers to prevent or delay process of rejection (110), and that each commands a formidable literature. The reader should also be aware of the considerable effort which has been directed towards a phenomenological and theoretical understanding of the solution and swelling properties of synthetic hydrogel and polyelectrolyte polymers (111–116).

3 Simple Coacervation

Several water insoluble polyacrylates have been used to microencapsulate live cells using simple coacervation or precipitation techniques. The polymer dissolved in an appropriate organic solvent (diethyl phthalate or polyethylene glycol 200) was coextruded along with a cell suspension through a coaxial needle assembly (Fig. 9a), into a third fluid stream. Air was used for EUDRAGIT RL and DMAEMA-MMA; hexadecane was used for HEMA-MMA. Flow against the air or hexadecane increased the shear force on the

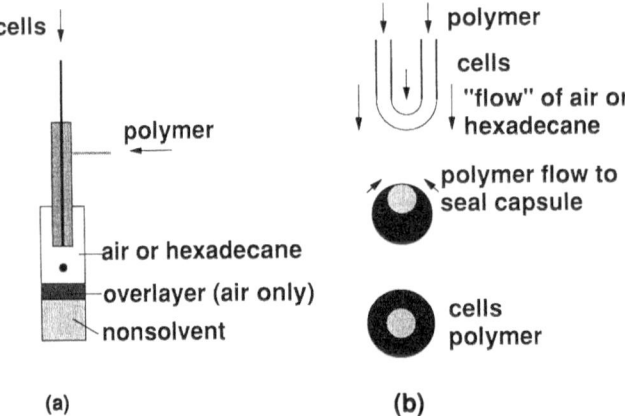

Fig. 9a, b. Schematic illustration of (a) coaxial extrusion assembly (b) capsule formation

capsule resulting in droplet formation at a controlled diameter (117). The resulting droplet, consisting of cells surrounded, more or less, by polymer solution fell through a hydrophobic oil phase (corn oil/mineral oil mixtures or hexadecane) to enable the polymer solution to completely flow around the capsules (Fig. 9b). Depending on the density of the cell suspension, better centered capsules were obtained as a result of this "capsule sealing" process. Capsules then crossed the interface from this oil phase into the precipitation bath, with or without the aid of agitation. In the case of EUDRAGIT RL and DMAEMA-MMA the precipitation bath was another oil phase of different composition to the upper layer and because the interfacial tension was so low, no agitation and no surfactant was needed in the precipitation bath. For HEMA-MMA, the precipitation bath was an isotonic aqueous solution (e.g., phosphate buffered saline) and crossing the interface was slow even with agitation and a surfactant rich bath. The physics of this interface and why crossing the interface is so slow, remains unclear.

3.1 Cationic Polyacrylates

Our initial work was with EUDRAGIT RL as a 12.0% w/v solution in diethyl phthalate. The very first studies (118) with EUDRAGIT RL involved making capsules using red blood cells emulsified in an oil phase. Beyond showing that capsules could be made and that at least a few encapsulated cells retained functioning hemoglobin this process was not practical for cells other than those (like erythrocytes) which could be subjected to high shear rates without lysis. Nevertheless, we were suitably encouraged to devise an alternative methodology that did not subject the cells to such high shears. Hence coextrusion has been found to be the method of choice in all subsequent studies (119).

Briefly, syringe pumps were used to pump (typically at the same rate, 1–5 mL/h) a cell suspension through the inner barrel and polymer solution through the annular space of a coextrusion assembly. Droplets were blown off the end of the needle by a coaxial air stream. The droplets were caught in a Teflon jar containing a nonsolvent. Typically the nonsolvent was overlayed with a small volume of an immiscible nonsolvent to control penetration of the miscible nonsolvent phase. After extrusion was complete, the jar was gently shaken for 20–30 min to extract all the solvent. The capsules were allowed to settle, removed from the jar by pipette and transferred to a wash solution to remove traces of solvents and non-solvents. EUDRAGIT RL capsules were washed free of oil and diethyl phthalate using the dialysed supernatant after $(NH_4)_2SO_4$ precipitation of outdated human plasma); this solution was found serendipitously to be an excellent means to transfer the capsules from organic to aqueous media.

A rat pancreatic islet microencapsulated in EUDRAGIT RL (120) is shown in Fig. 10. Capsules were generally spherical, approximately 500 μm in diameter with a 10–50 μm thick wall. Encapsulated islets were maintained in tissue

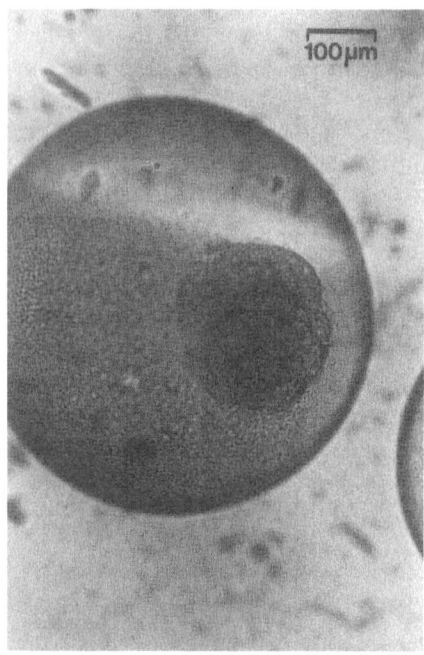

Fig. 10. A rat pancreatic islet in a EUDRAGIT RL microcapsule on the day after encapsulation (Reproduced with permission of ISAO press)

culture for several months, during which time most islets remained intact, at least as visible through a light microscope. It was clear from the static glucose challenges that the islets were viable inside the capsules and their viability was preserved in vitro to the same extent that control unencapsulated islets were maintained in vitro. However, perifusion studies indicated that not all batches of capsules responded appropriately and that there was a time lag in the response of the islets to increased glucose levels. It seemed apparent then, that the EUDRAGIT capsule wall was only slightly permeable to insulin and that the capsule wall placed a significant diffusion limitation on the release of insulin. Its limited permeability was also noted in the failure of CHO cells to grow in intact EUDRAGIT capsules (121). In addition to having a limited insulin permeability, EUDRAGIT RL, as expected, was not biocompatible. Within 2–3 weeks after subcutaneous implantation of a EUDRAGIT RL film in rats, a thick fibrous capsule developed around the film presumably because of its cationic nature. Clearly, EUDRAGIT RL is not appropriate for islet encapsulation. Nevertheless, its use as a model polymer did demonstrate that interfacial precipitation and organic solvents can be used to microencapsulate islets without danger to the islet itself.

Human diploid fibroblasts (HDF) were encapsulated within EUDRAGIT RL as above (122). The encapsulated cells were viable after encapsulation but grew only if collagen was coencapsulated to act as a growth scaffold for the anchorage dependent cells. Since collagen was awkward to encapsulate because of its limited solubility and encapsulated cells contracted the collagen gel, it was thought desirable to find a polymer which could be used as a substrate for HDF attachment and growth, without collagen supplements. Hence work was initiated to encapsulate cells using a water insoluble cationic polyacrylate copolymer which supported HDF growth (123). Capsules were prepared from some of the copolymers using the same method as for EUDRAGIT RL. Because of their ability to support HDF growth as films, 16% DMAEMA/MMA and 17/2.2 DMAEMA/MAA/MMA were used to encapsulate HDF. The 17/4.4 DMAEMA/MAA/MMA copolymer was also used in one or two encapsulations because of its higher water uptake. Using a 10% solution of 16% DMAEMA/MMA in diethyl phthalate, capsules were produced in reasonable yield and with an average diameter of 540 μm. Cells grew in approximately one-third of the capsules, with most of the remainder showing cell spreading but no growth. Since cell growth was also observed on the bottom of the culture dish in the former case and since coencapsulated high molecular weight FITC-dextran was found to leak out of the capsules with growing cells, it was presumed that these capsules were broken or had leaks or pinholes. In the other intact capsules cell growth did not occur presumably because of diffusion limitations associated with the capsule wall. Since cells would grow on the bottom of the culture dish when released from intact capsules immediately after encapsulation (by deliberately breaking the capsule), it appeared that encapsulation did not damage the cells. Rather the polymer although providing an appropriate surface charge was (presumably because of its lower water content, 10% in PBS) too impermeable to one or more critical nutrients or metabolites (glucose, lactate and others).

Encapsulations with DMAEMA/MAA/MMA terpolymers were also fairly unsuccessful. The increased water uptake (to 27%) resulted in a small improvement in cell viability, but the cells still did not grow in the majority of capsules. Permeability to glucose was enhanced, but either not enough with 2.2% MAA or the MAA containing terpolymer (4.4% MAA) was not as good a substrate for HDF growth. It remains to be seen whether an optimum terpolymer composition lies in the narrow band between these two very similar materials.

3.2 HEMA-MMA

3.2.1 Capsule Preparation

The encapsulation experience with EUDRAGIT RL lead to the conclusion that EUDRAGIT RL, while appropriate for demonstrating in a general sense that water insoluble polyacrylates could be used to microencapsulate live mammalian cells, was not suitable for the particular problem of encapsulating cells because of its limited biocompatibility and permeability. To address both of these limitations (and especially the former one), a series of noncrosslinked

HEMA/MMA copolymers was synthesized (61), one of which (containing a mole fraction of ~75% HEMA in the monomer mixture) has been used for encapsulation. HEMA-MMA was preferred over EUDRAGIT RL because of the former's biocompatibility and greater water uptake/permeability and mechanical durability. HEMA-MMA capsules were soft, elastic and tough. Unlike EUDRAGIT RL capsules which broke easily on contact with hard objects or when probed with forceps, HEMA-MMA capsules had to be torn apart or shredded with pairs of forceps or a scalpel to release the intracapsule contents. The water uptake of HEMA-MMA is greater than that of EUDRAGIT RL consistent with the observed difference in glucose permeability (124). As expected, capsules made from the copolymers used here elicited a very benign response when implanted subcutaneously or intraperitoneally in rats or mice. Upon explanation at 4 weeks, for example, the capsules were covered with only a very thin (1–4 cell layer thickness) fibrous capsule.

However, the mole fraction of 75% HEMA was not soluble in diethyl phthalate. The best solvent was found to be polyethylene glycol 200 since it was tolerated at least to a limited extent by the cells and since PEG was water soluble, enabling an aqueous salt solution to be used as the nonsolvent. While advantageous in the sense that the difficult to remove, hydrophobic oils and phthalate were no longer necessary, the choice of PEG 200 created other problems. Because of the higher viscosity of the HEMA-MMA/PEG 200 solution (~1 Pa-s vs 0.2 Pa-s for EUDRAGIT RL in diethyl phthalate) the coaxial air stream was replaced with an inert liquid (hexadecane) to increase the shear force (i.e., by increasing the viscosity). To avoid the complexities of pumping hexadecane, the hexadecane was kept stationary while the needle was moved to exert the shear force (Fig. 11) (117). We used a "submerged jet": i.e., an extrusion process with a submerged nozzle leading to droplet formation within another liquid phase.

In the submerged jet method the coextrusion needle was lifted out of a layer of hexadecane thereby shearing the capsule droplet off the needle at the hexadecane/air interface. The droplet (polymer solution plus cells) fell through the hexadecane layer into a precipitation bath consisting of PBS with 50 ppm of Pluronic L101 surfactant to aid droplet penetration. The PBS extracted the solvent, leaving the precipitated polymer behind as a capsule wall around the cells. The HEMA/MMA was used as a 9 or 10% solution in pure PEG 200. Cells were used as the core material, at a ratio of 1:1.5 or 1:2 polymer solution (v/v). After 30 min, capsules were washed in fresh PBS (no L101) for a further 30 minutes and then stored in RPMI with FCS and antibiotics at 37 °C, 5% CO_2. The process has been further modified to circumvent the technical problems associated with encapsulating Matrigel (for liver cells) and Cytodex beads (used for certain fibroblasts) which are used as cell growth substrates; HEMA-MMA is not a substrate upon which anchorage dependent cells will grow. For Matrigel which gels at room temperature, the cell suspension containing Matrigel was cooled prior to the extrusion needle. Cytodex beads with their preattached cells were kept in suspension using a micro-stirring bar in the feed syringe.

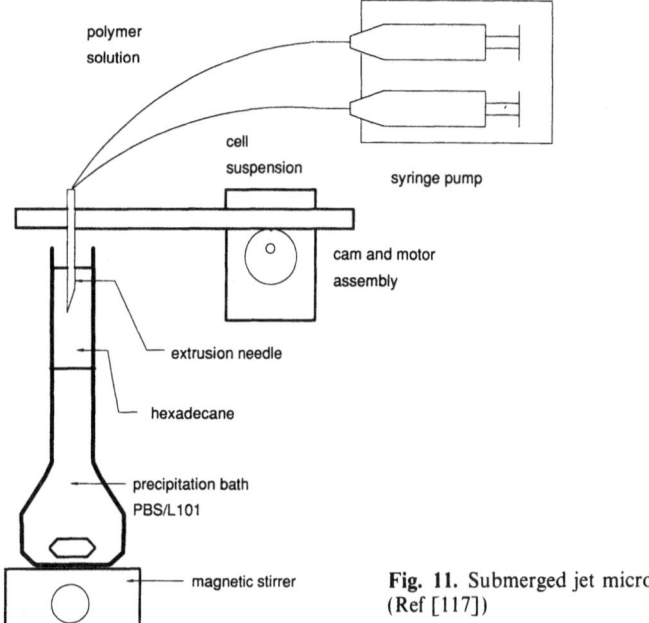

Fig. 11. Submerged jet microencapsulation apparatus, (Ref [117])

The capsule membrane appeared to consist of an outer skin, a thin macroporous layer and a very thick dense membrane, with an overall (mean) thickness of ~ 90 μm (Fig. 12). The capsule diameter was ~ 900 μm after 7 days. Smaller capsules can be made with a variation of this method in which the needle is held stationary and the hexadecane pumped past the end of the needle; the hexadecane is recirculated and care must be exercised to ensure that capsules are not entrained with the recirculated hexadecane (Fig. 13). The effect of hexadecane flowrate on capsule size (as it leaves the needle) in an early prototype is shown in Fig. 14; capsules shrink to approximately half their initial diameter as solvent is extracted. Note that capsules as small as 300 μm can be produced. As a consequence of the thin skin it is possible to damage the capsules through mishandling: forceps with serrated edges or forcing capsules through narrow bore needles are avoided as a result.

Because of concerns over the permeability of the capsule wall, we sought to modulate the capsule wall structure by changing the composition of the polymer solvent and precipitate bath and thereby changing the rate at which the polymer precipitated. Figure 15 shows scanning electron micrographs of capsules prepared using different conditions: 0.3 M glycerol precipitation bath (Fig. 15a), 15% water/85% PEG 200 as solvent (Fig. 15b) and the combination of 15% water and 0.3 M glycerol as precipitant (Fig. 15c). Although the capsule wall thicknesses were not much different, the modified capsule walls were more macroporous. The addition of precipitant (water) to the polymer solution

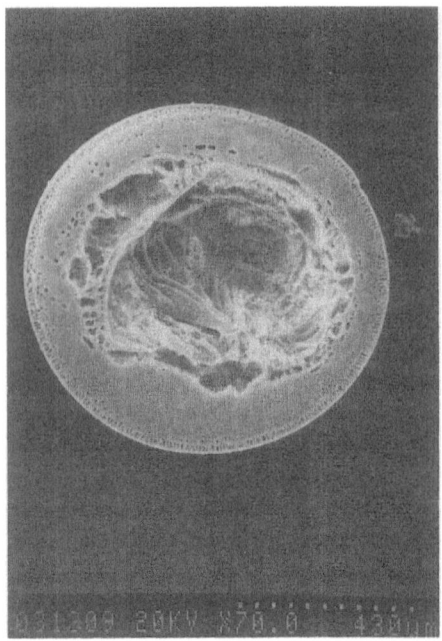

Fig. 12. SEM of a typical ∼ 900 μm HEMA-MMA capsule

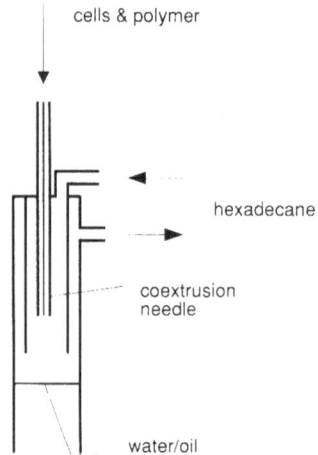

cells & polymer

hexadecane

coextrusion
needle

water/oil
interface

Fig. 13. Schematic illustration of extrusion assembly for producing small diameter capsules

placed the original system closer to the point of precipitation (the cloud point). Therefore, only a small amount of precipitant needed to diffuse into the nascent capsule wall before phase separation occurred. For the isotonic glycerol bath, improved solvent-precipitant compatibility due to the absence of salts in the precipitant was presumed to have promoted macrovoid formation. The mass

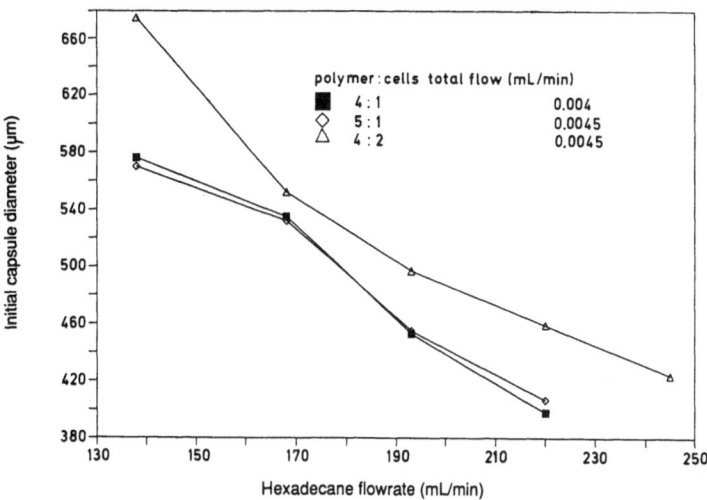

Fig. 14. Effect of hexadecane flowrate on capsule diameter (as formed; prior to shrinkage) in the initial prototype small diameter apparatus

transfer coefficients in glycerol precipitated capsules were only double those in capsules precipitated in PBS, consistent with the more macroporous character of the capsule wall, although the extent of macroporosity would imply even greater permeability changes. However the mass transfer coefficients in capsules made with 15% H_2O/85% PEG 200 as the solvent were not much different than "normal" capsules, despite the apparent difference in structure. It thus appeared that the capsule permeability was defined by the properties of the skin layer and that the dense sublayer was only a minor resistance to diffusion; i.e., the sublayer in normal PBS precipitated capsules appears more permeable than it looks from the SEM.

The microcapsule wall membranes, following the work of Strathmann and others (125–127), involves a ternary system of a polymer, a solvent and a miscible non-solvent. During immersion precipitation, the polymer solution undergoes an exchange of solvent for nonsolvent; as the nonsolvent content of the polymer solution increases, the system becomes thermodynamically unstable and phase separation occurs (Fig. 16). The kinetics of phase separation is also important for these demixing phenomena. In Fig. 16, the initial polymer solution composition is represented by the point A. Exchange of solvent for nonsolvent results in a composition pathway leading to point B where the ternary system becomes unstable and phase separation occurs. Further non-solvent-solvent exchange occurs until the final membrane composition lies at point C. At C the membrane consists of two phases: swollen polymer of composition given by point D and almost pure precipitant. The locus of points ABC may be shifted, so that point C is shifted along the polymer-precipitant

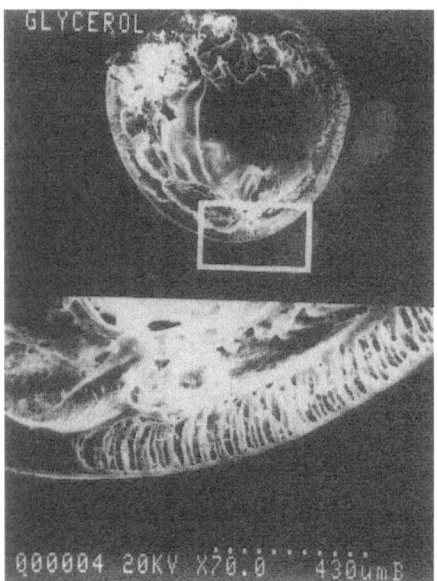
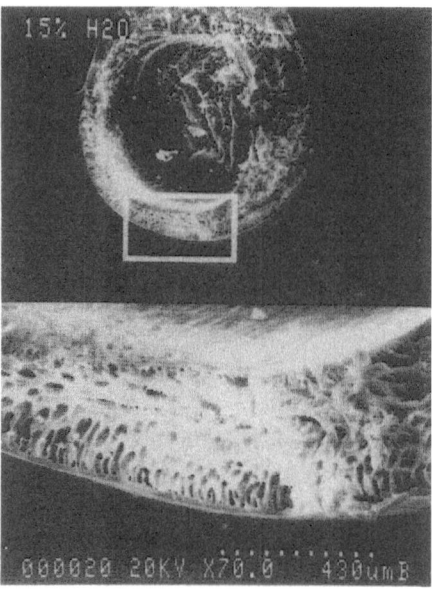
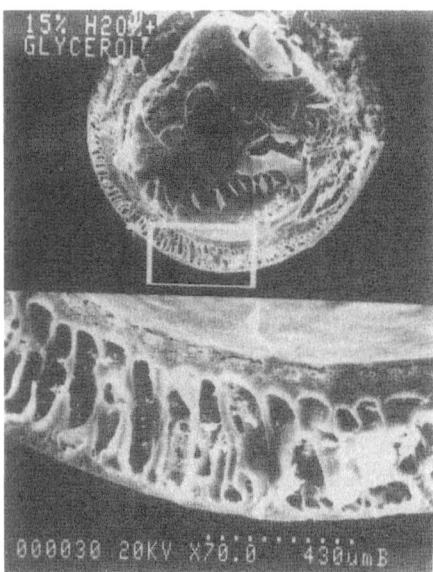

Fig. 15a–c. Scanning electron micrographs of capsules prepared using (**a**) 0.3 M glycerol as precipitant (**b**) 15% H₂O/85% PEG 200 as solvent (**c**) both 15% water, 0.3 M glycerol. (Reproduced with permission of J. Wiley & Sons, Inc. 1990)

axis, depending on formation parameters. The greater the precipitant content of the overall membrane, the greater the percentage of void spaces (which are filled with precipitant). Thus the degree of porosity and the pore size distribution in the membrane can vary greatly.

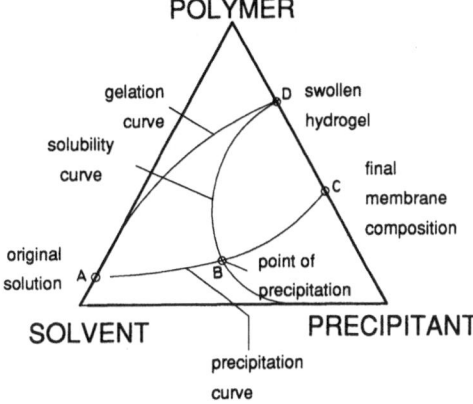

Fig. 16. Schematic illustration of ternary phase diagram of the polymer-solvent-precipitant system showing the precipitation pathway (*ABC*) and gelation pathway (*AD*) during membrane formation, (Ref. [117])

The skin and porous sublayer are presumed to be formed by different mechanisms: the skin by homogeneous gelation and the underlying porosity by liquid-liquid phase separation ("precipitation") (125). Immediately after immersion, there is a very rapid depletion of solvent from the polymer film surface and a relatively small penetration of precipitant. This causes the polymer concentration at the film/precipitant bath interface to increase, crossing the gel boundary and forming a dense gel layer, the membrane skin (pathway AD in Fig. 16). The skin thus formed presents a resistance to the out-diffusion of solvent and controls the subsequent sublayer formation. Precipitant continues to enter the forming membrane such that the sublayer composition follows pathway ABC. Once the solubility curve is crossed, precipitation occurs by phase separation followed by nucleation and growth of each phase. The better the compatibility of polymer and solvent, the slower the removal of the solvent from the polymer structure and hence the precipitation of the membrane is slower, leading to a dense, symmetric membrane. Conversely, closely matched solubility parameters of solvent and precipitant increase the rate of precipitation leading to a more open macroporous structure.

Capsules shrink slowly after they are made, reaching an "equilibrium" size within 2–3 days (~60% reduction in diameter). Some of the size reduction is due to loss of solvent, but analysis of the PEG content of capsules (largely the core solution) by HPLC (Fig. 17) suggests that most of the PEG is removed within the first hour of preparation (the washing stage). The remainder of the size reduction likely reflects a restructuring of the wall (the denser sublayer) as the polymer collapses on itself to minimize the internal surface area within this portion of the wall. Micrographs are available to document the progressive loss of internal surface (and porosity) in at least early forms of these capsules. A consequence of these changes in wall structure is that fresh capsules are much more porous than 4 or 7 day old capsules, even allowing encapsulated 2 million molecular weight dextran to leak out of the capsules within a few hours. Complement components can also permeate the capsules to lyse sensitized

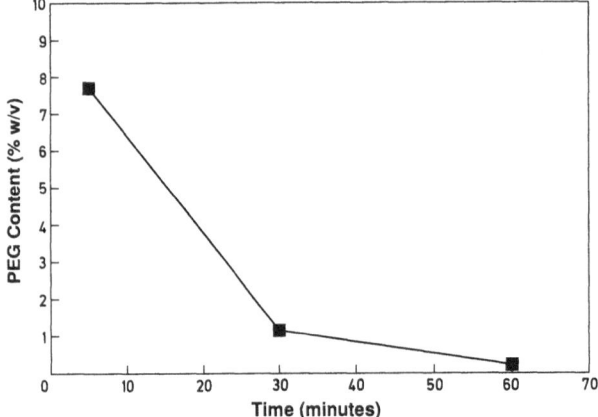

Fig. 17. PEG content of capsules immediately after preparation. PEG content measured by HPLC from an aliquot of crushed capsules. (Results provided by courtesy of Dr. B.Kuhn, Hana Biologics)

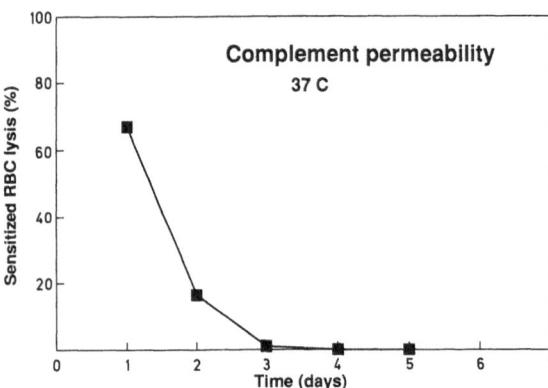

Fig. 18. Lysis of encapsulated sensitized sheep erythrocytes by exposure to complement (in the external medium) as a function of time after encapsulation. (Results provided by courtesy of Dr. A. Morgan, Hana Biologics)

sheep red cells for two days after encapsulation (Fig. 18) but no lysis (implying no permeation at least to molecules > 400 kDa) is detected at three days or more. The capsules are also tackier for the first few days after encapsulation and this can make handling more awkward.

Capsule permeability at room temperature was measured by incubating 4 day old capsules in a test solute solution for 3 days and then monitoring release by taking samples from the extracapsular supernatant at periodic intervals (117). ^{3}H-glucose and ^{3}H-insulin concentrations were measured using liquid scintillation counting, while albumin (MW $\approx$ 69 kDa) and alcohol dehydrogenase (ADH, MW $\approx$ 150 kDa) were measured with a spectrophotometer. The overall mass transfer coefficients (U) determined from the rate of increase of extracapsule concentration of 4 different solutes are shown in Fig. 19. Calculation of U involved no assumptions regarding the uniformity of the wall or the

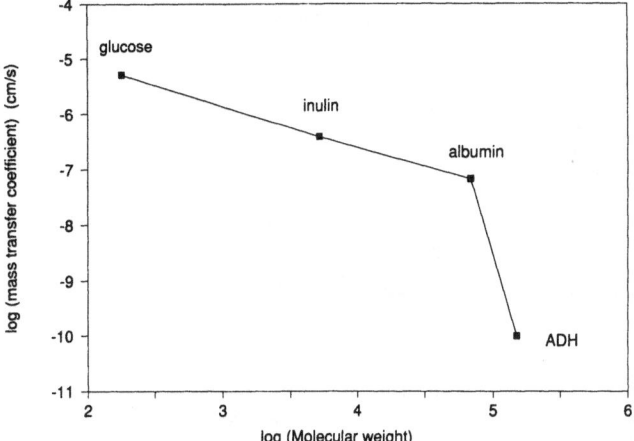

Fig. 19. Effect of solute molecular weight on the mass transfer coefficient (U) for PBS precipitated capsules. U is similar to the physiologist's permeability. (Reproduced with permission of Technomic Publishing Co.)

presence of internal mass transfer resistance. The external mass transfer resistance was presumed to be negligible since shaking at lower speeds had no significant effect on U. HEMA/MMA capsules are clearly not absolutely impermeable to molecules of molecular weight comparable to immunoglobins. The permeability however is extremely low. Only by making measurements over a long time (2–3 days), was it possible to detect such release from the capsule. Examination of the log-log plot of mass transfer coefficient against molecular weight (Fig. 19), suggests a change in the relationship between these two between albumin and ADH, suggesting that the molecular weight cutoff of the membrane is in the range of 70–150 kDa; i.e., on the order of 100 kDa.

3.2.2 Encapsulated Cell Behavior

Encapsulated islets secreted insulin into the extracapsule medium (here αMEM with 50 mg/dL glucose; 24 h static incubation) at a rate that appeared to be one-third that of control islets (Fig. 20) based on the number of islets added to the encapsulation syringe (typically 100–150) for up to 30 days after encapsulation. Unfortunately the capsules were opaque so that it was impossible to determine the actual number of islets in a batch of capsules until the end of an experiment. According to a vital stain (alcian blue), 70–80% of the islets (after 4 weeks) inside the capsule were intact and viable. However the yield of encapsulated islets was ~ 30% relative to the number added to the syringe. Hence the insulin secretion values, based on the actual number of islets in the capsules, was comparable to that for the control. Encapsulated rat islets also responded to

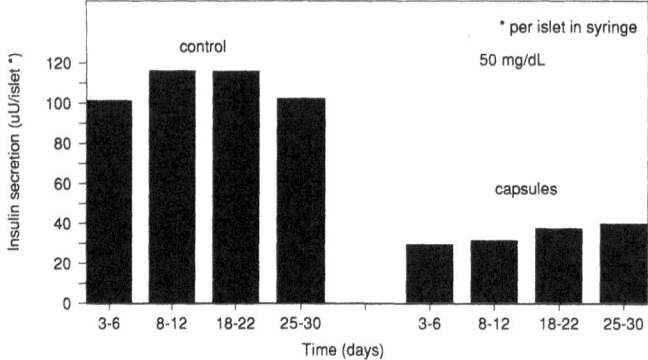

Fig. 20. Insulin secretion from microencapsulated islets. 24 h incubation in 50 mg/dL glucose in αMEM. Results reported per islet in syringe, not per islet in capsule (see-text). (Reproduced with permission of Technomic Publishing Co.)

increases in glucose concentration in a static glucose challenge assay. Both control, unencapsulated and encapsulated islets increased their production rate by comparable relative amounts on exposure to a higher glucose medium (128).

Using a variety of other cells we have shown that cells survive encapsulation and will grow or function afterwards for periods from 2 to > 6 weeks in vitro (128). Encapsulated cell counts and MTT activity (129) increase with time although not necessarily to the same extent since the mitochondrial activity (MTT signal/cell) appears to decrease at high cell density. Since the MTT signal could be measured on single capsules, we were able to show that there was a large variation in the formazan production among individual capsules within each batch (130). This variation appears to be related to the efficiency of polymer solution to entrap the cell mass in the aqueous core and was minimized through adjustment of the cell/polymer flow rate ratio.

Dopamine (PC12 cells/depolarization conditions), IL2 (MLA 144 cells), EGF (transfected FR3T3 cells), and various proteins (HepG2 cells) were all released from the different encapsulated cells. Dopamine release under depolarization conditions from microencapsulated PC12 cells, 2 weeks after microencapsulation is shown in Fig. 21 for two cell densities. The time course of formazan production (MTT activity) for 4 weeks was followed to assess PC12 survival and proliferation in microcapsules (Fig. 22). Experiments were carried out using blank capsules and capsules for which the core solution cell concentrations (cells/ml) were 4×10^5 (low density) and 4×10^6 (high density). Similarly encapsulated MLA 144 cells secreted detectable quantities of IL2 into the surrounding medium, as measured by the ability of CTL.L cells in separate culture to grow in aliquots of the extracapsule medium (Fig. 23). Approximately 300 capsules were incubated for more than 2 weeks in 2 ml of complete RPMI, with replacement of the medium every 2–3 days. Significant levels of IL2, measured in terms of cpm of ^{3}H thymidine incorporated into CTL.L cells

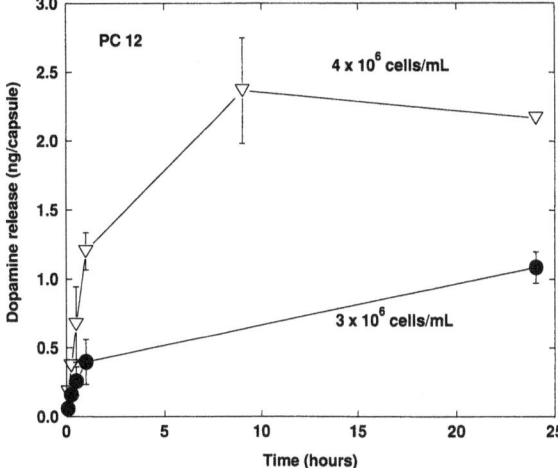

Fig. 21. Dopamine release by encapsulated PC12 cells at two cell densities. Capsules incubated in high K^+ release medium for 24 hours. $\pm$ sd, n = 3 (Reproduced with permission of Elsevier Publishing Co.)

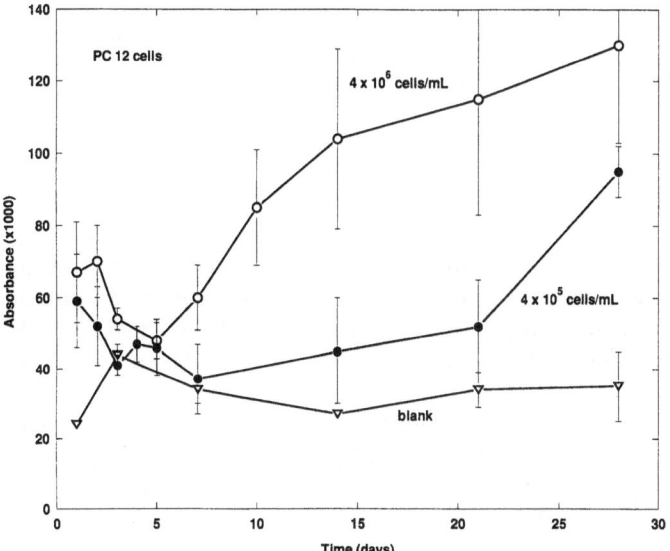

Fig. 22. Formazan absorbance (MTT signal) from blank, low and high density PC12 cell capsules. (mean $\pm$ sd, n = 3 batches) (Reproduced with permission of Elsevier Publishing Co.)

grown on the extracapsule medium, were detected throughout this time; IL2 free medium corresponds to a detection limit of $\sim$ 1000 cpm. Unfortunately comparison of the performance here with unencapsulated MLA 144 is complicated by the high intracapsule density (cells were encapsulated at 10^7 cells/mL) but low density relative to the volume of external medium (3.3×10^5 cells/mL). We are now trying to correlate IL2 production with the MTT signal.

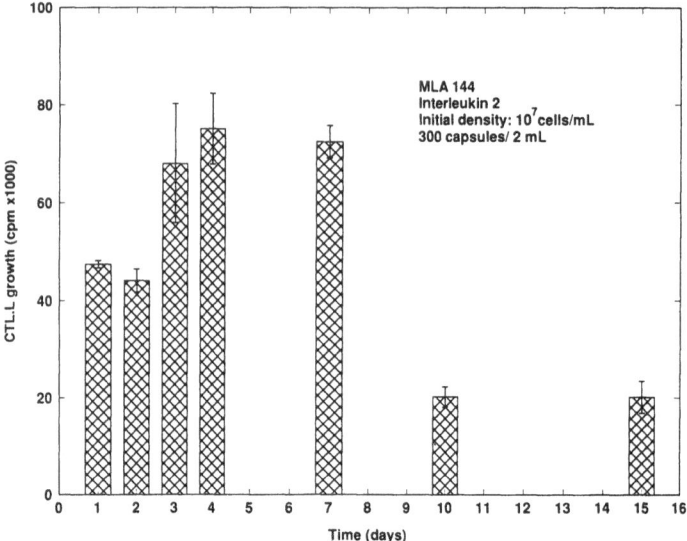

Fig. 23. Interleukin 2 production from microencapsulated MLA 144 cells. IL2 assayed by its effect on CTL.L cells to grow and incorporate ^{3}H-thymidine, when grown on the MLA 144 supernatant. (Reproduced with permission of Technomic Publishing Co.)

Histological assays by light and electron microscopy (by De Boni) have shown that, regardless of culture time (1–6 weeks), a peripheral layer ($\sim 100\ \mu m$) of PC12 cells intimately apposed to the interior aspect of the capsule, appear viable and intact, as assessed by morphological criteria similar to tumor spheroids (45). The cells comprising this layer are identical (by TEM) to cells grown without encapsulation. In contrast, the cells closer to the center show distinct and progressive changes as a function of time in vitro. Capsules maintained for 1 to 2 weeks contain cells in their center which exhibit progressive lysis with changes indicative of hypoxia (e.g., aggregation of cristae to the peripheral aspects of mitochondria and vacuolization of cytoplasm). By six weeks the centrally located cells are represented only as an amorphous mass surrounded, however, by normal appearing cells. There is some increase in the thickness of viable, normal cells along the capsule periphery, presumably resulting from division of the most viable cells routinely found at this locus. It remains to be seen how long beyond 6 weeks such a "steady state" in viable cell behavior can be maintained. A similar picture is seen for HepG2 cells in large capsules; in contrast early results from small ($\sim 400\ \mu m$ OD, $200\ \mu m$ ID) capsules indicate that viable HepG2 cells fill the entire capsule interior without central necrosis. Because of the better use of internal volume we expect to concentrate our attention on these small capsules. Mammalian cells may be encapsulated within synthetic polymeric membranes and survive to grow or function afterwards, despite the exposure to organic solvents and to extrusion processing. However it is not yet clear how the capsule wall affects subsequent cell behavior nor do we yet know what happens in vivo. These are the subjects of current research.

4 Complex Coacervation

4.1 Background and Initial Development

As an alternative to simple coacervation, microcapsules may also be formed through complex coacervation of water soluble polymers around mammalian cells. By this method we obviate the need for a non aqueous polymer solvent and produce a capsule wall with a higher water uptake, and presumably a higher permeability, than obtainable through use of a water insoluble polymer. Unfortunately, this higher water uptake is accompanied by a structural instability which must be understood and controlled in order to optimize this process for tissue transplantation purposes.

Insoluble polyelectrolyte complex may be formed when dissolved acidic and basic polyelectrolyte polymers are brought into intimate contact (131). Complex formation is generally agreed to be driven by the increase in entropy associated with the loss of small counterions into the bulk of the solution (132). Polyelectrolyte complex from concentrated solutions of strongly acidic and basic homopolymers has been shown to form sufficiently rapidly to produce a 20–30 nm thick membrane at the solution interface, as was found through reaction of dissolved poly(vinylbenzyl trimethylammonium chloride) with sodium poly (styrene sulfonate) (132).

In theory, cell containing microcapsules may be prepared by adding droplets containing cells and one of these polymers to a solution containing the other polymer, to form the capsule wall by complex coacervation of the polymers around the cell containing droplet. In practice, structural modifications have to be made in order to lessen the toxicity of the dissolved polymers, and to thicken, strengthen and improve the permeability of the capsule membrane. These improvements were achieved through use of methacrylate based polyelectrolyte co- and ter-polymers containing methacrylic acid (MAA) with others containing dimethylaminoethyl methacrylate (DMAEMA) (133, 134).

Lacking a data base to correlate polymer structure with complex forming ability, we were forced to proceed on an ad hoc basis. Approximately 30 acidic and basic copolymers spanning the full range of charge content and hydrophilicity were prepared (66, 67, 133, 134) and tested in pairs for complex forming ability in distilled water, physiological buffer, and buffered saline (133, 134). Most polymer pairs would not produce capsules. In contrast, a few combinations produced capsule membranes best described as ranging from thin and osmotically unstable to relatively thick and mechanically resilient.

Table 2. Selected water soluble copolymer characteristics

Polymer	mole fraction[a] %	mole fraction[a] %	mole fraction[a,b] %	mole fraction[a] %	Intrinsic Viscosity[c]
	HEMA	MMA	MAA	DMAEMA	(cc/g)
3	25	53	22	xx	56.6
10a	91.2	xx	8.8	xx	72.0
10b	91.5	xx	8.5	xx	114
12	22	40	xx	38	18.8
20a	91.8	xx	xx	8.2	26.2 (25.8[d])
20b	91.6	xx	xx	8.4	66.8 (53.0[d])

Key: HEMA: 2 - Hydroxyethyl methacrylate, MMA: Methyl methacrylate, MAA: Methacrylic acid,
     DMAEMA: Dimethylaminoethyl methacrylate
[a] By proton NMR. spectroscopy
[b] By acid-base titration
[c] in dimethyl formamide containing 0.2% Wt./Vol. LiBr.
[d] After quaternization.

Capsules with a good overall balance of properties were prepared from combinations of polymers containing moderate (Nos. 3 and 12) and low (Nos. 10 and 20) levels of charge. Structural characteristics of these polymers are listed in Table 2.

Potentiometric titrations of the basic polymers 12 (a mole fraction of 38% DMAEMA) and 20a (a mole fraction of 8.2% DMAEMA) in distilled water (Fig. 24) indicated that effective base strength was a weak function of alpha (the extent of ionization) at these moderate to low levels of functionality. As expected, polymer 20a proved the stronger base due its lower charge content. Titration of the acidic polymers in distilled water (Fig. 25) revealed a similar but stronger correlation between acid strength and charge content. Titration of the acidic polymers in isotonic saline revealed that acid strength increased in that medium, probably due to shielding of developing charges by the added salt. Other experiments (not shown) indicated that base strength increased also in saline (133, 134). This has important consequences in that effective charge content, and, therefore, stoichiometry of complexation becomes a function of ionic strength, thus affording a plausible explanation for the efficient complexation of polymer No.3 [a mole fraction of 22% MAA completely ionized in saline at pH 7.4] with polymer No.12 [a mole fraction of 38% DMAEMA, about 60% ionized (a mole fraction of 23%) under the same conditions], by supposing a similar effective charge content in the two polymers.

4.2 Encapsulation of Erythrocytes

To form the capsules, the polymers were first dissolved in isotonic buffered saline which was readjusted to pH 7.4 by titration with concentrated acid or base. Droplets containing the acidic polymer, with or without added cells, were

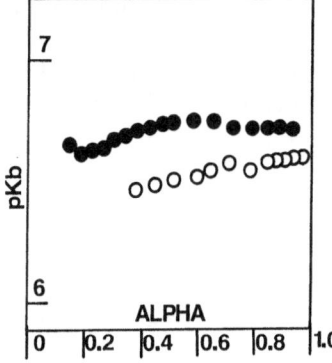

Fig. 24. pKb versus alpha curves. Polymer 12 (mole fractions of 38% DMAEMA, 22% HEMA, 40% MMA) in distilled water (*open circles*). Polymer 20a (mole fractions of 8.2% DMAEMA, 91.8% HEMA) in distilled water (*filled circles*)

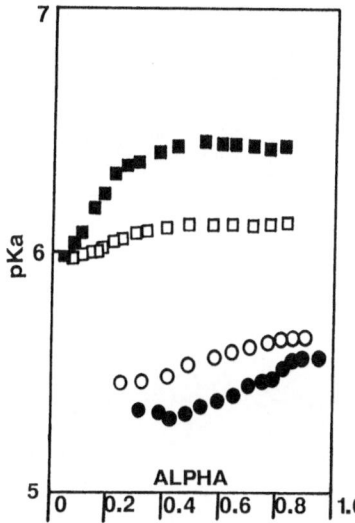

Fig. 25. pKa versus alpha curves. Polymer 3 (mole fractions of 22% MAA, 25% HEMA, 53% MMA) in distilled water (*filled squares*) and in 0.15M NaCl (*open squares*). Polymer 10a (mole fractions of 8.8% MAA, 91.2% HEMA) in distilled water (*open circles*) and in 0.15M NaCl (*filled circles*)

introduced to a receiving bath containing the basic polymer to form capsules which were transferred to polymer free buffered saline for long term storage.

We chose the acidic polymer as the interior member of the capsule forming pair due to the expected compatibility with cells of acidic polysaccharides such as alginate and carboxymethyl cellulose. For this reason, droplet viscosity and, therefore, stability was improved by synthesizing the acidic polymers as higher molecular weight analogues of the corresponding polybases.

A typical microcapsule containing 10% by volume of guinea pig red blood cells and prepared from polymers 3 and 12 is illustrated in Fig. 26 (I & II). The capsule wall was elastic and resisted puncture with a fine needle. [Such capsules (without cells) were stored intact for over six months in our laboratory]. The erythrocytes remained bright red over an 8 day storage period and retained the biconcave morphology of control cells [Fig. 26 (III)] after liberation from the capsules. [Fig. 26 (IV)]

A typical pear shaped capsule from polymers 10a and 20a is illustrated in Fig. 27 (I). Such capsules were prepared in moderate (40–60)% yield from these

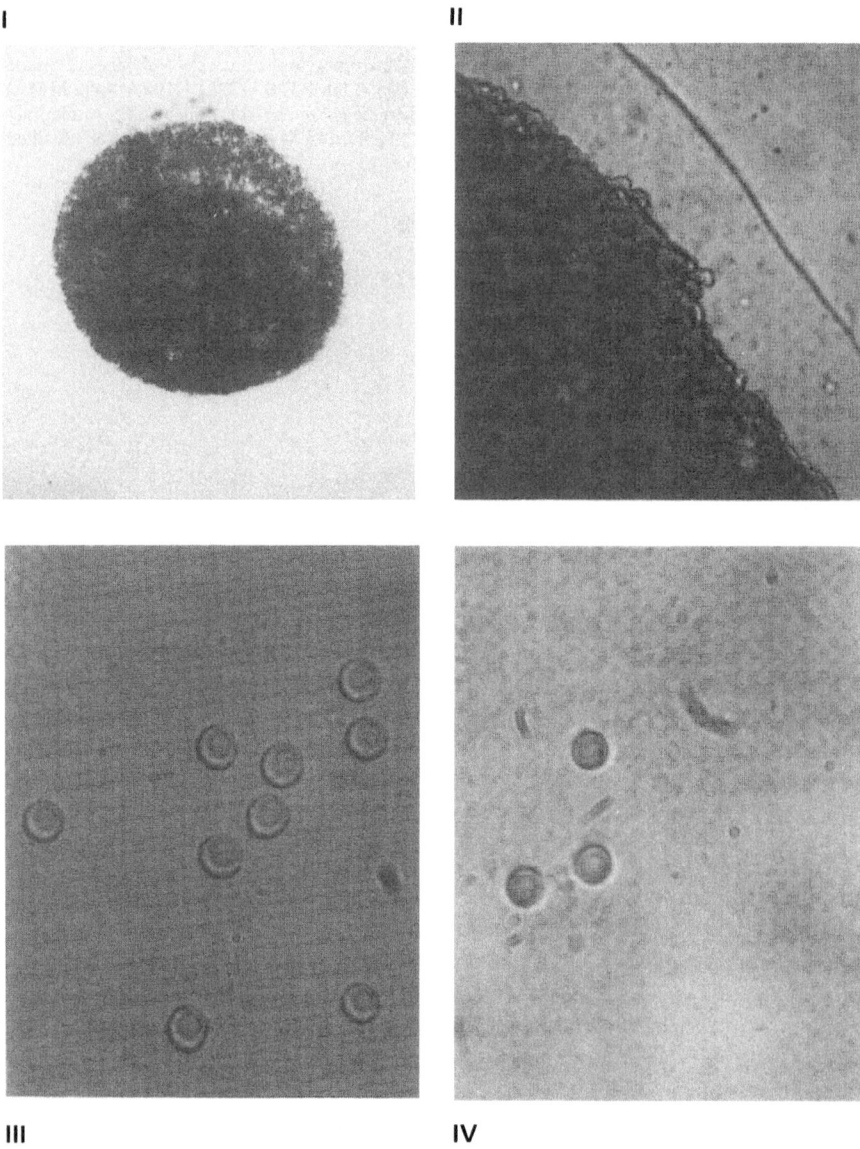

Fig. 26. (**I**) Capsule from Polymers 3 and 12 (both 5%) containing 10% guinea pig red blood cells (RBCs). (**II**) Close up of (I). (**III**) RBCs before encapsulation. (**IV**) RBCs removed from capsule 8 days after encapsulation. All solutions: Phosphate buffered saline. Long axes of photographs: (**I**) 1.4 mm, (**II**) 0.35 mm, (**III**) & (**IV**) 0.084 mm

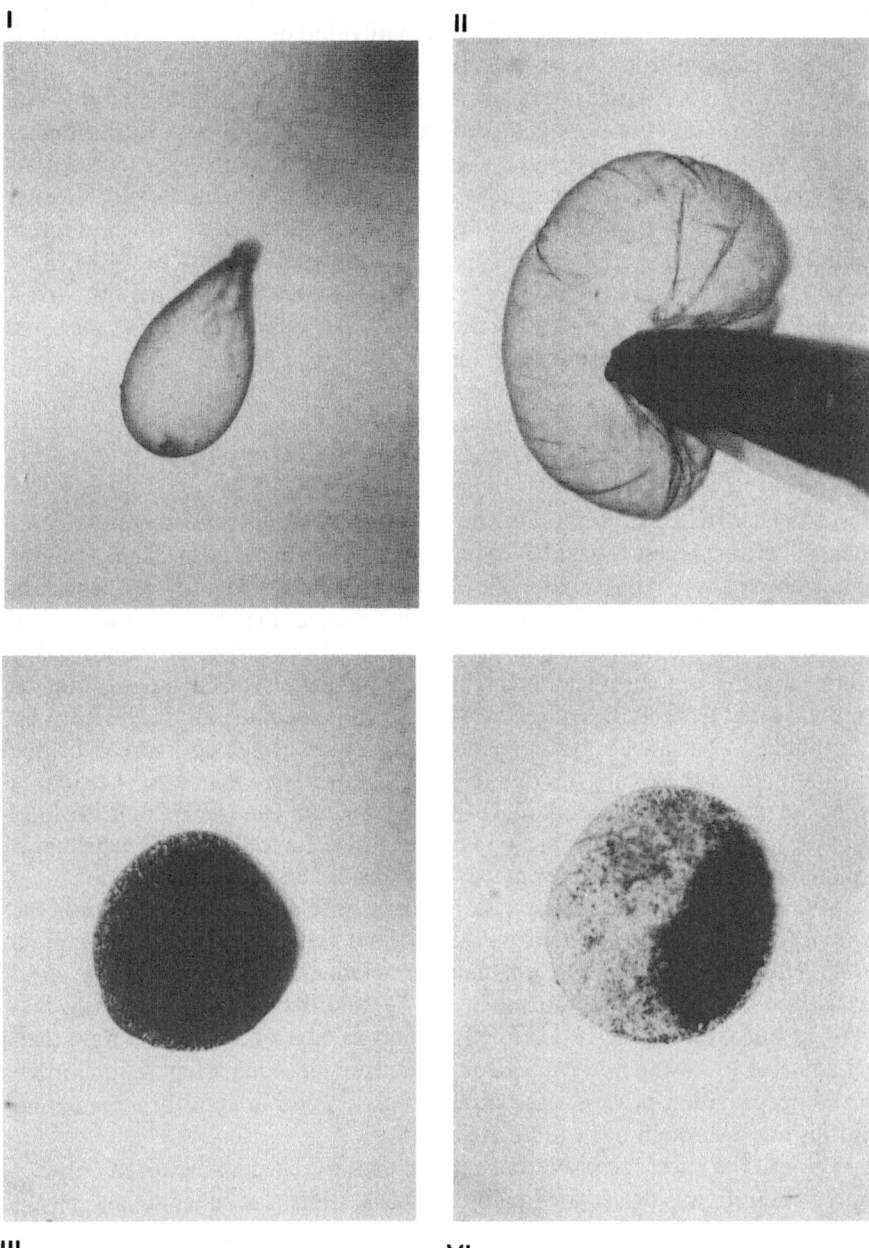

Fig. 27. Capsules by complex coacervation. (**I**) Capsule from polymers 10a and 20a, both mass fractions of 7.5% in PBS. (**II**) Capsule with inserted needle from polymers 10b (mass fraction 5%) and 20b-Q (mass fraction 3%) both in PBS. (**III**) Capsules as in (II) but containing 10% evenly dispersed haematocrit. (**IV**) Capsules as in (III) but containing sedimented RBCs. Long axes of photographs: (**I**)–(**IV**) 3.1 mm

polymers. A higher (near 100%) yield of capsules was achieved by use of polymers 10a and 20a, quaternized with methyl iodide as described in Sect. 2. A typical capsule from those polymers is indicated in Fig. 27 (II), resisting puncture with a fine gauge needle. Further improvement in capsule shape, yield, and long term stability were achieved through use of polymers with a higher molecular weight, namely 10b and a quaternized version of 20b. A representative capsule containing 10% guinea pig hematocrit is illustrated in Fig. 27 (III). A capsule (as opposed to gel) morphology was confirmed by observation of sedimented cells as in Fig. 27 (IV). Application of Drabkins test (133, 135) indicated that the survival of encapsulated cells over a 4 day period corresponded to about 41% of a control sample of cells stored at 4 °C over the same time period in phosphate buffered saline.

4.3 Encapsulation of Nucleated Cells

Erythrocytes are useful probes for the determination and adjustment of microencapsulation process variables to minimize cell trauma. Visual and microscopic observation of the bright red cells will often help to identify deficiencies in a capsule forming system. For example, color changes may be linked to oxygen concentrations and shape changes to well defined physiological trauma (136). Unfortunately, lacking nuclei and unable to divide, erythrocytes may not be used in longer term studies of cell survival; for this, we must use nucleated cells. It is unlikely that anchorage dependent cells could attach and spread on the inner surface of these capsules due to their high water content (88). Anchorage independent cells were, therefore, chosen by default. The transparent capsules allowed the use of visual microscopy to examine for cell division which was equated with a biocompatible environment inside the capsule.

Human Burkitt Lymphoma ("Raji") cells were obtained frozen from the American Type Culture Collection; thawed, and cultured, in RPMI containing 10% Fetal Bovine Serum. The polymers for microencapsulation were dissolved in distilled water with pH adjustment to 7.4, and dialysed extensively against distilled water through a 10^4 Da cut off dialysis bag. The polymers were then pH adjusted to 7.4 and freeze dried to constant weight. Final preparation for the microencapsulation process was limited to dissolution in appropriate medium and filtration through a 0.4 μm cartridge.

Prior to the microencapsulation experiments, the effect of dissolved polymer on cell viability was determined by the simple expedient of subculturing cells in medium contaminated with polymer. Relevance to the encapsulation process stems from the short term exposure of the cells to a high concentration and a longer term exposure to a residual concentration of dissolved interior polyelectrolyte polymer. Raji cells were subcultured in the presence of 0.5% (Wt./Vol) of polymers 3 and 12; no attempt was made to compensate for changes in solution osmolarity which accompany the addition of those polymers to medium (133, 134). The results of these experiments are reported in Fig. 28.

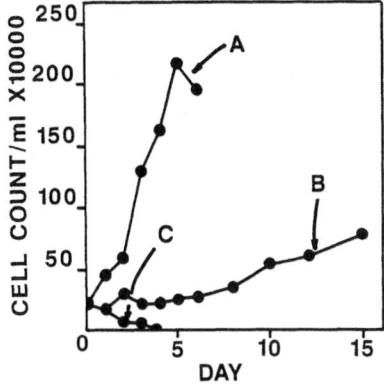

Fig. 28. Growth of Human Burkitt Lymphoma cells in RPMI containing 10% fetal bovine serum. (*A*) Control medium, (*B*) Control medium + 0.5% (Wt./Vol.) polymer 3. (*C*) Control medium + 0.5% (Wt./Vol.) polymer 12. Measurement by hemocytometer

As expected, the control cells (A) divided rapidly to a maximum solution concentration of about 10^8/ml. Cells in the presence of polymer 3 (28B) divided more slowly but steadily, and no signs of cell death were observed when the experiment was terminated after 15 days. In contrast, dissolved cationic polymer (28C) proved toxic and promoted rapid necrosis of the cells. We were indeed fortuitous in our initial choice of acidic polymer as the interior member of the capsule forming pair.

Raji cells were next immobilized in capsules prepared from 5% (Wt./Vol.) solutions of polymers 3 and 12. A batch of cells were harvested in the growth phase, divided, and encapsulated using phosphate buffered saline or RPMI + FBS as the vehicle for both polymers at an initial cell density of about 1.1×10^6/ml internal solution. Capsules, 1 day after preparation in PBS, are depicted in Fig. 29A. The individual cells are represented as dots in the capsule. After 3 days in PBS the cells have clumped together (Fig. 29B), a promising sign that some viability is retained. Cells entrapped with RPMI + FBS clump within 24 h of encapsulation (Fig. 29C), and divide continuously through day 6 (Fig. 29D) to fill the capsule at day 15 (Fig. 29E). The close-up of the filled capsule (Fig. 29F and others not shown) illustrates the thin capsule wall.

The following conclusions may be drawn from these studies. For example, the process of encapsulation is sufficiently mild that entrapped cells retain the ability to divide and multiply. The cells must also be substantially protected from the cationic polymer during encapsulation. The capsule wall must, therefore, be permeable to required nutrients including high molecular weight components of FBS and oxygen, and allow for a rapid removal of cell waste products and for equilibration of osmolarity and pH with the capsule exterior.

These, and similar results obtained through use of polymers 10 and 20, indicate that capsules by complex coacervation may prove useful vehicles for islet transplantation. Before this may be achieved, however, perifusion studies must be undertaken to determine if encapsulated islets of Langerhans retain sufficient viability to supply insulin "on demand"; and if that insulin can be delivered through the capsule wall quickly enough to restore normoglycemia in

A   B   C

D   E   F

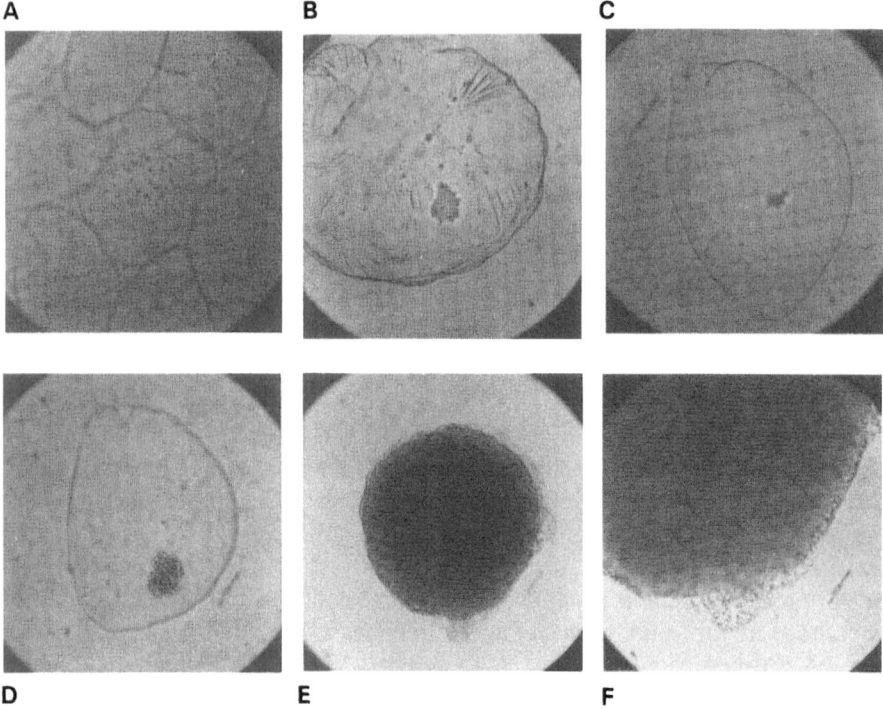

Fig. 29A–F. Capsules from polymers 3 and 12 and containing "Raji" cells. (**A**) Capsules in PBS, day 1. (**B**) Capsules as in (A), day three. (**C**) Capsules prepared in RPMI containing fetal bovine serum, day 1. (**D**) Capsules as in (C), day 6. (**E**) Capsules as in (C), day 15. (**F**) close-up, as in (C), day 15. Long axes of photographs, **A–E:** 1.68 mm, **F:** 0.84 mm

a timely fashion. Implantation studies must also be performed to determine if these capsules are, or can be made, sufficiently strong and bio-inert to survive the mechanical and physiological trauma of implantation.

5 Hybrid Methods

The drive towards microencapsulation systems based on the use of synthetic hydrophilic methacrylate based polymers is fueled by their proven biocompatibility, (56) hydrolytic stability, (57) ease of synthesis (66, 67) and enormous structural diversity made possible through copolymerization. In contrast, interest in polysaccharide gel formers such as alginate is founded upon the relative ease of capsule formation under physiological conditions. It would seem inevitable that attempts be made to combine the host biocompatibility and stability of methacrylate based polymers with the ease of capsule formation associated with the use of alginate. We present here the results of three preliminary studies which indicate that such an approach has merit.

5.1 Template Polymerization Around Alginate

5.1.1 Background and Method Development

Entrapment of mammalian cells in calcium alginate gel may be achieved by mixing the cells with dissolved sodium alginate under physiological conditions and extruding the mixture into a receiving bath containing calcium ion. Guluronic acid blocks in the alginate react with calcium to form the well known "eggbox" complex which crosslinks the polysaccharaide into a gel (137). Providing cell division is not expected, cells may be cultured in vivo in calcium alginate. Unfortunately, calcium alginate is unstable in vitro, and allows entry of immunoglobulins and other components of the hosts immune system. Permeability has been controlled and in vivo stability improved by forming a very thin polyelectrolyte complex with polylysine at the capsule surface, then liquefying the interior alginate by reaction with sodium citrate (138). Even so, questions remain as to the biocompatibility and physical strength of this fragile polysaccharide-protein complex.

The in vivo stability of a "natural" polyelectrolyte complex membrane, such as is formed between alginate and polylysine, (or even between synthetic polyelectrolytes) must never be assumed due to slow site-by-site displacement reactions which may occur with high molecular weight polymers (proteins, etc.) present in body fluids, and to processes of hydrolysis, enzymatically promoted or otherwise which may disrupt the membrane.

In brief, a water soluble polyhydric alcohol (a copolymer of HEMA with MAA) is mixed with the sodium alginate and cell containing solution and extruded into a receiving bath containing Ca^{+2} and a polyepoxide polymer [a copolymer of glycidyl methacrylate (GMA) with n-vinyl pyrrolidinone (NVP)]. The alginate gells, trapping the cells and establishing a framework around which the hydrogel coating may be formed by slower epoxy/alcohol etherification. Polymer structures are represented in Fig. 30 and an analysis of GMA-co-NVP polymers 52 and 53 made in Table 3.

A number of hydroxylated water soluble polymers were examined as coreactants with polymer 52 in the absence of calcium alginate, and were judged on the basis of the rate of gel formation and the physical properties of the gel. These polymers included sodium alginate, polyvinyl alcohol, and copolymers of HEMA with MAA. Of the polymers tested, best results were obtained with polymer 10a, a copolymer of HEMA with a mole fraction of about 10% MAA, which rapidly produced an elastic gel on exposure to polymer 52 in solution. Simple condensation of the carboxyls in polymer 10a with the epoxide functionality was ruled out as a competing reaction due to the measurable but slow reaction between polymer 52 and poly methacrylic acid. It is, therefore, likely

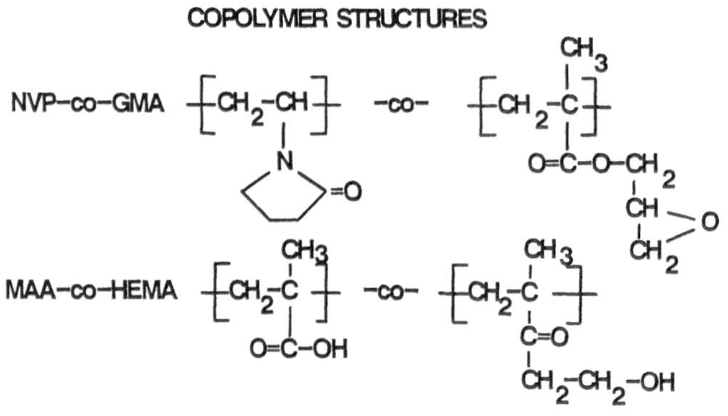

COPOLYMER STRUCTURES

Fig. 30. Monomer residues in copolymers used to prepare the hydrogel coating: *n*-vinyl pyrrolidin-one (NVP) *-co-* glycidyl methacrylate (GMA) & methacrylic acid (MAA) *-co-* 2-hydroxythyl methacrylate (HEMA)

Table 3. Characteristics of copolymers containing glycidyl methacrylate (GMA) and n-vinyl pyrrolidinone

Polymer	mole fraction(%) GMA[a]		mole fraction(%) GMA[b]
	In Feed	Conversion	In Polymer
52	10	23.7	40
53	2.25	20.2	9.6

Key: [a] 70°C for 1 h, Recovered by coagulation into petroleum ether. Yield expressed as a percentage of the monomer charge.
[b] Measured by Proton NMR Spectroscopy in d1 chloroform.

that the MAA serves primarily as a vehicle to solubilize the HEMA residues in polymer 10a and other such copolymers.

5.1.2 Capsule Preparation and Morphology

We have developed a rationale to describe the processes operative in the preparation of these hydrogel coated capsules, primarily through a visual comparison of lyophilized coated and uncoated alginate gel by Scanning Electron Microscopy (SEM).

Uncoated calcium alginate gel beads were prepared for examination by dripping a solution of 1% (Wt./Vol.) Kelko LV sodium alginate into a TRIS buffered isotonic solution of calcium chloride. Examination of these uncoated calcium alginate gel beads reveals a smooth featureless surface (Fig. 31a & b). This surface layer was very fragile and may be disrupted by simple handling (as in Fig. 34A). The interior of the gel (Fig. 31c & d) had a cellular, almost sponge like morphology.

Coated alginate beads and capsules were prepared by extruding an isotonic buffered aqueous mixture containing 1.3% sodium alginate and 2.5% polymer 10a (as described in Sect. 4) into an 5 Wt.% solution of polymer 52 in isotonic buffer. After 30 min, the beads were transferred to a calcium free solution of polymer 52 for 24 h before examination. It must be emphasized that exposure to this heavily epoxidized polymer for a prolonged time was used as a means of generating free standing hydrogel coatings suitable for examination by SEM, and that such treatment regimes should not be confused with milder versions used to coat cell containing capsules.

An intact coated gel bead, after a 30 min treatment with sodium citrate to remove surface entrained alginate, is pictured in Fig. 32a, b, and c at low and successively higher magnifications respectively. It would appear from these observations that the hydrogel coating forms by reaction of the two polymers in the alginate void space, hence the label "template polymerization" as applied to this process.

a　　　　　　　　　　　　　　　　b

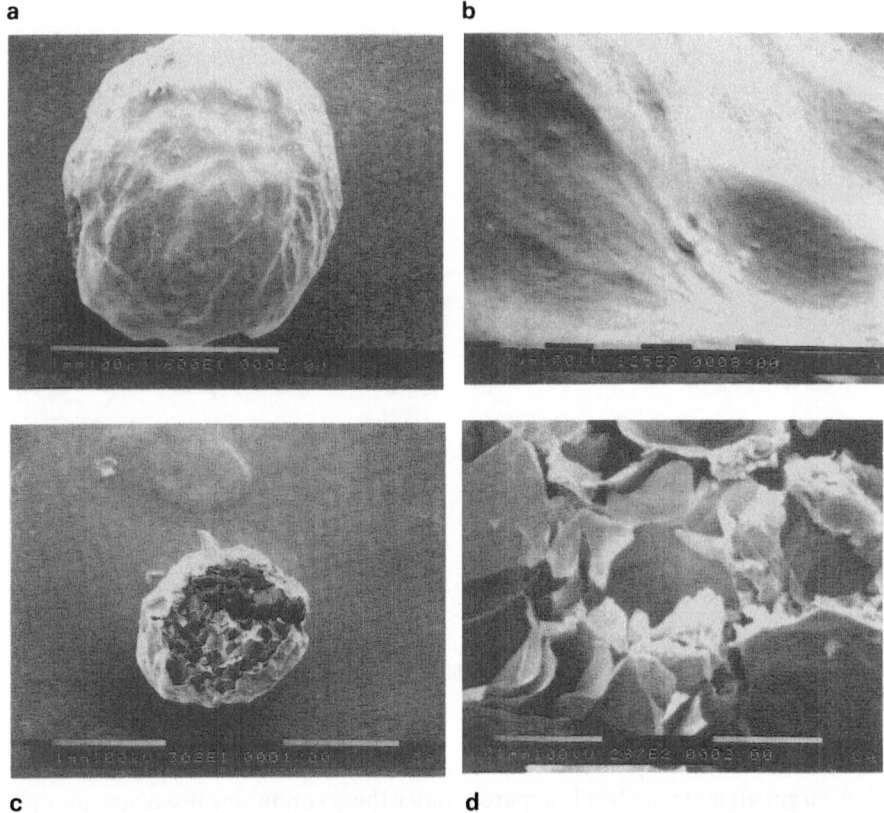

c　　　　　　　　　　　　　　　　d

Fig. 31a–d. SEMs of calcium alginate beads, magnifications as indicated by white spacebar. (**a**) Intact bead at low magnification. (**b**) Bead surface, featureless at higher magnification. (**c**) Fractured alginate bead at low magnification showing cellular interior. (**d**) Close-up of fracture surface

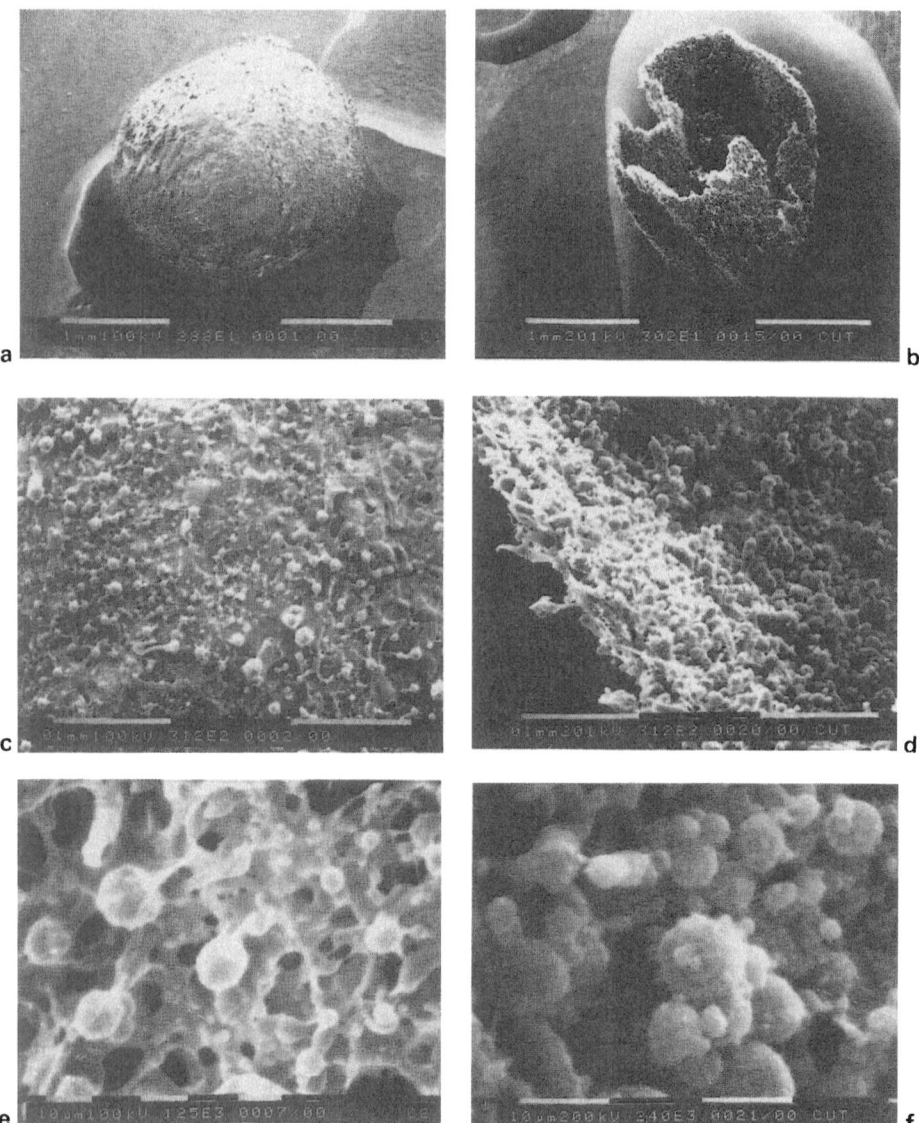

Fig. 32a–f. SEMs of hydrogel coated alginate gel bead, after treatment with sodium citrate [**a, b & c** – 30 min: **d, e & f** – 24 h], magnifications as indicated by white spacebar. (**a**) Intact bead at low magnification. (**b**) & (**c**) Surface of intact bead at higher magnification. (**d**) Fractured capsule at low magnification. (**e**) & (**f**) Cross section and interior surface, respectively, of fractured capsule at higher magnification

A virgin alginate gel bead prepared under these conditions dissolved after a 30 min treatment with isotonic sodium citrate. In contrast, examination of a subsequently fractured coated bead after the same treatment indicated that the internal alginate gel morphology was retained; strong circumstantial evidence

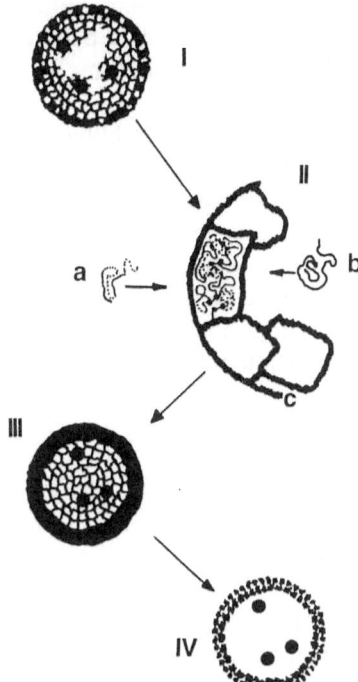

Fig. 33. Schematic representation of development of coating morphology. (**I**) Simultaneous ionotropic gelation of the alginate with synthetic hydrogel formation at the capsule surface; *filled cavities* at the bead surface denote hydrogel formation, *filled circles* denote cells. (**II**) Schematic of reaction of (*a*) Epoxidized copolymer with (*b*) Hydroxylated copolymer in alginate void space near surface of gel bead. (**III**) Hydrogel coated alginate gel bead. (**IV**) Capsule after liquification of interior alginate and extraction of surface alginate in citrate solution

for the formation of a continuous coating. A 24 h exposure to citrate liquefied the capsule interior as is evident from examination of the fractured capsule in Fig. 32d. Although a measurably thick (about 0.1 mm) coating is achieved, the process is shown to be self limiting. Examination of a cross section of the hydrogel and the interior surface (Figure 32e & f, respectively) indicates that the surface morphology is repeated throughout.

Hydrogel formation may be best rationalized, with reference to Fig. 33. The rapid formation of calcium alginate is accompanied by a slower reaction of polymer 10a (a) and 52 (b) at the surface of the gel (Fig. 33I & II). The hydrogel coating eventually becomes a diffusional barrier to further reaction (Fig. 33 III) and can be recovered as an inverse of the alginate matrix by treatment with citrate ion (Fig. 33 IV).

Citrate washed hydrogel coatings were examined by Fourier Transform Infrared (FTIR) spectroscopy with spectral subtraction and were found to consist of a reaction product of polymers 10a and 52 and to be free of alginate.

5.1.3 Encapsulation of Erythrocytes

Guinea Pig red blood cells (10% hematocrit) were encapsulated in isotonic TRIS buffered saline containing 1.35% alginate and 2.5% polymer 10a. The cell containing solution was exposed to isotonic TRIS buffered calcium chloride containing 5% of polymer 53 for 30 min, transferred to an isotonic buffered, calcium free solution containing 5% of polymer 53 for 4 h, then to isotonic TRIS buffered saline for long term storage.

The survival of encapsulated RBC's was determined by Drabkins test. Approximately 71% of encapsulated cells survived for 4 days as opposed to 96% of a control sample isolated at the same time and stored in TRIS buffered saline. Although these results indicate that the process of encapsulation followed by a short storage period did not result in undue necrosis, caution must be used in any extrapolations toward the behavior of nucleated cells. Even so, these indefinitely stable and mechanically resilient coatings hold promise as an alternative to polylysine treatment for the post stabilization of calcium alginate gel prior to transplantation.

5.2 Alginate Stabilized Methacrylate Emulsion

5.2.1 Background

High energy radiation is a preferred method for the initiation of polymerization under certain circumstances. The use of high energy radiation can also lead to a unique product unobtainable by other methods. We have shown that using a weak gamma source (for example, 800 min/M Rad) certain hydrophilic methacrylate based monomers may be polymerized in aqueous solution in the presence of dissolved sodium alginate to yield stable hydrogel emulsions (139). Over a certain compositional range, these may be destabilized as a cohesive precipitate in the presence of calcium ions. Evidence has been presented to the effect that the growing hydrogel nuclei are stabilized by a block or graft copolymer of the alginate with the vinyl monomer which precipitates onto its surface to stabilize against coagulation by electrostatic repulsion. We have also shown that this effect may be exploited to form hollow cell containing microcapsules under mild conditions.

5.2.2 Preparation and Examination of the Capsules

Based on the results of preliminary investigations it was concluded that capsules could be formed from emulsions containing about 2–10% (mass fraction) monomer and 0.5–2% (mass fraction) alginate subjected to radiation doses greater than about 7.5×10^{-2} M Rad to reduce residual monomer levels below toxic limits. Emulsions may be prepared in distilled water, saline, or buffered saline, but not in calcium containing solution.

Capsules were prepared from emulsion containing 1% sodium alginate and 5% PHEMA and coagulated in isotonic calcium chloride to form large capsules for examination by SEM. The surface of a large capsule is compared in Fig. 34B

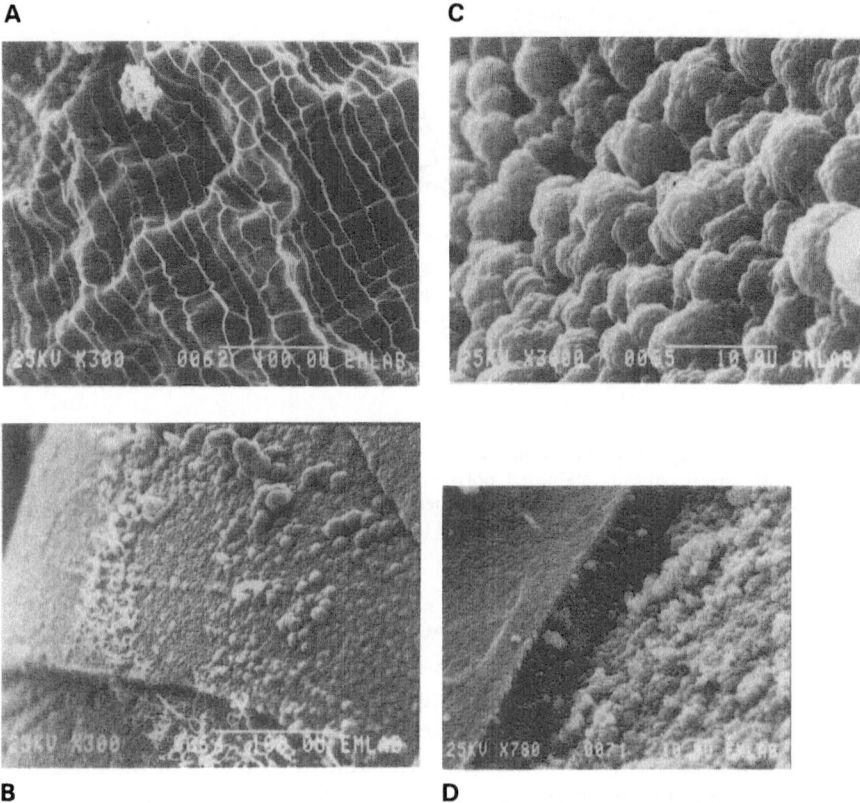

Fig. 34A–D. High magnification SEMs of exterior of capsules from **(A)** native (1%) calcium alginate with exterior surface "sloughed off" with handling. **(B & C)** A graft copolymer emulsion (1% alginate, 5% HEMA). **(D)** Graft copolymer capsule fragment showing interior wall surface (1% alginate, 5% HEMA)

with that of a calcium alginate gel bead (Fig. 34A). Unlike the alginate which displays a characteristic cellular morphology, the surface of the capsule is shown at higher magnification in Fig. 34C to be composed of cemented together particles of emulsion. A cross section of capsule wall (Fig. 34D) reveals that the inner surface is composed of smaller particles than the outer surface. This may be rationalized on the basis that the least stable large particles of emulsion will precipitate first of all to form the outer layers of the capsule wall. More stable components (i.e., smaller particles) will precipitate as calcium diffuses through the outer layers and into the capsule. The capsules are hollow and free of the internal calcium alginate meshwork morphology, an effect attributed to depletion of free alginate to levels below which it is able to gel in the presence of calcium ion (139).

The formation of this cohesive precipitate is pictured in Fig. 35. On the left are particles of emulsion stabilized by entrained or grafted alginate. The alginate

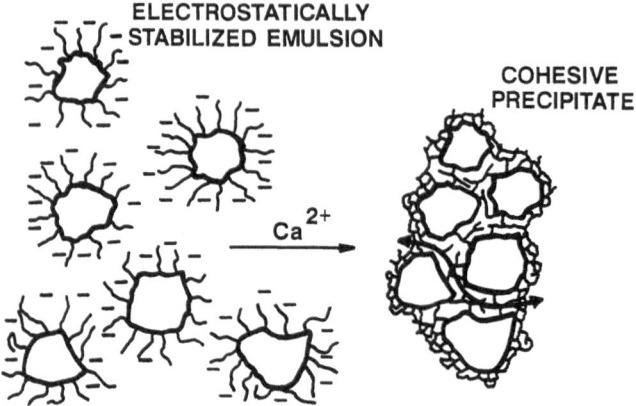

Fig. 35. Schematic representation of formation of "cohesive precipitate". Shown (*left*) is the emulsion with electrostatically stabilized polymer microparticles with extended surface bound sodium alginate chains. Shown (*right*) is the "cohesive precipitate" containing destabilized microparticles with collapsed surfactant and with some (assumed) ionotropic bridging chains

chains extend far into the solution as would be expected of a surface bound polymer in the "brush" regime (140). On addition of calcium the alginate chains collapse and self complex to destabilize the emulsion which precipitates. Some of the alginate will form bridges with surfactant on other particles of emulsion to form the cohesive precipitate. It seems reasonable to assume that the preferred method of mass transfer across the capsule wall will be around rather than through the particulate phase as is indicated in Fig. 35.

5.2.3 Encapsulation of Erythrocytes

A preliminary investigation was made of the toxicity of the graft copolymer emulsion using human erythrocytes as a model cell line. First of all the toxicity of an emulsion prepared from 5% HEMA and 1% alginate in phosphate buffered saline was estimated by examination of a mixture with 10% hematocrit after 1 h storage. No lysis could be observed and the cells could be separated and disrupted in distilled water. We then encapsulated 10% hematocrit in the same emulsion and stored the capsules in isotonic saline for one week. The cells were recovered by disruption of the capsules with citrate and were perceived as healthy as per the simple tests outlined above.

It would appear that these graft copolymer emulsions constitute a novel method of capsule preparation with some promise as a vehicle for cell transplantation purposes. Before such may be achieved, it must be ascertained if the surface bound alginate elicits an inflammatory response from the mammalian host, and if the capsules themselves are sufficiently permeable to allow for the efficient transport of nutrients and cell products whilst excluding components of the hosts immune system.

5.3 Precipitation of a Cationic Methacrylate Emulsion onto Calcium Alginate

Dilute aqueous polymeric emulsions are commonly stabilized through the use of polymeric surfactants. If the stabilizer is uncharged, the emulsion is stabilized entropically by segmental exclusion. In most instances, however, stabilization is a by product of coulombic repulsions generated by a polyelectrolyte surfactant. In a few instances the polymer itself is able to act as surfactant. For example, Eudragit RL, a commercially available partially quaternized cationic methacrylate based polymer, is able to form indefinitely stable emulsions in distilled water or buffered saline (141). These emulsions are prepared by adding polymer to boiling solution and are presumably stabilized by concentration of cationic functionality at the particle surface.

These emulsions of Eudragit RL become destabilized in the vicinity of calcium alginate gel, around which they precipitate to form a dense coating about 60 μm thick, bound to the negatively charged alginate by electrostatic attractions. The strength of this coating was measured by compressive testing and was found to increase with the concentration of alginate in the gel bead and the coating time, and to decrease with the ionic strength of the coating solution (141).

Coatings of Eudragit RL100 were shown to offer only minimum diffusional resistance to cyanomethemoglobin which was released from lysed red blood cells. Encapsulated red blood cells remained viable after coating, as measured by

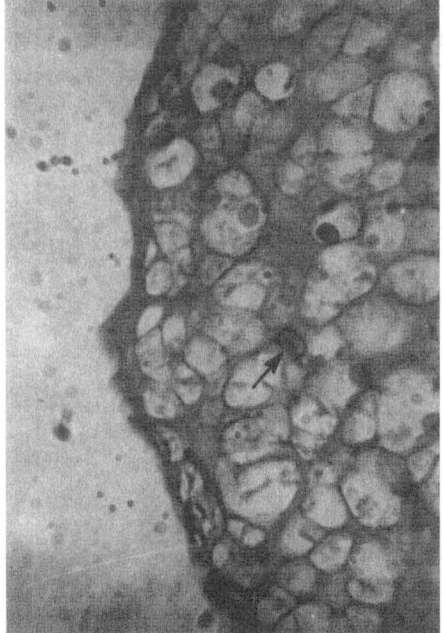

Fig. 36. Histological sections of Eudragit RL coated alginate beads containing immobilized erythrocytes stained with hematoxylin eosin (3% alginate, 60 min). (Reproduced with permission of Elsevier Publishing Co.)

glucose consumption, except for those at or near the surface of the bead and adjacent to the coating. In Fig. 36 are histological sections of Eudragit RL100 coated calcium alginate containing immobilized red blood cells. Staining by hematoxylin-eosin reveals the erythrocytes as biconcave disks randomly immobilized in the alginate network, and the capsule wall as a band of irregular thickness around the alginate. Some of the RBC's appeared spiky, presumably due to high calcium concentrations in the gel (142).

A methodology is available for the coating of calcium alginate gel beads with a methacrylate based cationic emulsion. Although subsequent implantation studies revealed that Eudragit RL100 coated calcium alginate elicited a pronounced inflammatory reaction in the mammalian host, these results indicate that this method of surface stabilizing calcium alginate, through reaction with cationic emulsion, may play a role in the development of future microencapsulation systems, provided that Eudragit RL may be substituted by a more biocompatible polymer. Thermodynamic analysis of this problem with a view to optimizing the polymer has been undertaken (143).

6 Conclusion

We have outlined the diversity and procedural flexibility afforded through the use of appropriately designed synthetic polymers for mammalian cell immobilization. Having selected the acrylate family of monomers because of their diversity, we have shown that mammalian cells may be microencapsulated in uncharged and polyelectrolyte polymer, in polyelectrolyte complexes and inside a cohesive precipitate from a destabilized emulsion; all without significant loss of viability. Through this chemical diversity and inherent biocompatibility, these systems hold forth the possibility of improved transplantation therapies for a wide variety of cellular diseases.

Acknowledgements. The work summarized here was supported by the Natural Sciences and Engineering Research Council, Medical Research Council, the Canadian Diabetes Association, the American Diabetes Association (WTKS), the Sedgewick Country mill levy fund (WTKS), the Ontario Centre for Materials Research (OCMR) and NIH (DK 29689).

7 References

1. Chang TMS (1972) Artificial cells. C.C. Thomas, Springfield, Illinois
2. Lim F, Sun AM (1980) Science 210:908
3. O'Shea GM, Goosen MFA, Sun AM (1984) BBA 804:133–136

4. Sun AM, O'Shea GM (1985) J Contr Rel 2:137
5. Ar'Rajab A, Bengmark S, Ahren B (1991) Transpl 51:570–574
6. O'Shea GM, Sun SM (1986) Diabetes 35:943–946
7. Fan M-Y, Lum Z-P, Fu X-W, Levesque L, Tai IT, Sun AM (1990) Diabetes 39:519–522
8. Sun Y, Vecek I, Sun AM (1991) Artif Org 15: Abstract 22
9. Chicheportiche D, Darquy S, Lepeintre J, Capron F, Halban PA, Reach G (1990) Diabetolog 33:457–461
10. Sutherland DER, Moudry-Munns KC (1990) Transpl Proc 22:571–574
11. Tze WJ, Tai J (1982) Transpl Proc 14:714–723
12. Lafferty KJ (1988) In: van Schilfgarde R, Hardy MA (eds) Transplantation of the endocrine pancreas in diabetes mellitus. Elsevier, Amsterdam, p 279
13. Scharp DM, Lacy PE, Santiago JV, et al. (1990) Diabetes 39:515
14. Warnock GL, Kneterman NM, Ryan EA, Evans MG, Seelis REA, Halloran PF, Rabinovitch A, Rajotte RV (1989) Lancet 8662:570–572
15. Hering BJ, Bretzel RG, Federlin K (1988) Horm Metabol Res 20:537–545
16. Gray DWR, Morris PJ (1987) Transp 43:321–331
17. Ricordi C, Finke E, Lacy PE (1986) 35:649–653
18. Rajotte RV, Warnock GL, Coulombe MG (1988) In: Transplantation of the Endocrine Pancreas in Diabetes Millitus, van Schilfgarde R, Hardy MA, (eds) Elsevier 125–135
19. Colton CK, Avgoustiniatos ES (1991) J Biomech Eng 113:152–170
20. Tze WJ, Wong FC, Cheng LM (1979) Diabetolog 16:247
21. Sun A, Parisus W, Macmorine H, Sefton MV, Stone R (1980) Artif Org 4:276
22. Maki T, Ubhi CS, Sanchez-Farpon H, Sullivan SJ, Borland K, Muller TE, Solomon BA, Chick WL, Monaco AP (1991) Transpl 51:43–51
23. Reach G, Jaffrin MY, Desjeux JF (1984) Diabetes 33:752–761
24. Altman JJ, Houlbert D, Chollier A, Leduc A, McMillan P, Galletti PM (1984) Trans ASAIO 30:382–386
25. Lanza RP, Butler DH, Borland KM, Staruk JE, Faustman DL, Solomon BA, Muller TE, Rupp RG, Maki T, Monaco AP, Chick WL (1991) Proc NAS 88:11100–11104
26. Warnock GL, Rajotte RV (1988) Diabetes 37:467
27. Eisenbarth GS, Eng N (1986) J Med 314:1360–1368
28. Sibley RK, Sutherland DE, Goetz F, Michael AF (1985) Lab Invest 53:132–144
29. Naji A, Silvers WK, Kimura H, Anderson AO, Barker CF (1983) Metabolism 32:62
30. Sangan J (1991) Am Soc Artif Int Org Chicago, IL
31. Sanberg P (1991) Am Soc Artif Int Org Chicago, IL
32. Sun AM, Cai Z, Shi Z, Ma F, O'Shea GM (1987) Biomat Artif Cells Artif Org 15:483–496
33. Wong H, Chang TMS (1988) Biomat Artif Cells Artif Org 16:731–739
34. Bruni S, Chang TMS (1989) Biomat Artif Cells Artif Org 17:403–411
35. Darquy S, Sun AM (1987) Trans ASAIO 33:356–358
36. Romi MM, Lum ZP, Sun AM, Chang PL (1991) Miami biotechnology winter symposium proceedings p 109
37. Aebischer P, Winn SR, Tresco PA, Jaeger CB, Greene LA, J Biomech Eng (in press)
38. Aebischer P, Tresco PA, Winn SR, Greene LA, Jaeger CB (1991) Exp Neurol 111:269–275
39. Aebischer P, Winn SR, Galletti PM (1988) Brain Research 448:364–368
40. Christensen L, Aebischer P, Galletti PM (1988) Trans ASAIO, 34:681–686
41. Aebischer P, Wahlberg L, Tresco PA, Winn SR (1991) Biomaterials 12:50–56
42. Clicheportiche S and Reach G (1988) Diabetolog 31:54
43. Dionne KE, Colton CK, Yarmush ML (1989) Trans ASAIO 35:739–741
44. Weir GC, Bonner-Weir S, Leahy JL (1990) 39:401–405
45. Sutherland RM (1988) Science 240:177–184
46. Williams DF (ed) (1987) Definitions in biomaterials. Elsevier, Amsterdam
47. Gin H, Dupuy B, Bonnemaison-Bourignon D, Bordenave L, Bareille R, Latapie MJ, Ch Baquey, Bezian JH, Ducassou D (1990) Biomat Artif Cells Artif Org 18:25–42
48. Ricker AT, Stockberger SM, Halban PA, Eisenbarth GS, Bonner-Weir S (1986) In: Immunology of Diabetes, Jaworski MA et al. (eds) Elsevier, 193–200
49. Mazaheri R, Atkison P, Stiller C, Dupre J, Vose J, O'Shea G (1991) Transpl 51:750–754
50. Weber CJ, Zabinski S, Koschitzky T, Wicker L, Rajotte R, D'Agati V, Peterson L, Norton J, Reemtsma K (1990) Transpl. 49:396–404
51. Calafiore R, Calcinaro F, Basta G, Pietropoalo M, Falorni A, Piermattei M, Brunetti P (1990) 39:175–181

52. Soon-Shiong P, Skjak-Braek G, Smidsrod O, Espevik T, Otterlei M, Heintz R, Sawhney A, Hubbell J (1991) Third Int'l Congress on Pancreatic and Islet Transplantation, Lyons, France, Abstracts p 193
53. Ma X, Vecek I, Sun A (1991) Artif Org 15 (4) Abstract #16
54. Sawhney AS, Soon-Shiong P, Hebbell JA (1991) Trans Soc Biomater 14:156
55. Ratner BD, Hoffman AS (1976) ACS Symposium Series 31:1
56. Peppas NA, Moynihan HJ (1987) In: Peppas NA (ed) 2, CRC Press p49
57. Bevington JC, Eaves DE, Vale RL (1958) J Polym Sci 33:317
58. Gregonis DE, Chen CM, Andrade JD (1976) ACS Symposium Series 31:88
59. Simon GP (1991) Ion Exchange Training Manual, Van Nostrand Reinhold, NY
60. Williams DF (1987) J Mat Sci 22:3421
61. Stevenson WTK, Evangelista RA, Broughton RL, Sefton MV (1987) J Appl Polym Sci 34:65
62. Macret M, Hild G (1982) Polymer 23:81
63. Flory PJ (1953) Principles of polymer chemistry, Cornell University Press, Ithaca, NY
64. Fort RJ, Polyzoidis TM (1976) Eur Polym J 12:685
65. Askill IN, Gilding DK (1981) Polymer 22:342
66. Shao Wen, Yin Xiaonan, Stevenson WTK (1991) J Appl Polym Sci 42:1399
67. Shao Wen, Yin Xiaonan, Stevenson WTK (1991) J Appl Polym Sci 43:205
68. Shao Wen, Yin Xiaonan, Stevenson WTK, Polymer International (In Press)
69. Greenley RZ (1975) In: Brandrup J, Immergut H (eds) Polymer handbook, 2nd edn, Wiley, NY
70. Kelen T, Tudos F (1975) J Macromol Sci Chem A9(1), 1
71. Salomon JC, Watterson AC, Quach L, Raheja MK (1985) ACS Polymer Preprints, 26(1), 196
72. Shao Wen, Stevenson WTK, Manuscript submitted to Colloid and Interface Science
73. Webster OW, Sogah DY, p163 In: Comprehensive Polymer Science, Vol.4, Eastmond GC, Ledwith A, Russo S, Sigwald P (eds)., (1989) Pergammon Press, London. Schubert W, Bandermann F (1989) Makromol Chem 190, 2721. Schubert W, Sitz HD, Bandermann F, ibid, 190:2193
74. Gergonis DE, Russel GA, Andrade AD, deVisser AC (1978) Polymer, 19:1279
75. Bovey FA, Mirau PA (1990) Makromol Chem Makromol Symp 34:1
76. Dusek K, Bohdanecky M, Prokopova E (1974) Eur Polym J 10:239
77. Jeon SI, Jhon MS (1984) J Polym Sci Polym Chem edn., 22:3555
78. Bovey FA (1969) Polymer Conformation and Configuration, Academic NY
79. Hummel DO (1966) Polym Rev 14
80. Stevenson WTK, Sefton MV (1988) J Appl Polym Sci 36:1541
81. Heacock JK (1963) J Appl Polym Sci 7:2319
82. Smyth G, Quinn FX, McBrierty V (1988) Macromolecules, 21:3198
83. Rabin Y, Cohen J, Priel Z (1988) J Polym Sci Part C: Polym Lett 26:397
84. Pratt KJ, Williams SK, Jarrell BE (1989) J Biomed Mater Res 23:1131–1148
85. Rapoza RJ, Horbett TA (1990) J Biomed Mater Res 24:1263–1288
86. Andrade JD, Smith LM, Gregonis DE (1985) In: Surface and Interfracial Aspects of Biomedical Polymers, 1:249–292 Plenum NY
87. Ratner BD (1985) In: Polymers in Medicine II, Biomedical and Pharmaceutical Applications, Chiellini E, Guisti P, Migliaresi C, Nicolais L (eds) Plenum NY, 13
88. Lydon MJ (1986) Br Polym J 18(1), 22
89. Smetana K, Vacik J, Souckova D, Krcova Z, Sulc J (1990) J Biomed Mat Res 24:463
90. Ratner BD, Miller IF (1972) J Polym Sci Part A-1, 10:2425
91. Migliarisi C, Nicodimo L, Nicolais L, Passerini P, Stol M, Hrouz J, Cefelin P (1984) J Biomed Mat Res 18:137
92. Armstrong RW, Strauss UP (1969) Polyelectrolytes In: Encyclopedia of polymer science and technology vol 10, Wiley J, NY, p 78
93. Philip Molyneux, Chapter 1 In: Water Soluble Synthetic Polymers, Volume I: Properties and Behavior, CRC Press, Boca Raton, FL (1985)
94. Brannon-Peppas L, Peppas NA (1990), Biomaterials, 11:635
95. Kiler J, Peppas NA (1989) Proceed Adv Biomed Polym 1, 107 also Proceed Intern Symp Contr Rel Bioact Mater 16:101
96. Salomon JC, Watterson AC, Quach L, Reheja MK (1985) ACS Polymer Preprints, 26(1), 196
97. Tinland B, Mazet J, Rinaudo M (1988) Makromol Chem Rapid Commun 9:69
98. Aqueous Size Exclusion Chromatography, Dubin PL (ed) (1988) Lesevier, NY
99. Posner GH (1972) Org React 19:1
100. Noma K, Niwa M, Lida S, Nakazato Y (1975) Kobunshu Ronbunshu, 32:189

101. Kurihara M, Kamachi M, Stille JK (1973) J Polym Sci, Polym Chem Ed Vol. II, 587
102. Hata T, Chong K-J (1971) Bull Chem Soc Japan, 45(2):654
103. Kitano H, Hasegawa M, Kaku T, Ise N (1990) J Appl Polym Sci 39:241
104. Leemans L, Fayt R, Teyssie PH (1990) J Polym Sci: Part A: Polym Chem Ed 28:1255
105. Lee H, Neville K (1967) Handbook of Epoxy Resins, McGraw-Hill, London
106. Menschutkin N (1890) Z Phys Chem 5:589
107. Arcus CL, Hall WA (1963) J Chem Soc 4199. (1964) J Chem Soc 1157
108. Shao Wen, Yin Xiaonan, Stevenson WTK (1991) Biomaterials 12:479
109. Nicholson JW, Brookman PJ, Lacy OM, Sayers GS, Wilson AD (1988) J Biomed Mat Res 22:623
110. Parker S, Braden M (1989) Biomaterials 10:91
111. Fuoss RM (1951) Disc Faraday Soc 11:125
112. Nagasawa M, Holtzer A (1964) J Am Chem Soc 86(4):531
113. Manning GS, J Chem Phys 51(3), 924. J Chem Phys 51(3):934
114. Dunlap PN, Leal LG (1984) Rheological Acta, 23:238
115. Davis RM, Russel WB (1986) J Polym Sci: Part B: Polym Phys 24:511
116. Ritger PL, Peppas NA (1987) J Controlled Release 5:37
117. Crooks CA, Douglas JA, Broughton RL, Sefton MV (1990) J Biomed Mater Res 24:1241–1262
118. Sefton MV, Broughton RL (1983) 717:473–477
119. Sefton MV, Dawson RM, Broughton RL, Blysniuk J, Sugamori ME (1987) Biotech Bioeng 29:1135–1143
120. Sugamori M, Sefton MV (1984) Trans ASAIO, 35:791–799
121. Broughton RL, Sefton MV (1989) Biomaterials 10:462–465
122. Boag AH, Sefton MV (1987) Biotech Bioeng 30:954–962
123. Mallabone CL, Crooks CA, Sefton MV (1989) Biomaterials 10:380–386
124. Douglas JA, Sefton MV (1990) Biotech Bioeng 36:656–664
125. Strathmann H (1986) In: Bungay PM Lonsdale HK de Pinho MN (eds) synthetic membranes: scinece, engineering and applications. D. Reidel, Dordrecht, p 1
126. Formmer MA, Messalem RM (1973) Ind Eng Chem Prod Res Develop 12:328–333
127. Strathmann H, Scheible P, Baker RW (1971) J App Polym Sci, 15:811–828
128. Sefton MV, Kharlip L, Horvath V, Roberts T, J Contr Rel (in press)
129. Uludag H, Sefton MV (1990) Biomaterials 11:708–712
130. Uludag H, Sefton MV, Biotech Bioeng (in press)
131. Fuoss RM, Sadak H (1949) Science 110:552
132. Michaels AS, Miekka RG (1961) J Phys Chem 69:1765
133. Wen S, Xiaonan Y, Stevenson WTK (1991) Biomaterials 12:374
134. Wen S, Xiaonan Y, Stevenson WTK (1991) Biomaterials 12:479
135. Ferguson PL, Mills TE, Vonder Muehall E, Thompson M (1973) Clinical Hematology, Interpretations and Techniques, Wiley, NY, USA
136. Haradin AR, Weed RI, Reed CF (1969) Transfusion 9:229
137. Clark AH (1987) Structure and mechanical properties of biopolymer gels. In: Advances in Polymer Science, Vol 60, Springer, Berlin Heidelberg New York
138. Goosen MFA, O'Shea GM, Gharapetian HM, Sun AM (1985) Biotech and Bioeng Vol. XXVII, 146
139. Stevenson WTK, Sefton MV (1987) Biomaterials 8:449
140. deGennes PG (1987) Adv Colloid and Interface Sci 27:189
141. Lamberti FV, Sefton MV (1983) Biochimica et Biophysica Acta 759(1–2), 81
142. Weed RI, LaCelle PL, Merril EW (1975) J Clin Invest 48:795
143. Lamberti FV, Sefton MV (1989) J Coll Interfac Sci 130:1–13

Cellulose Derivatives

E. Doelker
School of Pharmacy, University of Geneva, Quai Ernest-Ansermet 30,
CH-1211 Geneva 4, Switzerland

The fundamental and derived properties of cellulose derivatives are presented concomitantly with applications in various life sciences (pharmaceutics, cosmetics, food, packaging). Emphasis is placed on drug delivery systems. Because most applications are related to the solubility of the materials, the subject is reviewed with regard to this parameter: 1) derivatives soluble in water (swelling, thermal gelation, cloud point, liquid crystal formation, bioadhesion); 2) derivatives soluble in organic solvents (manufacture of aqueous nanodispersions, permeation properties, solute osmotic delivery, dissolution-controlled drug delivery, phase separation); 3) derivatives soluble in nonacidic media and organic solvents (enteric coatings and pH-sensitive gels); 4) derivatives insoluble in water and organic solvents (low-substituted and crosslinked tablet disintegrating agents, crosslinked water absorbents).
    The data are presented on a comparative basis in order to emphasize the differences between similar derivatives.

List of Symbols and Abbreviations

Symbols for cellulose derivatives are listed in Table 1

Roman Symbols

A	surface area
BP	British Pharmacopoeia
c_*, c_0	concentrations
c^*	polymer volume fraction
C_*	dimensionless concentration at the polymer-gel interface
c_s	solute solubility
CA	Codex Alimentarius
CTFA	The Cosmetic, Toiletry and Fragrance Association, Inc.
CP, CP_0	cloud points
D	diffusion coefficient
D_s	diffusion coefficient of the solvent
D_{eff}	effective diffusion coefficient
D_{iw}	diffusion coefficient in water
$(DEB)_D$	diffusional Deborah number
DM	dilatometry
DMA	dynamic mechanical analysis
DR	degree of reaction
DS	degree of substitution
DSC	differential scanning calorimetry
DTA	differential thermal analysis
E	elasticity modulus
EP	European Pharmacopoeia
g_1, g_2	surface energy ratios
H	hysteresis of the polymer-water free energy of adhesion
IGT	incipient gelation temperature
IPT	incipient precipitation temperature
k	kinetic constant
k_d	dissolution rate constant
k_s	softening point depression coefficient
K_∞	infinite dilution partition coefficient
K_{CP}	salting out constant
L, L_0	lengths, thicknesses
LCST	lower critical solution temperature
L_p	mechanical water permeability
m	molar concentration
M	molar mass
$\bar{M}_c$	number-average molar mass per crosslinked unit

$\bar{M}_n$	number-average molar mass
$\bar{M}_w$	mass-average molar mass
$\bar{M}_w/\bar{M}_n$	polydispersity index
M_t	mass of species dissolved or released at time t
$\bar{M}_v$	viscosity-average molar mass
MC	moisture content
MS	molar substitution
n	kinetic exponent
N	mole fraction
NF	National Formulary
P	hydrostatic pressure
PC	permeability coefficient
Q_{max}	maximum swelling coefficient
r_0	radius of the unswollen polymer particle
S	solubility coefficient
SI	substitution index
t, t_s, t_{max}	times
T	temperature
T_g	glass transition temperature
T_s	softening temperature
TBA	torsional braid analysis
TMA	thermomechanical analysis
TSDCT	thermally stimulated depolarization current technique
USP	United States Pharmacopeia
$\bar{v}$	specific volume of polymer
V	molar volume
w, w_0^*	weight fraction
W_a	work of adhesion
W_c	work of cohesion
W_R	reduced coefficient of spreading
W_s	work or coefficient of spreading
WVTR	water vapor transmission rate
x_*	position of the polymer-gel interface
x_d	fractional non-polarity (dispersion)
x_p	fractional polarity
X	crosslinking ratio
x_2, x_3	normalized volume corrections

Greek Symbols

α_1	linear expansion coefficient
β	contribution of the penetrant to the expansion coefficient
γ	surface tension, surface free energy, interfacial tension
δ	solubility parameter

ε   diffusivity ratio
η   viscosity
η_{sp} specific viscosity
θ   contact angle
θ_a advancing contact angle
θ_D characteristic diffusion time
θ_r receding contact angle
λ_m characteristic relaxation time
ξ_* dimensionless position of polymer-gel interface
π   osmotic pressure
ρ   density
ρ_x crosslinking density
σ   reflexion coefficient
σ_{max} maximum strength
σ_0 adhesive strength extrapolated to zero film thickness
τ   dimensionless time
$\phi_{2\infty}$ solubility of water in polymer particles
Φ   interaction parameter according to Girifalco and Good [67]
χ   Flory–Huggins interaction parameter
χ_c critical interaction parameter

1 Introduction and Synthetic Aspects

The chemical modifications of cellulose considered here are partial or total esterification and etherification of the three hydroxyls present on the anhydro-glucose unit and subsidiarily crosslinking and grafting of the backbone. These transformations were carried out most of the time to improve the processability and performance of cellulose and more specifically to obtain derivatives which are soluble in water and in organic solvents (cellulose itself, although very hydrophilic, is not soluble in water due to extensive hydrogen bonding between the hydroxyl groups of the macromolecular chains).

Approximately 2% of the 5×10^{11} metric tons of cellulose generated yearly by biosynthesis throughout the world is recovered industrially and of these 10^8 tons, about 2% is transformed into various esters (3/4) and ethers (1/4). Historically, cellulose nitrate is the oldest (inorganic) ester synthetized, but it has found almost no applications as a biopolymer. The first organic ester and which remains widely used in the field of life sciences is cellulose acetate (1865). Ethers are more recent, since methylation was first described in 1905.

To illustrate our purpose better and for the sake of clarity fundamentals and applications will be presented concomitantly throughout the sections.

The synthesis and the general properties of these compounds have been reviewed extensively [1–15] and they will be recalled only when necessary to explain recent data. In this review, emphasis will be laid on papers dealing with cellulose derivatives used in the field of pharmacy, medicine, cosmetics and food, but applications will essentially be related to drug delivery. This presentation will be on a comparative mode between cellulose derivatives. The first two sections are concerned with all types of cellulose derivatives, while the following are each devoted to a special solubility-class of derivatives.

1.1 Classification of Cellulose Derivatives

Derivatives of cellulose are usually classified according to the type of chemical reaction taking place on the hydroxyl groups [16]:

1. Substitution of the hydrogen atom in the hydroxyl groups
   (a) esterification
   (b) etherification
2. Selective and controlled oxidation of the cellulose hydroxyls
3. Substitution of the hydroxyls
   (a) intramolecular reactions
   (b) intermolecular reactions
4. Reactions involving radical additions
5. Reactions involving electrophilic substitutions
6. Interaction of cellulose with organometallic compounds
7. Synthesis of block or graft copolymers of cellulose

Item 1, and accessorily items 2, 3 and 7 are concerned with the modified celluloses described here. Additionally, esters and ethers are frequently divided into single-substituted and mixed compounds, according to the number of different substituents involved. They can also be classified as ionic or nonionic derivatives. Table 1 lists the main esters and ethers cited in the specialized literature and which are commercially available.

Table 1. Main esters and ethers used as biopolymers together with abbreviations

Single substituent	Mixed[1]

A. ESTERS

Cellulose nitrate (CN)	Cellulose acetate butyrate (CAB)
Cellulose acetate (CA)	Cellulose acetate propionate (CAPr)[3]
Cellulose triacetate (CTA)	Cellulose acetate phthalate (CAP)
	Cellulose acetate trimellilate (CAT)

B. ETHERS

Methylcellulose (MC)	Methylethylcellulose (MEC)
Ethylcellulose (EC)	Methylhydroxyethylcellulose (MHEC)
Hydroxyethylcellulose (HEC)	Hydroxypropylmethylcellulose (HPMC)
Hydroxypropylcellulose (HPC)	Hydroxybutylmethylcellulose (HBMC)
Sodium carboxymethylcellulose[2] (NaCMC, CMC)	Ethylhydroxyethylcellulose (EHEC)
Calcium carboxymethylcellulose (CaCMC)	Sodium carboxymethylhydroxyethyl- cellulose (NaCMHEC, CMHEC)
	Carboxymethylethylcellulose (CMEC)

C. ETHER ESTERS

Hydroxypropylmethylcellulose phthalate (HPMCP)
Hydroxypropylmethylcellulose acetate succinate (HPMCAS)
Hydroxypropylcellulose acetate (HPCA)
Esters of hydroxyethylcellulose and diallyldimethylammonium chloride (Polyquaternium-4)[3]
Esters of hydroxyethylcellulose and 2-hydroxypropyltrimethylammonium chloride (Polyquaternium-10)[4]
Esters of hydroxyethylcellulose and a lauryldimethylammonium substituted epoxide (Polyquaternium-24)[4]

D. CROSSLINKED DERIVATIVES

Crosslinked sodium carboxymethylcellulose (croscarmellose sodium)

E. GRAFTED DERIVATIVES

Hydroxyethylcellulose grafted with alkyl C_{12}–C_{24} chains (g-HEC)

---

[1] For mixed ethers, the order for substituents may either be alphabetical or follow the respective degree of substitution. Thus, hydroxypropylmethylcellulose (U.S. Pharmacopeia, USP) is also known as methylhydroxypropylcellulose (European Pharmacopoeia, EP).
[2] NaCMC is frequently designated as sodium cellulose glycolate and under a purified form as cellulose gum. An acidic form (non-neutralized) of CMC is also commercially available.
[3] Cellulose acetate propionate is sometimes abbreviated CAP which most commonly designates cellulose acetate phthalate.
[4] CFTA (The Cosmetic, Toiletry and Fragrance Association, Washington, D.C.) adopted names.

This chemical classification is of little use with respect to the various applications as biopolymers, in which most of the time interactions with solvents are involved. This is illustrated by the fact that the solubility of a given compound may change completely with the degree of substitution and/or with the molar substitution. This is for example the case of HPC and NaCMC which are commercially available as insoluble and water-soluble products. We therefore use a classification of cellulose-based biopolymers made according to their behavior in water and to their ionic character (Table 2).

Cellulose esters and ethers of various types and viscosities are characterized by their *degree of substitution* (DS), i.e. the average number of substituted hydroxyl groups (the maximum value is 3, see also Table 3). Furthermore, ethers or ether esters with hydroxyalkyl groups attached are characterized by the *degree of reaction* (DR), also frequently named *molar substitution* (MS), i.e. the average number of molecules of reagent (alkylene oxide) reacted with each

Table 2. Classification of cellulose biopolymers based on interaction with solvent and ionic character

Soluble in water	Soluble in organic solvents	Soluble in non-acidic media and organic solvents	Insoluble in water and organic solvents
Nonionic	*Nonionic*	*Ionized above pK_a*	*Nonionic*
MC	EC	CAP	Oxidized cellulose[3]
MEC	EHEC[2]	CAT	L-HPC[4]
MHEC	HBMC	HPMCP	
HPMC	CN	HPMCAS	
HEC	CA		
HPC[1]	CTA		
EHEC	CAB		
g-HEC	CAPr		
	HPCA		
Anionic			*Anionic*
NaCMC			Oxidized cellulose[3]
CMHEC (sodium)			L-NaCMC[4]
CMEC (acid)			CaCMC
			Croscarmellose sodium
Cationic			
Polyquaternium-4			
Polyquaternium-10			
Polyquaternium-24			

[1] HPCs are also soluble in a few polar organic solvents.
[2] Highly substituted EHECs are soluble in organic solvents.
[3] The oxidized groups formed may include aldehyde and/or carboxyl groups.
[4] Low-substituted derivatives.

anhydroglucose unit (this value can exceed 3). Thus, the ratio MS/DS expresses the average length of the pendant chain (Table 3).

Most often, the extent of substitution and of reaction are expressed on a weight percentage basis. In some cases, the substitution index (SI, the percentage of substituted anhydroglucose units) has been determined [17, 18].

Table 3. Structure of a section of two anhydroglucose units (cellobiose) in a cellulose derivative

Cellulose derivative	Substituent R (other than H)

Esters	
CN	$-ONO_2$
CA, CTA[1]	$-COCH_3$
CAPr	$-COCH_3$, $-COCH_2CH_3$
CAB	$-COCH_3$, $-COCH_2CH_2CH_3$
CAP	$-COCH_3$, $-COC_6H_4COOH$
CAT	$-COCH_3$, $-COC_6H_3(COOH)_2$
Ethers	
MC	$-CH_3$
EC	$-CH_2CH_3$
HEC	$-(CH_2-CH_2O)_nH$
HPC	$-(CH_2CH(CH_3)O)_nH$
MEC	$-CH_3$, $-CH_2CH_3$
MHEC	$-CH_3$, $-CH_2CH_2OH$
HPMC	$-CH_2CH(CH_3)OH$, $-CH_3$
HBMC	$-CH_2CH_2CH(CH_3)OH$, $-CH_3$
EHEC	$-CH_2CH_3$, $-CH_2CH_2OH$
NaCMC	$-CH_2COO^- Na^+$
CMHEC (sodium)	$-CH_2COO^- Na^+$, $-(CH_2CH_2O)_nH$
CaCMC	$(-CH_2COO^-)_2 Ca^{2+}$
Ether esters	
HPMCP	$CH_2CH(CH_3)OH$, $-CH_3$, $-COC_6H_4COOH$
HPMCAS	$-CH_3CH_2CH(CH_3)OH$, $-COCH_3$
	$-COCH_2CH_2COOH$, $-CH_2CH(CH_3)OCOCH_3$,
	$-CH_2CH(CH_3)OCOCH_2CH_2COOH$
HPCA	$-CH_2CH(CH_3)OH$, $-COCH_3Cl^-$
Polyquaternium-4	$-(CH_2CH_2O)_n R'N^+(CH_3)_2(CH_2CHCH_2)_2$
Polyquaternium-10	$-(CH_2CH_2O)_n R'N^+(CH_3)_3Cl^-$
Polyquaternium-24	$-(CH_2CH_2)_x-(CH_2CH(OH)CH_2N^+(CH_3)_3R'C)_y$

[1] At least 92% of esterified hydroxyl groups.

When considering a cellulose derivative, the following chemical characteristics are likely to influence its properties:

1. Nature of the substituent group
2. Proportion of substituted hydroxyl groups
3. Uniformity of substitution in the repeating unit and along the polymer chain
4. Average length and molecular mass distribution (viscosity).

Some data will illustrate these points in relation to cellulose ethers, but most of the conclusions remain valid for esters.

Properties are highly dependent upon the polarity of the substituent. Water-solubility is obtained at a DS value of about 1 for nonionic ethers. If hydrophobic ether groups predominate, the water solubility may disappear at a DS levels above 2 [5]. Methylcellulose is in fact soluble up to about 1.8, whereas ethylcellulose only up to 1.2–1.4 (not available commercially). Introducing polar groups (carboxyls, hydroxyls) of course strongly increases hydrophilicity.

Industrially, esterification and etherification of the starting material (cellulose from cotton linters or wood pulp) are performed heterogeneously (topochemical reaction). Because of restricted access to a part of the macromolecules, due to intra- and intermolecular hydrogen bonding, the normally expected increased reactivity of the primary hydroxyl groups is not observed. In cellulose (isotactic β-1,4-polyacetal), the secondary hydroxyl of C-2 often shows a higher reactivity. The relative reaction rates differ according to the reactant involved and the starting material. However, etherification with alkylene oxides takes place by preference of the hydroxyl group of C-6. Consequently, nonuniformity of distribution occurs within the anhydroglucose unit and along the polymer chain, resulting in mixtures of fully substituted, irregularly substituted and unsubstituted cellulose molecules (Table 4). Accordingly, in the industrial alkali process a DS of 1.2 is necessary for ensuring water solubility, whereas an experimental solution method results in a more uniform substitution, with ethers being soluble at a DS of only 0.6.

Knowledge of the substituent distribution in hydroxyalkyl derivatives is complicated by the fact that the reactant (alkylene oxide) may either be added to

Table 4. Distribution of substitution in methylcellulose of DS 1.27 [8]

Monomeric species	Mol %	DS Share
Glucose	16	0
6-O-Methylglucose	11 ⎫	
2-O-Methylglucose	34 ⎬47	0.47
3-O-Methylglucose	2 ⎭	
3,6-Di-O-Methylglucose	2 ⎫	
2,6-Di-O-Methylglucose	24 ⎬31	0.62
2,3-Di-O-Methylglucose	5 ⎭	
2,3,6-Tri-O-Methylglucose	6	0.18

unsubstituted hydroxyl groups of the cellulose or to the growing polyalkylene oxide side chains. The substitution pattern is also dependent on the alkylene oxide used. Thus, HPCs significantly differ from HECs, because of the formation of secondary rather than primary hydroxyls when propylene oxyde is used instead of ethylene oxide. The consequence is that for instance at a MS level of approximately 4, the content of unsubstituted anhydroglucose in HPC is only one tenth of that in HEC [18].

Chain length and molecular mass distribution have of course also strong implications for the performance of modified cellulose. For ethers, the degree of polymerization, DP, is usually between 50 and 2000 (300 for esters), as compared with 100 to 3000 for cotton cellulose and 600 to 1000 for wood cellulose. The relatively high polydispersity of the molecular mass originates from the starting material.

An interesting example of the variation of substitution, average molecular mass and distribution of seven batches of HPMC 2208 from two different manufacturers is provided by Dahl et al. [19]. HPMC 2208 is a well-defined USP type of HPMC with a nominal methoxyl and hydroxypropyl content of 22 and 8%, respectively. Here all the batches were of 15 000 mPa s viscosity grade (2% in water).

All HPMC samples had similar viscosity, except batch 7 which is outside USP specifications (75–140% of the nominal value 15 000). Methoxyl contents are uniformly high and three batches fall outside the USP specifications 19–24%. Hydroxypropyl contents, although within the specified limits (4–10%), vary relatively more than methoxyl contents, with batch 5 showing the lowest value. When measuring the release of the drug naproxen from compressed matrix tablets, the authors observed in vitro a strong dependence of the release rate (expressed in $h^{-0.5}$) on the hydroxypropyl content of the HPMC (Fig. 1).

It can be concluded that simply stating the USP type can be insufficient for some borderline cases and that each batch even from the same manufacturer should be carefully controlled. Alternately, based on experimental evidence, the specifications of the USP (and other pharmacopoeas) should be reinforced. The

Table 5. Physico-chemical characteristics of various batches of HPMC 2208 15 000 mPa s (adapted from [19])

Batch Number	Viscosity (mPa s)	Methoxyl content (%)	Hydroxypropyl content (%)	$\bar{M}_n$	$\bar{M}_w$	$\bar{M}_w/\bar{M}_n$
1[1]	15 200	23.7	8.7	143 953	608 274	4.24
2[1]	14 000	22.5	10.9	161 338	613 714	3.80
3[1]	14 200	25.9	11.1	211 649	622 861	2.94
4[1]	15 000	26.4	7.2	187 060	613 351	3.28
5[1]	15 000	25.5	5.3	119 901	568 659	4.74
6[2]	15 600	22.7	10.7	167 805	633 932	3.78
7[2]	12 491	23.4	9.5	158 695	603 797	3.80

[1] From manufacturer A.
[2] From manufacturer B.

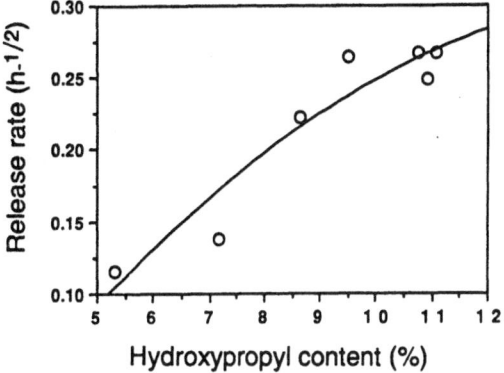

Fig. 1. Influence of HPMC 2208 hydroxypropyl content on naproxen release from matrix tablets [19]

seven batches examined by size exclusion chromatography also varied with respect to their average molecular mass and polydispersity index, but it must be borne in mind that these parameters are strongly influenced by differences in substitution. Anyway, the polydispersity indices, for example, are much lower than those obtained for various batches of 50 mPa s HPMC, ranging from 18.1 to 341 [21]. The latter figures are exceptional when considering polydispersity indices for other low molecular weight derivatives [21].

For technological reasons it is sometimes desirable to use a definite particle size fraction of a given compound. A priori, it is expected that the molecular mass and polydispersity do not depend on the particle size. Actually, gel permeation chromatography on a fractionated 10 mPa s EC sample showed some effect [22]: the average molecular mass and the polydispersity index of the various size fractions ranged from ca. 18 800 to 24 800 and from 3.1 to 3.4, respectively.

1.2 Crosslinking of Cellulose Derivatives

Since they are hydroxyl-containing polymers, cellulose ethers and possibly esters can be reacted covalently with many species, both mono- and polyfunctional, in order to stabilize and insolubilize their structure. The intermolecular reactions include [23]:

1. formation of acetals with monofunctional aldehydes (formaldehyde, acetaldehyde)

   $2 \text{ Cell–OH} + \text{RCHO} \rightarrow \text{Cell–O–CH(R)–O–Cell}$

2. formation of hemiacetals with dialdehydes (glyoxal, glutaraldehyde)

   $2 \text{ Cell–OH} + \text{OHC–R–CHO} \rightarrow \text{Cell–O–CH(OH)–R–CH(OH)–O–Cell},$

3. formation of acetals with dialdehydes

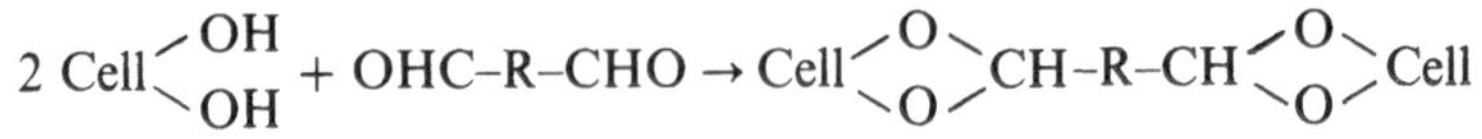

$$2 \text{ Cell} \underset{OH}{\overset{OH}{<}} + \text{OHC–R–CHO} \rightarrow \text{Cell} \underset{O}{\overset{O}{<}} \text{CH–R–CH} \underset{O}{\overset{O}{>}} \text{Cell}$$

4. formation of ether or methylene links with reagents containing methylol groups

$$2 \text{ Cell–OH} + 2 \text{ HOH}_2\text{C–R–CH}_2\text{OH} \rightarrow$$

$$\text{Cell–O–CH}_2\text{–R–CH}_2\text{–O–CH}_2\text{–R–CH}_2\text{–O–Cell}$$

or

$$\text{Cell–O–CH}_2\text{–R–CH}_2\text{–R–O–Cell}$$

5. formation of ether links with epoxides

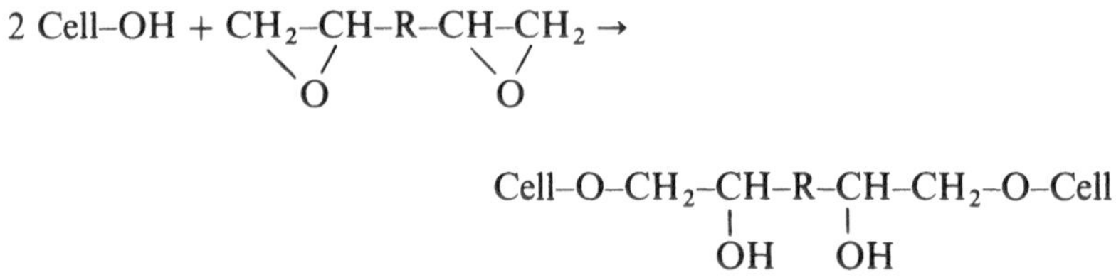

$$2 \text{ Cell–OH} + \text{CH}_2\text{–CH–R–CH–CH}_2 \rightarrow$$

with epoxide O bridges

$$\text{Cell–O–CH}_2\text{–CH–R–CH–CH}_2\text{–O–Cell}$$
$$\underset{OH}{|} \qquad \underset{OH}{|}$$

6. formation of ether links with ethylene imine derivatives

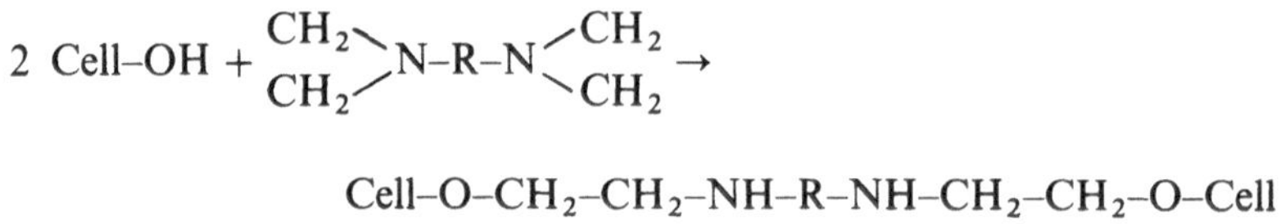

$$2 \text{ Cell–OH} + \underset{CH_2}{\overset{CH_2}{>}} \text{N–R–N} \underset{CH_2}{\overset{CH_2}{<}} \rightarrow$$

$$\text{Cell–O–CH}_2\text{–CH}_2\text{–NH–R–NH–CH}_2\text{–CH}_2\text{–O–Cell}$$

7. formation of ether links with sulfones

$$2 \text{ Cell–OH} + \text{HOCH}_2\text{–CH}_2\text{–SO}_2\text{–CH}_2\text{–CH}_2\text{OH}$$

$$(\text{or } \text{CH}_2\text{=CH–SO}_2\text{–CH=CH}_2)$$

$$\rightarrow \text{Cell–O–CH}_2\text{–CH}_2\text{–SO}_2\text{–CH}_2\text{–CH}_2\text{–O–Cell}$$

8. formation of ether links with labile chlorine compounds

$$2 \text{ Cell–OH} + \text{Cl–R–Cl} \rightarrow \text{Cell–O–R–O–Cell}$$

Under inadequate conditions (e.g. in very dilute solutions) some of these reactions may proceed intramolecularly. Crosslinking by esterification is also possible when carboxylic groups are present (NaCMC) or when using crosslinkers with hydroxyl groups.

Crosslinking through esterification is also possible without further additive, i.e. between the carboxyl and hydroxyl groups of the cellulose derivative chains. A striking example is the insoluble pharmaceutical tablet disintegrating agent croscarmellose sodium, which is produced by internal crosslinking mediated via pH and temperature control (see Sect. 6).

The crosslinking index is usually expressed as moles of crosslinking agent per mole of repeating unit, X, or moles of crosslinked units per g of polymer, $1/\bar{M}_c$ ($\bar{M}_c$ = molecular weight per crosslinked unit), whereas the crosslink density ρ_x corresponds to $\rho_x = 1/\bar{v}\bar{M}_c$ ($\bar{v}$ = specific volume of the polymer).

1.3 Grafting of Cellulose Derivatives

Much effort has been devoted over the last few years to preparing branched cellulose or cellulose derivatives, by combining the cellulose backbone with a synthetic polymer which confers desirable properties. The length of these branches or grafts varies considerably, depending on the copolymerization conditions. The most widely used monomers are acrylic and vinyl monomers, with the following order of reactivity [16]: ethyl acrylate > methyl methacrylate > acrylonitrile > acrylamide > styrene.

As far as biopolymers other than plain cellulose are concerned, grafting has been reported for crosslinked cellulose, oxidized cellulose, carboxymethylated cellulose, cellulose acetate and even for crosslinked derivatives. The grafted material content is usually given as the ratio of the increase in weight to the initial cellulose weight.

Practical applications in life sciences are still rare. As an example, cellulose graft copolymers of poly (acrylic acid) or poly (methacrylic acid) cause coagulation of proteins and blood clotting in the form of their calcium salt, and they have bactericidal properties as silver and copper salts. Grafted celluloses have found application as membrane and water-sorbing materials. A nice example of the effect of several of these methods for modifying cellulose is given by Tabata and Ikada [24]. The surface of cellulose microspheres was chemically treated *in situ* using various etherifying, esterifying and crosslinking agents. Phagocytosis by macrophages was related to the charge (zeta potential) and hydrophobicity of the microsphere surface.

1.4 Compendium Specifications

When used as pharmaceutical, cosmetic or food additives, cellulose derivatives must compulsorily comply with official standards. Table 6 lists existing specifications in the main compendia.

From these figures, it can be concluded that a large number of derivatives are available as biopolymers but some of them are not well-defined in compendia or their specifications differ from one compendium to the other.

2 Basic Data for Intermolecular Interactions

All the applications concerned with cellulose derivatives imply an interaction with another body. The latter can be a liquid such as an organic solvent,

Table 6. Compendium specifications of cellulose derivatives

Material	Compendium[1]	Specification
Oxidized cellulose	USP/NF	16.0–24.0% carboxyl
	BP	16.0–22.0% carboxyl
Oxidized regenerated cellulose	USP/NF	18.0–24.0% carboxyl
Methylcellulose	USP/NF	27.5–31.5% methoxyl
	BP, EP	No limits are given
	CTFA, CA	No limits are given
Methylethylcellulose	CA	No limits are given
Methylhydroxyethylcellulose	BP, EP	No limits are given
Hydroxypropylmethylcellulose	USP/NF	
Type 1828		16.5–20.0% methoxyl, 23.0–32.0% hydroxypropoxyl
Type 2208		19.0–24.0% methoxyl, 4.0–12.0% hydroxypropoxyl
Type 2906		27.0–30.0% methoxyl, 4.0–7.5% hydroxypropoxyl
Type 2910		28.0–30.0% methoxyl, 7.0–12.0% hydroxypropoxyl
	BP[2], EP[3]	No limits are given
	CTFA, CA	No limits are given
Hydroxypropylcellulose	USP/NF	Not more than 8.5% hydroxypropoxyl
	BP, EP	No limits are given
	CTFA, CA	No limits are given
Hydroxyethylcellulose	USP/NF	No limits are given
	BP, EP	No limits are given
	CTFA	No limits are given
Ethylhydroxyethylcellulose	CTFA, CA	No limits are given
Hydroxybutylmethylcellulose	CTFA	No limits are given
Carboxymethylcellulose sodium	USP/NF	6.5–9.5% Na
Type 12		10.5–12.0% Na (DS = 1.15–1.45)
	BP[4], EP	6.5–10.8% Na
	CTFA[5], CA	No limits are given
Carboxymethylhydroxyethylcellulose	CTFA	No limits are given
Carboxymethylcellulose calcium	USP/NF	No limits are given
Ethylcellulose	USP/NF	44.0–51.0% ethoxyl
	CTFA, CA	No limits are given
Cellulose acetate[6]	USP/NF	29.0–44.8% acetyl
Cellulose nitrate[7]	BP	No limits are given
Cellulose acetate phthalate	USP/NF	21.5–26.0 acetyl, 30.0–36.0% phthalyl
	BP[8], EP	17.0–26.0 acetyl, 30.0–40.0% phthalyl
Hydroxypropylmethylcellulose phthalate	USP/NF	
Type 200731		18.0–22.0% methoxyl, 5.0–9.0% hydroxypropoxyl, 27.0–35.0% phthalyl
Type 220824		20.0–24.0% methoxyl, 6.0–9.0% hydroxypropoxyl, 21.0–27.0% phthalyl
	BP[9], EP[10]	20.0–35.0% phthalyl
Croscarmellose sodium	USP/NF	No limits are given
Polyquaternium-4	CTFA	No limits are given
Polyquaternium-10	CTFA	No limits are given
Polyquaternium-24	CTFA	No limits are given

a plasticizer (e.g. for film coating, casting) or water when in contact with physiological fluids. It can also be some solid (e.g. other polymers, diluents, pigments).

2.1 Glass Transition Temperature

The Flory–Huggin's polymer solution theory and the solubility parameter concept have been largely applied, together with measurements of glass transition temperature. It is therefore interesting to collect the constants used in predictive equations. The glass transition temperatures of cellulose derivatives have been investigated using various techniques including dilatometry (DM), differential scanning calorimetry (DSC), differential thermal analysis (DTA), thermomechanical analysis (TMA), torsional braid analysis (TBA), dynamic mechanical analysis (DMA) and thermally stimulated depolarization current technique (TSDCT).

As it is well known, different techniques yield different T_g values and only those values determined by the same method are comparable. Moreover, differences are caused not only by the nominal composition of the products (molecular mass, degree of substitution) but also by interbatch variability. It should also be mentioned, that some of the values obtained were derived from films and not from the starting materials, and that either the solvent composition used for casting or the residual solvent may have an influence on the transition values.

Furthermore, some of the techniques (DM, TMA, DMA) have shown themselves to be sufficiently sensitive to be also capable of detecting low-energy secondary transitions. Even with other techniques (e.g. DSC), it is probable that the low T_g values reported are only secondary transitions. The relevance of these low-temperature transitions is unknown but could be of significance in the diffusion of oxygen and water. Incidentally, no information is available in the

---

Notes for Table 6

[1] USP/NF: United States Pharmacopeia/National Formulary, BP: British Pharmacopoeia, EP: European Pharmacopoeia, CTFA: Cosmetic Toiletry and Fragrance Cosmetic Ingredient Dictionary, CA: Codex Alimentarius (F.A.O.).
[2] Named Hypromellose in the BP.
[3] Named Methylhydroxypropylcellulose in the EP.
[4] Named Carmellose sodium in the BP.
[5] Named Cellulose Gum in the CTFA Dictionary.
[6] Also named Diacetate and Triacetate.
[7] A nitrated cellulose dampened with not less than 25% of organic solvent is mentioned as Pyroxylin in USP XXI and BP. It is also known as Colloxylin. It serves to prepare collodions (solutions in ether and alcohol).
[8] Named Cellacephate in the BP.
[9] Named Hypromellose phthalate in the BP.
[10] Named Methylhydroxypropylcellulose phthalate in the EP.

Table 7. Reported glass transition temperatures for various cellulose derivatives

Material	Method	T_g (°C)	Ref.
Methylcellulose			
Methocel[R] A4M	DCS	196	15
Hydroxypropylmethylcellulose			
Type 2910			
Pharmacoat[R] 606	DSC	177	25
Methocel[R] E15	TMA	172–175[1]	26
Pharmacoat[R] 606	DSC	155	27
Pharmacoat[R] 606	DSC	180	28
Pharmacoat[R] 606	DTA	169–174	28
Pharmacoat[R] 606	TBA	153.5, 158.5	28
Pharmacoat[R] 606	DSC	155.8	29
Pharmacoat[R] 606	TMA	163.8, 174.4	29
Pharmacoat[R] 603	DMA	160	30
Pharmacoat[R] 606	DMA	170	30
Pharmacoat[R] 615	DMA	175	30
Pharmacoat[R] 606	DMA	154	31
Pharmacoat[R] 606	DMTA	170	31
Type 2208			
Methocel[R] K4M	DSC	184	15
Methocel[R] K4M	DSC	57	32
Hydroxyethylcellulose			
Natrosol[R] 250MR	DSC	106	15
Natrosol[R] 250L	DMA	120	30
Hydroxypropylcellulose			
Klucel[R] L	TBA	124	28
Not specified	DMA	80	33
Not specified	DMA	110	34
Klucel[R] EF	DMA	No primary transition	30
Sodium carboxymethylcellulose			
High viscosity grade	DSC	78	32
Ethylcellulose			
Ethylcellulose[R] N50	DSC	129	25
Ethylcellulose[R] N50	TMA	147–149[1]	26
Ethylcellulose[R] N50	DSC	133	28
Ethylcellulose[R] N50	DTA	133	28
Ethylcellulose[R] N50	TBA	131.5	28
Ethocel[R] STD 10 IND	DSC	128	35
Ethocel[R] STD 20 PG	DSC	133, 136	22
Cellulose acetate			
Triacetate	DM	105	36
Triacetate	DM	175	37
2,3-Diacetate	DM	120	36
Secondary acetate	DM	195	37
Eastman	TMA	164	38
Not specified	Not specified	191	39

Table 7. Contd.

Material	Method	T_g (°C)	Ref.
Cellulose acetate propionate			
CAP 482–20	DSC	147	32
Cellulose acetate phthalate			
Wako	DSC	18.5	40
Eastman	TMA	180[1]	26
Eastman	DSC	171	28
Eastman	DTA	170	28
Eastman	TBA	163	28
EP	TSDCT	23	41
Hydroxypropylmethylcellulose phthalate			
HP-55	TMA	153–165	26
HP-55	DSC	133	28
HP-55	DTA	133	28
HP-55	TBA	133	28
HP-55	TSDCT	28	41
HP-55	Not specified	135	42
HP-50	DSC	146	28
HP-50	DTA	146	28
HP-50	TBA	137	28

[1] The values obtained by TMA in the penetration mode have been reported by the authors as softening temperatures.

literature (except from the studies carried out by Rials et al. [33] and by Russel and van Kerpel [37]) concerning transitions below room temperature.

Grafting and crosslinking influence the transition temperature of polymers. As an example, it was observed that for cellulose triacetate or secondary cellulose acetate the glass transition increased with increasing molecular mass of the poly(methylmethacrylate) or poly(methacrylic acid) grafted chain [42].

2.2 Solubility Parameters, Surface Energies

As for every material, the behavior of cellulose derivatives is governed by equations in which solubility parameters or surface free energy terms are frequently included. For comparison purposes, it seemed opportune to collect recently published data (Table 8). Polarity is for most applications a relevant parameter and therefore fractional polarities x_p were calculated from solubility parameters or surface free energies, Eq. (4) Sect. 2.3.

From the Table, it can be seen that not many experimental data are actually available by comparison with calculated values, but mention should be made of the solubility parameter maps published for instance for HPMC [52], HPC [52], EC [52–55], CN [56, 57], CA and CTA [53, 56–58], CAB [56, 57] and CAP [58].

Table 8. Experimentally determined and calculated values of the solubility parameter δ, as well as fractional polarities x_p of various cellulose derivatives (subscript d is for dispersive, p for polar and h for hydrogen-bonding)[1]

Material	Method	Solubility parameter $(MPa)^{1/2}$					Ref.
		δ	$δ_d$	$δ_p$	$δ_h$	x_p (%)	
Methylcellulose							
DS = 2.0	Calc.[3]	21.3	14.1	6.3	14.6	56	45
DS = 1.8	Calc.[3]	21.6	13.6	6.7	15.3	60	14
Hydroxypropylmethylcellulose							
DS = 1.0 and 1.0[2]	Calc.[3]	22.8	14.4	5.8	16.7	60	45
HPMC 2910							
Not specified	Exptl.	23.15	—	—	—	—	46
Not specified	Calc.[3]	24.4	—	—	—	—	47
DS = 1.9 and 0.23	Calc.[3]	20.7	13.7	6.2	14.2	56	14
HPMC 2906	Calc.[3]	21.3	13.7	6.5	14.9	59	14
DS = 1.8 and 0.13							
HPMC 2208	Calc.[3]	22.5	13.7	7.0	16.7	63	14
DS = 1.4 and 0.21							
Hydroxyethylcellulose							
MS = 2.5	Calc.[3]	25.5	14.3	59	18.9	69	14
Hydroxypropylcellulose							
Not specified	Exptl.	21.7–23.6	—	—	—	—	48
Not specified	Calc.[3]	24.9	—	—	—	—	47
MS = 4.0	Calc.[3]	21.3	14.6	3.8	15.1	53	14
Ethylcellulose							
Not specified	Calc.[3]	20.6	16.7	2.9	11.7	34	47
49.9% Ethoxyl	Exptl.	19.0	—	—	—	—	49
47.9% Ethoxyl	Exptl.	19.4	—	—	—	—	49
44.8% Ethoxyl	Calc.	—	16.7	—	11.7	—	50, 51
49.7% Ethoxyl	Calc.	—	17.1	—	9.9	—	50, 51
Cellulose nitrate							
1/2 sec	Exptl.	22.1	—	—	—	—	51
Cellulose acetate							
CTA	Calc.	—	15.6	—	10.6	—	50, 51
CA	Calc.	25.1	19.6	12.7	11.0	45	50, 51
E-398	Calc.	—	15.6	—	11.9	—	50, 51
E-383	Calc.	—	15.5	—	12.5	—	50, 51
E-394	Calc.	—	15.6	—	12.1	—	50, 51
Cellulose acetate propionate							
CAPr	Calc.	—	15.8	—	10.1	—	50, 51
Cellulose acetate butyrate							
EAB-171	Calc.	—	15.8	—	10.1	—	50, 51
EAB-381	Calc.	—	15.9	—	9.9	—	50, 51
EAB-500	Calc.	—	16.0	—	8.7	—	50, 51

Table 8. Contd.

Material	Method	Solubility parameter (MPa)$^{1/2}$					Ref.
		δ	δ_d	δ_p	δ_h	x_p (%)	
Cellulose acetate phthalate							
CAP	Calc.	27.2	—	—	—	—	47
Hydroxypropylmethylcellulose phthalate							
HPMCP	Calc.[3]	26.4	—	—	—	—	47

[1] While this review article was in press, a complete set of partial solubility parameters was published by Archer [167] based on solubility data. These parameters concern MC, the four types of HPMC (Table 6) and several ethylcelluloses with various ethoxyl contents.
[2] Does not correspond to any commercially available product.
[3] Calculated using the group contribution method of Van Krevelen (44).

Available surface free energies for cellulose derivatives are even scarcer than solubility parameters (Table 9). The total surface free energy was generally assimilated to the surface tension of a polymer solution in the range of concentration independence, whereas the non-polar and polar contributions were obtained from contact angle measurements.

The critical surface tension γ_c as deduced by using Zisman's approach, is also usually considered as an approximation of the surface free energy of solids. Thus, γ_c values of 30.8 mN m^{-1} [63] and 39.0 mN m^{-1} [62] have been reported for MC and CA, respectively. However, these values are quite different from those reported by Rowe [59, 60], which makes questionable the comparison of values obtained by different techniques.

Surface tensions of 2% aqueous solutions of commercially used water-soluble cellulose ethers were determined for comparative purposes in the author's laboratory (Table 10). Five derivatives were selected, each with two different molecular weights (nominal viscosity grades).

Solubility parameters for polymers are usually obtained from viscosity or swelling experiments but it has been known for a long time that they also can be calculated from physical constants, such as thermal coefficients, critical pressure and more particularly surface tension, using semi-empirical equations. They generally give values of poor accuracy.

Table 9. Values of total, non-polar or dispersion (superscript d) and polar (superscript p) surface free energy for some cellulose derivatives

Derivative	γ (mN m^{-1})	γ^d (mN m^1)	γ^p (mN m^{-1})	x_p (%)	Ref.
Methylcellulose	50.0	22.0	28.0	56	59, 60
Hydroxypropylmethylcellulose	48.4	18.5	29.9	62	59, 60
Hydroxypropylcellulose	41.0	24.7	16.3	41	61
Cellulose acetate	46.6	31.2	15.4	33	62

Table 10. Surface tension γ of some cellulose derivatives measured at 25 °C (2% solutions)[1]

Material	DS or MS	Nominal viscosity at 2% (mPa s)	γ (mN m^{-1})
Methylcellulose	1.8	400	50.3
	1.8	4000	51.2
Hydroxypropylmethylcellulose			
Type 2910	1.9/0.23	4000	48.7
Type 2208	1.4/0.21	4000	50.5
Hydroxyethylcellulose	2.5	4500–6500	67.7
	2.5	100 000	66.0
Hydroxypropylcellulose	4.0	150–400	43.6
	4.0	4000–6500	46.2
Sodium carboxymethylcellulose	0.7	2500–4500[2]	71.6
	0.9	2500–4500[2]	69.0

[1] Unpublished data.
[2] Viscosity at 1%.

2.3 Interactions with Materials in the Solid State

In most applications, cellulose derivatives are not used alone but are admixed with materials, that may be either polymers, or plasticizers or various substrates (tabletted cores, fillers, lubricants, biological tissues). In compatibility studies, a systematic examination has gradually been substituted for trial-and-error approaches. Several experimental methods have been proposed for the evaluation of interaction phenomena in polymeric systems [64]. For the solid state (without "solvent"), the main techniques are:

– microcalorimetry
– differential scanning calorimetry and differential thermal analysis
– thermomechanical analysis
– infrared spectroscopy
– X-ray diffraction.

Polymer-polymer or polymer-plasticizer miscibility has frequently been evaluated from glass transition temperatures T_g (DSC, DTA) or from softening temperatures T_s (TMA). If the two materials are compatible, a single T_g or T_s value is observed lying between the T_g's of the individual materials. For partially compatible materials, two T_g's are detectable if the compatibility limit is exceeded: one for the blend and the other for the material in excess [64]. In pharmaceutics, this approach has been mostly used to investigate the interaction of polymers with plasticizers in coating for the purpose of lowering the temperature of film formation or of correlating with mechanical or permeability characteristics.

Thermomechanical analysis in the penetration mode has been used to check the effect of plasticizers, solvents and moisture on the softening temperature of three cellulose derivatives used as film coating agents [26]. Table 11 lists the

Table 11. Softening point depression coefficient k_s of plasticizers for EC, HPMC and HPMCP [26]

Plasticizer	EC	HPMC	HPMCP
Diethyl phthalate	2.34	2.30	1.59
Triacetin	2.04	2.40	1.63
Polyethylene glycol 400	1.00	8.86[1]	2.96
Propylene glycol	0.07	0.35	—
Castor oil	1.77	—	—

[1] PEG 4000 was used in this case.

softening point depression coefficients k_s obtained from the equation:

$$T_s = T_0 e^{-k_s N} \tag{1}$$

where T_s denotes the softening temperature of the plasticized films (in °C), T_0 that of the pure polymer and N the mole fraction of the plasticizer.

As expected, the efficacy of the various plasticizers differs for each film forming agent, but surprisingly, no real effect of propylene glycol, although generally recommended, was noted for HPMC. Because of the plasticizer present, some solvent may be retained by the film, which may also lower the softening points. Figure 2 highlights this effect for films of ethylcellulose complemented with propylene glycol, which is a poor plasticizer.

The authors also observed that the softening temperature of CAP, HPMC, HPMCP and EC films gradually dropped when stored at high relative humidity, thus demonstrating the plasticizing action of water, whereas no significant changes could be noted with most other plasticizers which therefore acted as protective agents againts moisture.

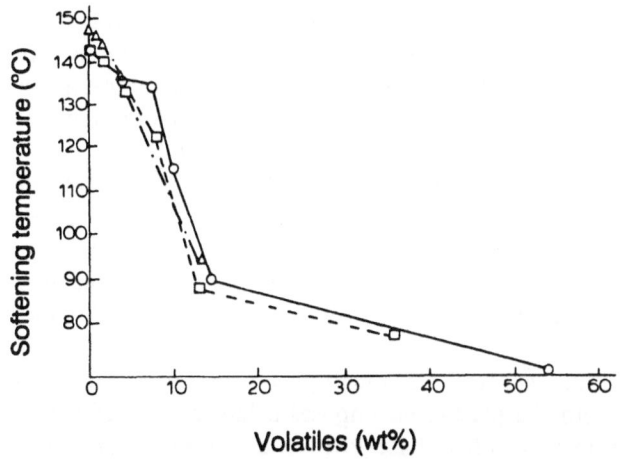

Fig. 2. Effect of solvent retention on the softening temperature of ethylcellulose films plasticized by 17% (△), 33% (□) and 50% (○) of propylene glycol [26]

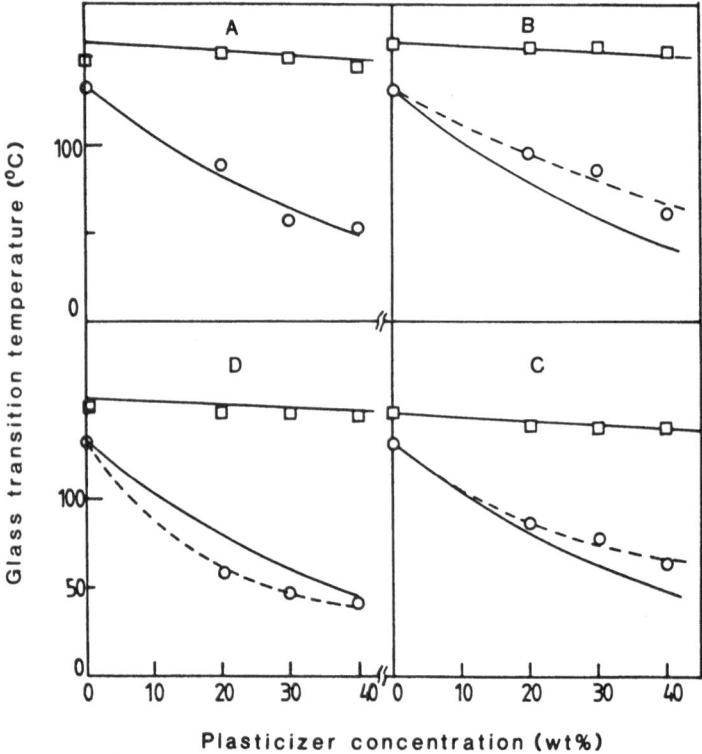

Fig. 3. Variation of the glass transition temperature of ethylcellulose/hydroxypropylmethylcellulose blends with blend composition and diethyl phthalate concentration: (A) 20/80, (B) 40/60, (C) 60/40, and (D) 80/20 w/w ethylcellulose/hydroxypropylmethylcellulose. (□) T_g of hydroxypropylmethylcellulose-rich phase; (○) T_g of ethylcellulose-rich phase. *Full lines* represent the behavior of the individual polymers [65]

Insoluble films most often contain porosigen additives such as hydrocolloids or low molecular weight solutes. EC-HPMC blends plasticized for example with diethyl phthalate are very popular. Unfortunately, diethyl phthalate plasticizes preferentially ethylcellulose, while hydroxypropylmethylcellulose remains almost unaffected, as demonstrated by torsional braid analyses (Fig. 3).

It was shown for individual polymers (full lines) and for EC-HPMC blends (dashed lines), that the glass transition temperature of the EC-rich phase gradually decreases while that of the HPMC-rich phase is only little affected. However, the behavior of the ethylcellulose-rich phase shows deviation from that of the plasticized polymer alone. The glass transition of the EC-rich phase either compares well (A) or is higher (B, C) than that of the ethylcellulose-diethyl phthalate system. The authors suggest that a high proportion of HPMC, with unfavorable interaction with the plasticizer, imposes a barrier to preferential diffusion of the plasticizer into the other phase. For blends containing only 20% HPMC, the glass transition of the EC-rich phase is shifted to lower temperatures due to increased partitioning of diethyl phthalate in this phase (D).

Some authors use the terms "glass transition temperature" and "softening temperature" interchangeably when reporting data and this may lead to erroneous conclusions, if one is not cautious about the technique which is used. Thus, no effect of lactose on the T_g of HPMC was observed using DSC, whereas the T_s decreased drastically ($k_s = 7.27$) with TMA. In contrast, a lowering of both T_g and T_s ($k_s = 7.52$) was seen with the two methods when incorporating ephedrine hydrochloride, a drug with good plasticizer efficiency [66].

The theoretical study of the compatibility between polymers (here cellulose derivatives) and additives has been performed most of the time by using the partial solubility parameters, the Flory–Huggins interaction parameter and the surface free energy approaches. Thus, Aulton et al. [46] concluded that none of the plasticizers screened (monomeric glycols and polyethylene glycols) were thermodynamically ideal for HPMC ($\delta = 23.15 \pm 2.45$ Pa$^{1/2}$ in strongly hydrogen bonded solvents). Although PEGs have δ values of 24.55 ± 5.15 Pa$^{1/2}$, values for χ ranged from 0.54 for PEG 200 to 1.19 for PEG 6000, indicating poor compatibility. More recently, partial solubility parameters and surface free energies have been used to calculate the values of an alternative interaction parameter Φ originally proposed by Girifalco and Good [67] and for which Wu [68] derived the equation:

$$\Phi = \left[\frac{x_d(A) \cdot x_d(B)}{x_d(A) \cdot g_1 + x_d(B) \cdot g_2} + \frac{x_p(A) \cdot x_p(B)}{x_p(A) \cdot g_1 + x_p(A) \cdot g_2} \right] \tag{2}$$

where x_d and x_p are the fractional non-polarity (dispersion) and the fractional polarity of either material A or material B.

Fractional non-polarity and polarity are given by Eqs. (3) and (4), respectively, i.e.:

$$x_d = \frac{\gamma_d}{\gamma} \quad \text{or} \quad x_d = \frac{\delta_d}{\gamma} \tag{3}$$

and

$$x_p = \frac{\gamma_p}{\gamma} \quad \text{or} \quad x_p = 1 - \left(\frac{\delta_d}{\delta} \right)^2 . \tag{4}$$

Constants g_1 and g_2 are defined as the ratios of their surface energies γ:

$$g_1 = \frac{\gamma(A)}{\gamma(B)} \quad \text{and} \quad g_2 = \frac{\gamma(B)}{\gamma(A)} . \tag{5}$$

Alternatively, γ and subsequently constant g_1 and g_2 can be deduced from the solubility parameters and molar volumes V:

$$g_1 = \frac{\delta^2(A) \cdot V^{1/3}(A)}{\delta^2(B) \cdot V^{1/3}(B)} \quad \text{and} \quad g_2 = \frac{1}{g_1} = \frac{\delta^2(B) \cdot V^{1/3}(B)}{\delta^2(A) \cdot V^{1/3}(A)} \tag{6}$$

This theoretical approach has been applied to the adhesion of film coatings to tablet surfaces and to the interaction in the granulation of substrates with polymeric binding agents [59, 60, 69–72]. In addition to the interaction coefficient Φ, the authors have calculated the work of cohesion W_C of each component, the work of adhesion between the two components W_a, the spreading coefficient W_s of each component on the other:

$$W_c(A) = 2\gamma(A) \quad \text{and} \quad W_c(B) = 2\gamma(B) \tag{7}$$

$$W_a = 4\left[\frac{\gamma^d(A)\gamma^d(B)}{\gamma^d(A) + \gamma^d(B)} + \frac{\gamma^p(A)\gamma^p(B)}{\gamma^p(A) + \gamma^p(B)}\right] \tag{8}$$

$$W_s(A/B) = W_a - W_c(A) \quad \text{and} \quad W_s(B/A) = W_a - W_c(B) . \tag{9}$$

Considering only the adhesion of material B on substrate A and the cohesion of material B, the ideal adhesive strength $\sigma_{max}(B/A)$ and cohesive strength $\sigma_{max}(B)$, are given by [70]:

$$\sigma_{max}(B/A) = 0.25\,\Phi \cdot \delta(B) \cdot \delta(A) \tag{10}$$

$$\sigma_{max}(B) = 0.25\gamma^2(B) \tag{11}$$

making possible the direct calculation of the reduced spreading coefficient W_R:

$$W_R = \frac{W_a}{W_c(B)} = \frac{\sigma_{max}(B/A)}{\sigma_{max}(B)} . \tag{12}$$

In a system (pharmaceutical tablet) composed of a compressed core material A (e.g. microcrystalline cellulose or anhydrous lactose) the value of the interaction parameter Φ provides a good prediction of the mode of failure of the system, i.e. interfacial or cohesive within the weaker component. Table 12 presents values for Φ and $\sigma_{max}(B/A)$.

In all cases, microcrystalline cellulose yielded a slightly higher interaction value and a maximum adhesive strength. Systematically cellulose acetate gave also higher values than EC. Should the interaction coefficient be equal or close to unity, an interfacial failure would not be possible.

A good example of correlation between the calculated works of adhesion (Eq. 8) and the measured adhesion of HPC to various substrates has been published by Johnson and Zografi [61] (Table 13). Films of HPC were either

Table 12. Predicted interaction parameters Φ and ideal butt adhesive strength σ_{max} (B/A) for various polymer film/substrate combination [70]

Film	Φ		σ_{max} (B/A) (MPa)	
	MCC[1]	AL[2]	MCC[1]	AL[2]
Cellulose acetate	0.82	0.80	202.2	200.3
Hydroxypropylcellulose	0.74	0.72	181.0	178.8
Ethylcellulose	0.64	0.62	129.5	126.2

[1] Microcrystalline cellulose.
[2] Anhydrous lactose.

Table 13. Correlation of extrapolated adhesive strength σ_0 with work of adhesion W_a for various HPC film-substrate combinations [61]

System[1]	σ_0 (kPa)	W_a (mN m^{-1})
2% Sprayed-PE	294	69
2% Cast-PE	314	69
2% Cast-PMMA	402	85
5% Cast-PET	412	88
2% Cast-PET	451	88
2% Sprayed-PMMA	471	85
2% Cast (ethanol)-PE	481	69

[1] Water was normally used as solvent.

Table 14. Calculated parameters for the various paracetamol (A)[1]-binder (B) combinations [71]

Parameter	HPMC	Acacia	PVP[2]	Starch
W_c(B) (mN m^{-1})	96.8	101.2	107.2	117.4
W_a (mN m^{-1})	102.4	105.9	110.0	115.1
W_s (A/B) (mN m^{-1})	10.6	−7.1	−3.0	2.1
W_s (B/A) (mN m^{-1})	5.6	4.7	2.3	−2.3
W_R	1.06	1.05	1.03	0.98
Φ	0.97	0.99	1.00	1.00

[1] The work of cohesion of paracetamol is 113.0 mN m^{-1}.
[2] Polyvinylpyrrolidone.

cast or sprayed and the three substrates used were poly(methyl methacrylate) (PMMA), poly(ethylene terephthalate) (PET), and polyethylene (PE). The experimental parameter was the adhesive strength of HPC films extrapolated to zero thickness. As predicted, the trend for adhesion intensity depended on the substrate, the concentration of the solution and the preparation method.

Granulation (agglomerating of a substrate) is best described using spreading coefficients. Rowe [71] has been able to correlate various tablet or granule properties of paracetamol prepared using four distinct binders with the spreading coefficient W_s(B/A). Table 14 reports the various calculated parameters and Fig. 4 illustrates the correlations.

Note that according to Rowe [59] there is a strongly adhering film of binder around the substrate when W_s(B/A) is positive, whereas when W_s(A/B) is positive, no film will be formed around the substrate and the resulting granules will be very porous and of lower strength.

The same author [69] has been able to show from calculations using data from nine drugs that there is a parabolic relationship between the reduced spreading coefficient and the fractional polarity of the substrate. The relationship, which is independent of the method used and specific for each binder (Fig. 5), gives an explanation for the apparently surprising rank ordering of polymers and is a tool for predicting the optimum binder for any particular substrate.

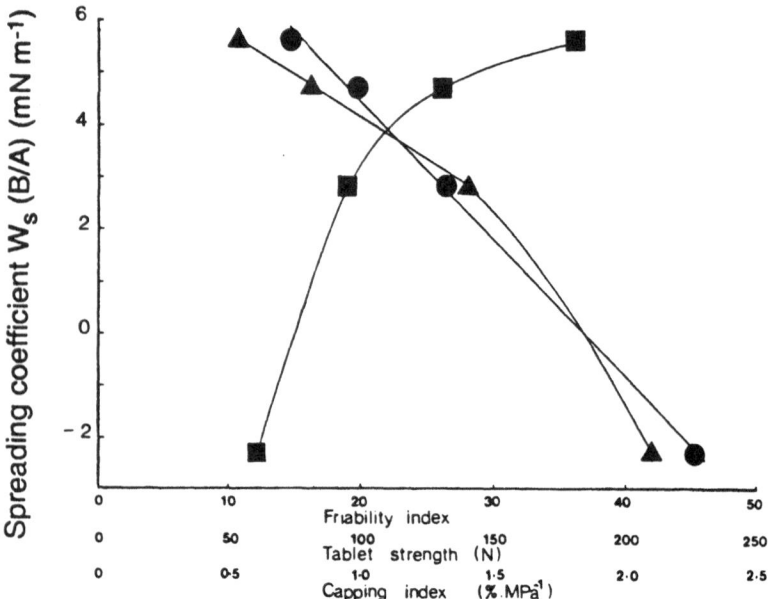

Fig. 4. Relationship between the calculated spreading coefficient $W_s(B/A)$ and granule friability (●), tablet crushing force (■) and tablet capping index (▲) [71]

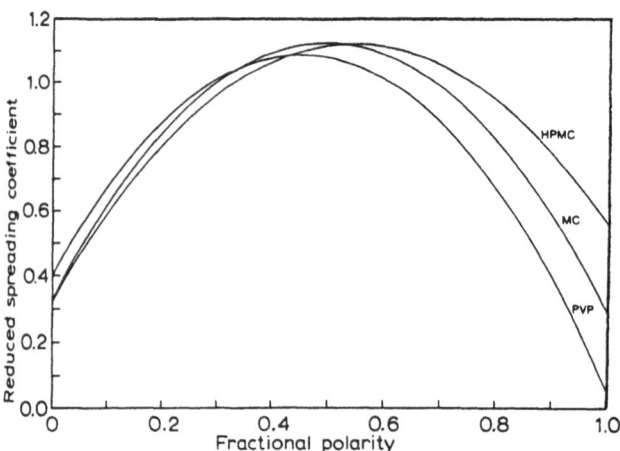

Fig. 5. Results of the quadratic curve fitting for the reduced spreading coefficient vs fractional polarity for HPMC, MC and PVP [69]

Solutions for the equations where $W_R = 1$ provide information on the fractional polarity ranges where film formation is likely to occur, HPMC being the most versatile binder:

$$\text{HPMC}: \quad x_p = 0.33 - 0.75, \quad \Delta x_p = 0.42$$

$$\text{MC} \quad : \quad x_p = 0.30 - 0.69, \quad \Delta x_p = 0.39$$

$$\text{PVP} \quad : \quad x_p = 0.29 - 0.61, \quad \Delta x_p = 0.32$$

These examples clearly show that these approaches based on measured or calculated polarities apply to a variety of situations and have a potential for use in product development. The analysis of bioadhesion phenomena will be addressed in Sect. 3, that will deal with specific properties of water-soluble cellulose derivative. The other three sections will concern derivatives soluble in organic solvents (Sect. 4), derivatives soluble in nonacidic media and organic solvents (Sect. 5) and derivatives insoluble both in water and organic solvents (Sect. 6). I think that is appropriate to make a distinction according to the solubility or insolubility in water and/or organic solvents as most technological and biological applications involve interactions with a given liquid.

3 Derivatives Soluble in Water

3.1 Moisture Content

Water-soluble derivatives are hygroscopic and at high relative humidities start to dissolve. Data are available on the effect of the degree of substitution on water absorption (see e.g. [18, 19, 21, 24]) but there is no direct comparison of derivatives commonly used today. Table 15 lists the moisture contents of a series of

Table 15. Moisture content MC of some water-soluble derivatives after 30 day storage at 74% relative humidity and hydrogen-bonding partial solubility parameter δ_h[1]

Material	DS or MS	Nominal viscosity at 2% (mPa s)	MC (%)	δ_h (MPa$^{1/2}$)
Methylcellulose	1.8	4000	12.6	15.3
Hydroxypropylmethylcellulose				
Type 2910	1.9/0.23	4000	11.8	14.2
Type 2906	1.8/0.13	4000	11.0	14.9
Type 2208	1.4/0.21	4000	14.4	16.7
		15000	15.0	16.7
		100000	14.6	16.7
Hydroxyethylcellulose	2.5	4500–6500	21.7	18.9
	2.5	100000	20.5	18.9
Hydroxypropylcellulose	4.0	4000–6500	10.9	15.1
	4.0	1500–3000[2]	10.9	15.1
Sodium carboxymethylcellulose	0.7	2500–4500[2]	30.6	—[3]
	0.9	2500–4500[2]	33.2	—[3]

[1] Unpublished data.
[2] Viscosity at 1%.
[3] The solubility parameter could not be calculated for the ionic NaCMC.

ethers after equilibration at 74% relative humidity (30 days at room temperature) together with the partial solubility parameters for hydrogen bonding as calculated previously [15]. Most of the derivatives are those described in Table 10.

Except for HPMC types 2910 and 2906, rank order of moisture contents follows strictly that of δ_h values calculated for the exact substitution degrees. Molecular weight has very little effect on water content at equilibrium.

Careful characterization of moisture interactions in cellulose derivatives has been reported for HPMC films, both unplasticized and plasticized. Using various techniques (a radiotracer technique involving the use of tritiated water, DSC and TGA), some authors [73] proposed the classification of the water present into three categories: the tightly bound water present in plain HPMC films was located in the ordered regions of the polymer and was absent in plasticized films. The moderately bound water (detectable by TGA) was in the rank order of the hydrophilicity of the polymer additives and this order was reversed for the calculated free water content.

3.2 Swelling and Water Penetration

Water-soluble (uncrosslinked) polymers absorb water, swell and concomitantly dissolve (erode). Not much has been published on water penetration kinetics and swelling of cellulose derivatives. Wan and Prasad [74] studied the linear swelling of MC films at 34, 37 and 40 °C. Concurrently, they were able to explain the slow disintegration of pharmaceutical tablets prepared using this binder at low temperature by the higher degree of swelling.

Water penetration is of utmost importance for drug release from hydrophilic films and from the so-called hydrophilic compressed matrix tablets [14, 75]. In the former case, the drug is incorporated by solubilizing or suspending it in the polymer solution, which is subsequently dried off. In the latter situation, the drug is mixed with the polymer and the powder blend is compressed in a die at an adequate pressure.

The first step necessary to release the drug is water penetration with formation of a gelified outer layer. For water-soluble cellulose derivatives, both water penetration and drug release are generally controlled by Fickian diffusion and are described by the so-called Higuchi's equations (see e.g. [14]). Recently, we have developed a model that can predict the solute (drug) and penetrant (water) concentration profiles with simultaneous swelling [76]:

$$\frac{1 - \xi_*}{\xi_*} = \frac{C_*(1 - \beta)}{2\varepsilon(1 - C_*)} \cdot \frac{1}{\sum_{n=1}^{\infty} \exp(-\lambda_n^2 \tau)} \tag{13}$$

where $\lambda_n = (2n - 1)\pi$. ξ_* is the dimensionless position (depth) of polymer-gel interface ($= x_*/L_0$); C_* is the dimensionless concentration at the polymer-gel interface; β is a dimensionless constant ($= M_s \cdot c_0 \cdot \rho^{-1}$); ε is the diffusivity ratio

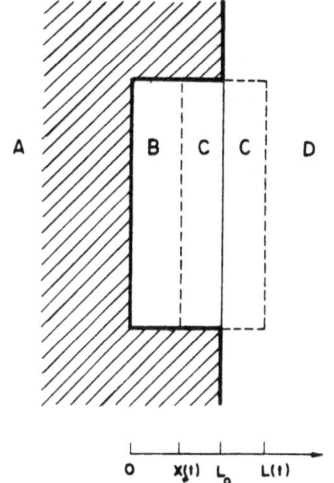

A   B   C   C   D

$0 \quad x_{(t)} \quad L_0 \quad L_{(t)}$ x

Fig. 6. Diagrammatic representation of drug diffusion in swellable polymer. *A*: inert support; *B*: solvent-free polymer; *C*: swollen polymeric gel; *D*: surrounding swelling agent [76]

of the drug in the solvent-free polymer (D_1) and in the gel phase (D_2); τ is a dimensionless time ($= D_1 \cdot t/L_0$); M_s is the molecular weight of the polymer (swelling agent), ρ its density and c_0 is the swelling agent concentration in the surrounding medium. Other symbols are apparent in the schematic representation of Fig. 6.

Figure 7 shows both experimental (symbols) and fitted (full lines) concentration profiles of KCl, water and HPMC 2208 in the gel phase after 4 and 6 h. The thickness of the swollen polymer changed from 1.75 mm at 4 h to 3 mm at 5 h and 5.25 mm at 6 h (initial thickness $L_0 = 2.0$ mm).

In this model and especially in the Higuchi equations, no allowance is made for a phenomenon occurring quite frequently with dosage forms made with cellulose derivatives, i.e. polymer dissolution. Under proper conditions, this erosion mechanism may lead to a zero order drug release, a goal often sought after. A model has in fact been proposed by Lee and Peppas [77] and Harland et al. [78] which accounts for concomitant polymer swelling and dissolution and has been checked experimentally for some cellulose derivatives, beside other hydrophilic polymers. In this model, two moving fronts, the swelling and the eroding fronts, are considered (Fig. 8).

In some conditions, the normalized gel layer thickness ($=(S - R)/a$) may reach a constant value because the synchronization of the swelling and eroding fronts generates pseudo-order release of a solute incorporated in the polymeric matrix. Experimental results concern mixes made of the drug diclofenac sodium and the excipient poly(vinyl alcohol), HPMC type 2208 or NaCMC (Fig. 9).

The synchronization of the two fronts is rapidly achieved with PVAL and soon produces a zero-order release. In the system containing HPMC, the synchronization is still not reached when, at approximately 300 min, the swelling front stops. Nevertheless, from that moment, drug release is constant because the system behaves as a conventional erodible system.

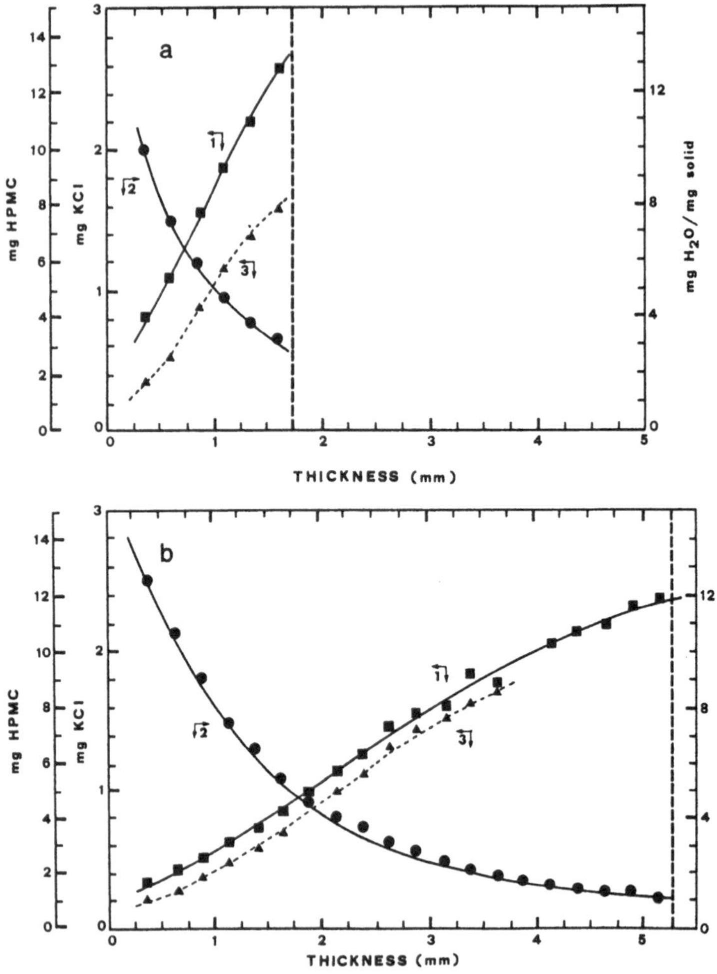

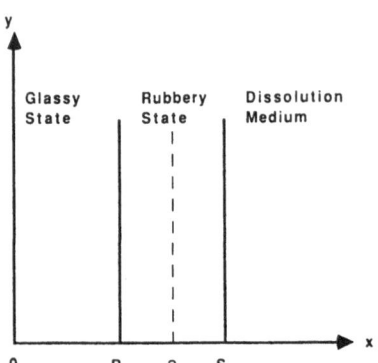

Fig. 7a, b. Concentration profiles of KCl (*1*), water (*2*) and HPMC 2208 (*3*) in the gel phase of the compressed tablets after 4 h (**a**), and 6 h (**b**) [76]. The *dashed line* corresponds to the swelling front

Fig. 8. Diagrammatic representation of a swelling polymeric system in contact with a penetrant indicating the initial position of the tablet surface *a*, the dissolution medium/swollen polymer front (eroding front) *S*, and the rubbery-glassy front (swelling front) *R*. The coordinate *O* is the center of the tablet and *x* is the distance coordinate [78]

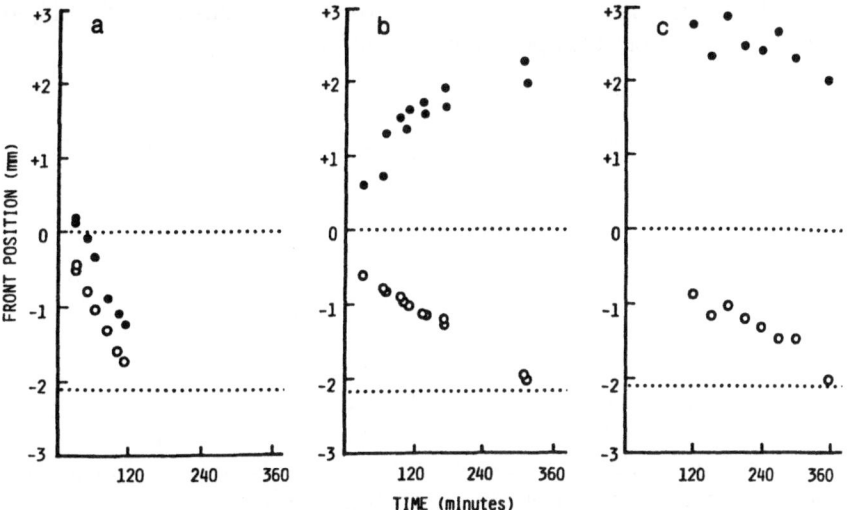

Fig. 9a–c. Swelling (○) and eroding (●) front movement of systems containing 50% diclofenac, 30% mannitol and 20% PVAL (a), HPMC 2208 (b) or NaCMC (c) [79]

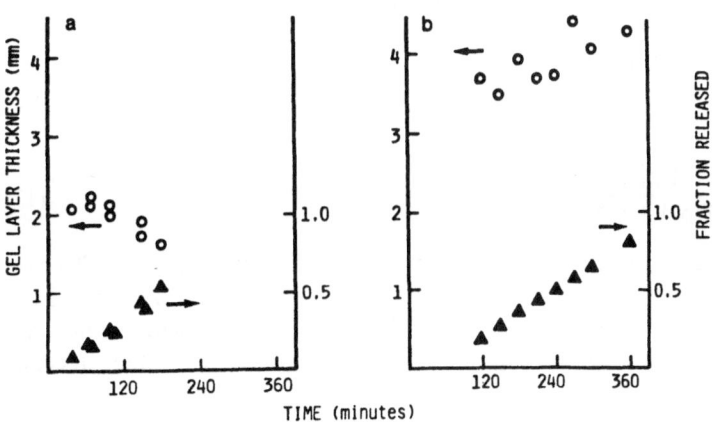

Fig. 10a, b. Gel layer thickness (○) and drug fraction release (▲) in systems containing diclofenac sodium, mannitol and NaCMC at (a) 10% and (b) 20% [79]

In the NaCMC system, the layer thickness first increases and then stabilizes. Constant release is observed only from about 100 min, but is achieved very soon when the polymer concentration is reduced to 10% (Fig. 10).

For various reasons, almost constant drug release has been observed by some authors and kinetics controlled by the swelling of the polymer have been postulated many times as an explanation. The author [15] has, in fact, proved theoretically that anomalous water transport – a necessary criteria for a non-Fickian *solute release* – is doubtful with these materials. Let us consider the diffusional Deborah number $(DEB)_D$ as proposed by Vrentas et al. [80, 81], to characterize the solvent transport in glassy amorphous polymers:

$$(DEB)_D = \lambda_m/\theta_D \tag{14}$$

where λ_m is a characteristic relaxation time and θ_D a characteristic diffusion time which is defined as:

$$\theta_D = L^2/D_s \tag{15}$$

with L being the thickness and D_s the diffusion coefficient of the solvent.

A close look at the scaling law proposed by Peppas and Lustig [82] shows that the polymer concentration in the swollen gel affects the $(DEB)_D$ value. Knowing that anomalous diffusion is observed when $(DEB)_D \simeq 1$, one can calculate the polymer volume fraction at the swelling interface c* from the glass transition temperature T_g. To estimate this parameter which has to be maximum, the weight fraction of penetrant w_0^*, is first calculated using the equation for the depression of polymer glass transition due to water:

$$w_0^* = \frac{T_g - T}{\beta/\alpha_1} \tag{16}$$

where T is the absolute temperature, β the contribution of the penetrant to the expansion coefficient (0.20 for many systems) and α_1 is the linear expansion coefficient of the polymer ($3.7 \cdot 10^{-4}$ K for most polymers).

The polymer volume fraction c* is then calculated from the polymer density ρ_p:

$$c* = \frac{1/\rho_p}{(1/\rho_p) + w_0^*} \tag{17}$$

Table 16 lists c* values calculated from selected T_g values at 37 °C. The case of poly (hydroxymethyl methacrylate) (PHEMA) is included for comparison. A high value of c* is observed only for PHEMA, a material with a low glass transition temperature, confirming that anomalous transport of water in cellulose derivatives is unlikely.

Table 16. Polymer volume fractions at the swelling interface c* calculated at 37 °C from selected T_g values [15]

Derivative		T_g (°C)	c*
MC		196	0.71
HPMC	type 2910	155	0.77
	type 2208	184	0.74
HEC		106	0.85
HPC		124	0.85
NaCMC		(78)[1]	0.89[1]
PHEMA		55	0.95

[1] Questionable value.

Mention should finally be made of the fact that swellable cellulose derivatives, often admixed with low-density lipid materials, have found applications in the so-called floating drug delivery tablets or capsules. These hydrodynamically balanced systems are intended to float on the gastric juice and therefore to undergo retarded gastric emptying. The only physical analysis published for these dosage forms was the measurement of the resultant force F acting vertically on the system [83].

3.3 Bioadhesion

One consequence of the swelling of hydrocolloids is that adhesion to substrates, in particular to biological tissues or mucosae, may occur. Advantage can be taken of this finding to retain a dosage form in a certain site where drug delivery or absorption is optimum.

Many pharmaceutically acceptable hydrophilic polymers have been proposed and tried for bioadhesive or muco-adhesive systems, and among them, water-soluble cellulose derivatives.

The mechanisms of bioadhesion have been reviewed by Kaelble and Moacanin [62] and by Peppas and Buri [84] and may be classified as chemical (electronic and adsorption theories) and physical (wetting, interpenetration and fracture theories).

According to the very popular wetting theory a material (bioadhesive, phase b) displaces the surrounding liquid (e.g. the gastric content, phase g) and adheres spontaneously on the mucus (tissue, phase t) when the spreading coefficient, $W_s(b)$, is positive, i.e.:

$$W_s(b) = \gamma(g, t) - \gamma(b, t) - \gamma(b, g) > 0 . \tag{18}$$

Accordingly, bioadhesion is favored either by a large value of the gastric content-mucus interfacial tension or by low values of the bioadhesive systems-mucus and bioadhesive system-gastric content interfacial tensions. The terms involving the bioadhesive polymer, $\gamma(b, t)$ and $\gamma(b, g)$, and which are to be minimized, can be defined [67] as:

$$\gamma(b, t) = \gamma(b) + \gamma(t) - 2\Phi[\gamma(b) \cdot \gamma(t)]^{1/2} \tag{19}$$

$$\gamma(b, g) = \gamma(b) + \gamma(g) - 2\Phi[\gamma(b) \cdot \gamma(g)]^{1/2} \tag{20}$$

where $\gamma(b)$, $\gamma(t)$ and $\gamma(g)$ are the surface tensions of the various phases and Φ are interaction parameters as defined previously in Eq. (2).

Alternatively, when taking into account the contributions of dispersion and polar forces for deeper examination, one may write [62]:

$$\gamma(b, t) = \{[\gamma^d(b)]^{1/2} - [\gamma^d(t)]^{1/2}\}^2 + \{[\gamma^p(b)]^{1/2} - [\gamma^p(t)]^{1/2}\}^2 \tag{21}$$

$$\gamma(b, g) = \{[\gamma^d(b)]^{1/2} - [\gamma^d(g)]^{1/2}\}^2 + \{[\gamma^p(b)]^{1/2} - [\gamma^p(g)]^{1/2}\}^2 . \tag{22}$$

Minimization of these two interfacial terms may be achieved by having the dispersion and the polar components of the bioadhesive system and of the gastric mucus as close as possible, i.e. the materials should have similar polarities.

Unfortunately, values of γ^d and γ^p for mucus and glycoproteins, the main components of the mucous layer, are not yet available, but from the fact that the best candidates for bioadhesion are hydroxyl- and carboxyl-containing polymers, one can infer that the polar component is the predominant factor. The experimental data relating cellulose derivatives (mostly by measuring bioadhesive force) generally only compare NaCMC to one or two nonionic derivatives and the former always proved to be superior to the latter, and equivalent to other carboxyl-containing polymers (see e.g. [85]).

Two recent studies provide more data on cellulose derivatives. Ranga Rao and Buri [86] measured *in situ* the adhesiveness of glass beads (as a model) and aspirin crystals coated with the polymers to be tested on the rat stomach or jejunum mucosae. The percentage of particles retained on the tissue upon washing by dilute HCl (for the stomach) or a phosphate buffer (for the jejunum) was considered to be an index of bioadhesion (Table 17).

The test proved to be more discriminating with denser glass beads. Binding of the coated particles to the intestine was less than (MC, HPMC, pectine) or equal to (NaCMC, polycarbophil) that to the stomach. As far as the cellulose derivatives are concerned, the adhesiveness to the stomach decreased in the order NaCMC > HPMC > MC (glass beads) and HPMC $\simeq$ NaCHC > MC (aspirin crystals). In the intestine, the order was NaCMC $\gg$ HPMC = MC (glass beads) and NaCMC > HPMC > MC (aspirin crystals). These results are in agreement with previous studies.

In an evaluation of adhesive patches for buccal administration, Anders and Merkle [86] compared, among others, several HECs and HPCs varying in their molecular mass (as characterized by the viscosity). The duration of mucosal adhesion of drug-free patches was measured in vivo (Fig. 11).

Generally, the duration of mucosal adhesion was much more affected by the molecular mass of the polymer than its nature. Furthermore, a very long duration was observed for the medium viscosity grades of HEC (250G, 250K), although they have the same surface energy characteristics as the other HECs. Again, the latter adhered as effectively as the less polar HPCs, for which

Table 17. Index of bioadhesion (%) of glass beads and aspirin crystals coated with various polymers on the stomach and intestinal mucosae [86]

Polymer	Glass beads		Aspirin crystals	
	Stomach	Intestine	Stomach	Intestine
None	12	0	72	79
MC	45	6	80	58
HPMC	54	6	98	75
NaCMC	72	74	96	96
Polycarbophil	93	89	100	100
Pectin	49	25	89	56

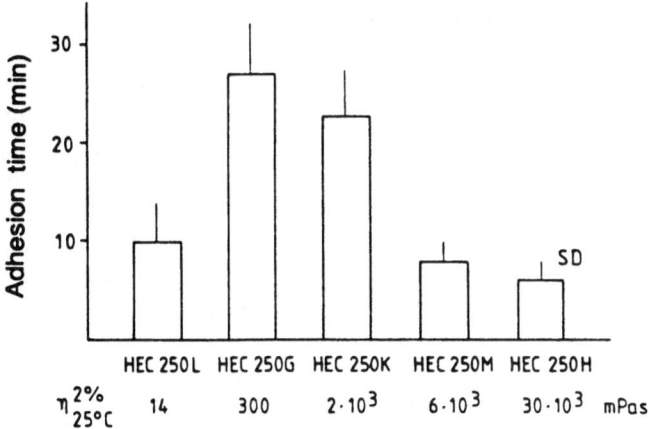

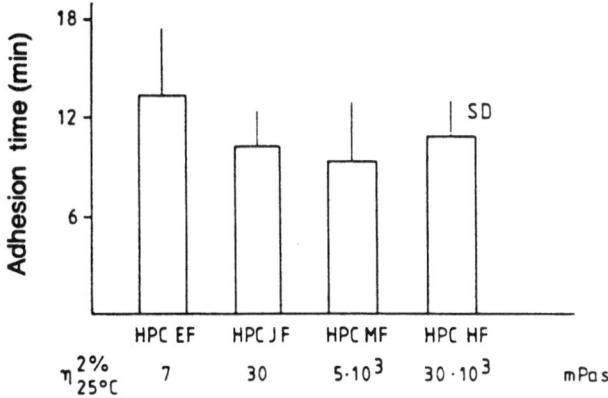

Fig. 11. Duration of mucosal adhesion of patches prepared with various viscosity grade HECs (*top*) and HPCs (*bottom*) [86]

a minimum was this time found in the adhesion duration vs viscosity grade relationship. This behavior is rather peculiar in that the adhesive strength usually increases as the macromolecule chains become longer, favoring their interpenetration and entanglement. However, beyond a certain critical value, the length of the chains became clearly detrimental to bioadhesive bonds.

These results, together with the fact that the anionic NaCMC is a strong bioadhesive polymer despite its high surface energy, give no real support to the wetting theory. In fact, according to Mikos and Peppas [87] thermodynamic (reversible) adhesion would be valid only for poor bioadhesives. Here, only the interfacial energy would be concerned and in particular bioadhesion would not be influenced by the pH of the mucus. For better bioadhesive systems, adhesion is not reversible and the fracture theory is more helpful.

3.4 Thermal Gelation and Cloud Point

One of the main features of nonionic water-soluble cellulose derivatives is that they exhibit, like some other polyethers, an inverse solubility-temperature behavior, i.e. there is phase separation on heating above the so-called lower critical solution temperature (LCST). The temperature at which a polymer-rich phase separates is normally referred to as the cloud point (CP). For ideal solutions, this temperature corresponds to the theta-temperature. Actually, for some derivatives, the cloud point may be preceded, if the concentration is not too low, by a sol-gel transformation with an increase in viscosity and possibly formation of liquid crystals (see Sect. 3.5). As it will be seen later, this reversible thermotropic behavior may be detrimental to the performance of the derivatives or can be advantageously utilized to develop applications.

Big differences in the data have been reported in the literature for a given product. The reason for this, besides real differences due to interbatch variations, is that the authors use varying process conditions and do not always observe the same transformation. In this respect, the most accurate work is that of Sarkar [88]. Three different temperatures were defined:

- the incipient gelation temperature (IGT) corresponds to the temperature at which the viscosity reaches a minimum (Fig. 12),
- the incipient precipitation temperature (IPT) is the temperature at which light transmission reaches 97.5%,
- the cloud point (CP) is the temperature at which light transmission is 50%.

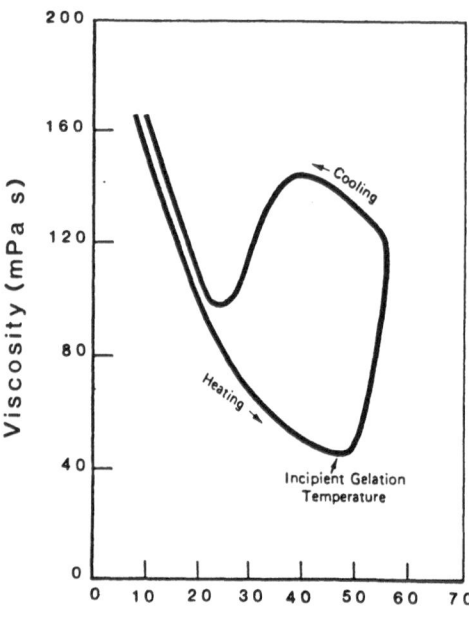

Fig. 12. Gelation of a 2% solution of a 50 mPa s viscosity grade methylcellulose (DS = 1.6–1.8; shear rate = 86 s^{-1}) [88]

The last two temperatures are frequently used interchangeably in the literature. Thus, the cloud point is sometimes defined as the lowest temperature for haze development, the temperature of precipitation being higher [10].

Note that a "melting point" has also been measured by some workers, as the temperature at which the viscosity of the gel begins to decrease rapidly.

The transition temperature depends on many factors:

- type, degree and uniformity of substitution,
- molecular mass and concentration of polymer,
- presence of additives (electrolytes, surfactants),
- experimental conditions (heating rate, stirring).

The type and the number of substituents are of course of utmost importance for the behavior of the derivatives (Fig. 13).

Only a monotonic decrease in viscosity upon heating is observed for NaCMC and HEC solutions due to a gradual loss of hydration water by the macromolecules. There is no gelation nor precipitation or cloud point at least under 100 °C. The drop in viscosity of the HPC solution is drastic and precipitation occurs at about 40–45 °C without any gelation. In contrast, gelation is evident for MC which forms a firm gel at 50–55 °C, before the cloud point. The behavior of HPMCs 2906 and 2910 is quite similar to that of MC, but being more hydrophilic, their temperature of gelation is higher (62–68 °C and 58–64 °C, respectively) and they form semi-firm gels. HPMC 2208 forms a mushy gel at 70–90 °C with no measurable strength. The thermal gelation points mentioned above have been reported for 2% solutions, but Sarkar [88] has shown that the values of IGT, IPT and CP vary considerably with polymer

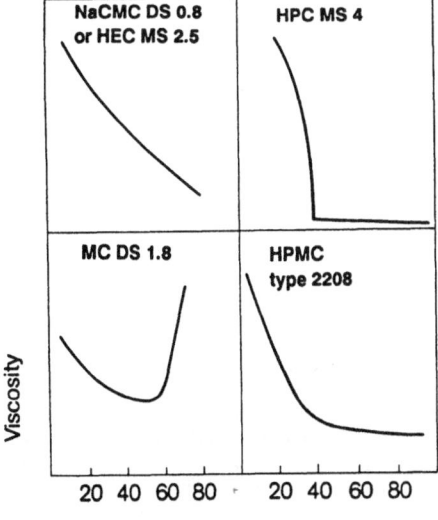

Fig. 13. Viscosity-temperature behavior of water-soluble cellulose ethers [11]

concentration. Figures 14 and 15 illustrate this effect for HPMC 2906. Both IPG and CP decrease initially with increasing concentration until a critical concentration is reached (Fig. 14). Figure 15 shows that there is no direct relationship between the precipitation temperature and the gelation temperature. Thus, up to a concentration of about 6.5% the solutions become turbid before gelation, whereas above that limit the solutions gel before any turbidity is visible.

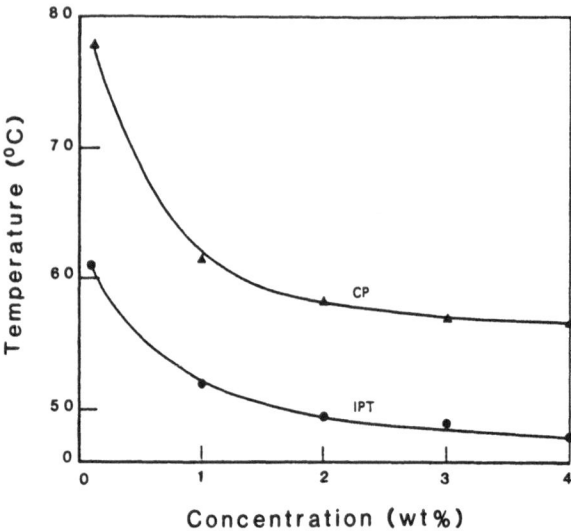

Fig. 14. Incipient precipitation temperature *IPT* and cloud point *CP* of HPMC 2906 (50 mPa s viscosity grade) as function of concentration [88]

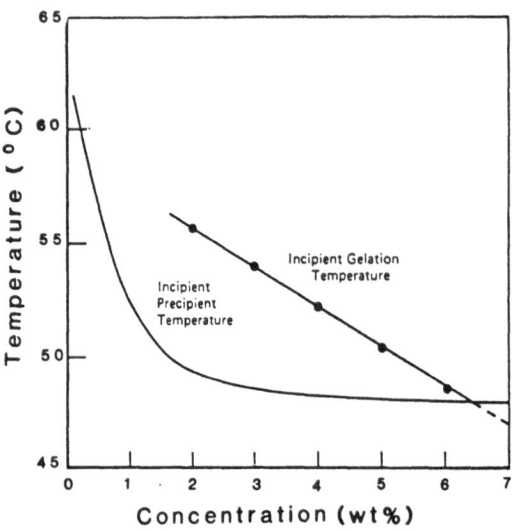

Fig. 15. Incipient gelation temperature IGT and incipient precipitation temperature IPT for aqueous solutions of HPMC 2906 (50 mPa s viscosity grade) as function of concentration [88]

These patterns are valid for heterogeneously etherified derivatives, but it should be recalled that, for instance, MC prepared in solution does not gel because of the reduced fraction of di- and tri-O-methylglucose residues. It is in fact known, that highly substituted regions form the junction zones. Sarkar [88] also demonstrated that the molecular mass had no significant effect on the IGT, IPT and CP values, in contrast to normal polymers where an increase in demixing temperature is expected with decreasing chain length. However, this conclusion differs from other findings [4].

Because of their great affinity for water, most solutes (in particular electrolytes, but also non-electrolytes like sugars) depress the transition temperatures by dehydrating the polymer and thus favor polymer–polymer interactions. This salting-out efficiency varies widely for the different cellulose derivatives and depends on the nature of both the cations and anions.

The effect of solutes is usually expressed as the depression in the gelation temperature or cloud point caused by a given concentration of additive or by the concentration of added solute tolerated by the polymer concentration. From the data available in many classical articles [1, 5–7, 9, 10, 12], the cation tolerance for nonionic ethers seems to follow the order:

$$Pb^{2+} > Zn^{2+} > Cu^{2+} > Fe^{3+} > NH_4^+ > Ca^{2+} > Ba^{2+} > K^+$$
$$> Mg^{2+} > Na^+ > Al^{3+}$$

and for anions:

$$I^- > CNS^- > borate > NO_3^- > CO_3^{2-} > Cl^- > acetate$$
$$> tartrate > SO_4^{2-} > PO_4^{3-} .$$

Conversely, some large ions (e.g. basic drugs) may rise the gelation temperature, and presumably the cloud point [89, 90]. The increase in hydration is considered to result from the adsorption of these ions carrying water onto the macromolecule. In this regard, Mitchell et al. [90] provide new data for two HPMCs 2208 (Table 18). The abilities of additives to depress the cloud point, or salting-out constants K_{CP}, were calculated from the relationship:

$$\log CP = \log CP_0 + K_{CP} m \tag{23}$$

where CP is the observed cloud point (similarly to Sarkar [88]), CP_0 is the theoretical cloud point in absence of additive and m is the molar concentration.

The salting-out constants for the two molecular mass grades, when available, were very similar. The nature of the anion plays a major role, with a high efficiency being offered by the various phosphates, which are salts commonly used in buffered media for drug release studies. In contrast, pH had no significant effect on the cloud point. Two of the drugs tested gave significantly positive K_{CP} values (salting-in effect).

Comprehensive recent data on the effect of electrolytes and surfactants and of their combinations can be found in the work of Carlsson et al. [91–93].

Table 18. Salting-out constants K_{CP} of additives derived from cloud points of 2% solutions of HPMCs 2208 (15 000 and 100 000 mPa s viscosity grades) [90]

Additive	Salting-out constant K_{CP}	
	15 000 mPa s grade	100 000 mPa s grade
NaH_2PO_4	−0.53	−0.63
KH_2PO_4		−0.57
Na_2HPO_4	−1.32	−1.30
Na_3PO_4	−2.32	−2.20
$Na_4P_2O_7$	−2.52	−2.68
K_2HPO_4		−1.20
$AlCl_3$		−0.30
NaCl		−0.22
KCl		−0.10
Potassium tartrate		−1.16
Na_2SO_4		−1.28
Theophylline	0.00	
Aminophylline	0.32	
Tetracycline HCl	0.35	
Quinidine bisulfate	0.00	

Increases and decreases in the cloud point have been observed. An interesting application of the lowering effect will be discussed later.

Anionic derivatives, like NaCMC, behave differently. They are much more resistant to dehydration, but chemical interaction is significant with multivalent cations. Precipitation, gelification and/or coagulation occur with trivalent cations due to crosslinking. NaCMC is also incompatible with positively charged polymers, e.g. gelatin, below the isoelectric point (see Sect. 3.6). Being a polycarboxylic acid, CMC is also very sensitive to pH and will precipitate in acidic media. Depending on the DS, the apparent pK value varies from 4 to 4.4, the intrinsic value pK_{int} (obtained by double extrapolation to zero ionic strength and zero titration index) amounting to 3.79 for a DS less than unity and increasing up to 4.30 for a DS of 2.84 [94].

The incidence of the ability of salts to reduce the hydration of compressed hydrophilic matrices has been investigated by some authors [90, 95]. The disintegration times of plain HPMC 2208 (15 000 mPa s grade) tablets in various solutions of electrolytes could be correlated with the cloud points (Table 19).

At an ionic strength of 0.9, the compressed matrix was not able to retain its integrity in phosphate media for two hours. In contrast, with NaCl, the matrix did not disintegrate until a ionic strength of 2.0. Above this value, the matrix remained intact but did not swell. Concurrently, Mitchell et al. [90] observed with tablets prepared with the same excipient but containing the drug pro-pranolol hydrochloride a burst release due to loss of integrity. We have noticed similar behavior many times: the compressed hydrophilic tablets swelled per-fectly in artificial gastric juice or dilute HCl, but disintegrated in intestinal juice or neutral buffers. This was not the case if the tablets were first placed in

Table 19. The effect of solutes and ionic strength on the disintegration of HPMC 2208 tablets (adapted from [90])

Additive	Concentration (mol l^{-1})	Ionic strength	Disintegration time (min)	CP (°C)	Matrix texture[1]
NaH$_2$PO$_4$	0.6	0.6	>120	29.2	RG
	0.9	0.9	52	18.9	SG
Na$_2$HPO$_4$	0.1	0.3	>120	52.0	RG
	0.3	0.9	23	28.5	SG
	0.8	2.4	>120	6.3	NG
NaCl	0.9	0.9	>120	43.7	RG
	1.5	1.5	>120	32.0	RG
	2.0	2.0	85	24.7	SG
	2.5	2.5	>120	19.0	NG

[1] RG, rapid gelling, matrix intact; SG, slow gelling, disintegration; NG, non gelling, matrix intact.

acidic medium because swelling was achieved before transfer into the phosphate buffer.

The sol-gel transformation may be used to create an *in situ* gelling system for drug delivery. Recently, the interaction between for instance EHEC at low concentrations and charged surfactants in aqueous solution was found to give rise to thermogels which are liquid at room temperature but highly viscous at body temperature. As the gelation temperature and cloud point are close and both imply change in chain conformation, the authors first examined the effect of electrolytes, surfactants and their combinations on the cloud point [91–93, 96] and the viscosity [91, 97, 98] of EHEC or other cellulose ethers. Interestingly, for some surfactants it was found that in the absence of salt the binding of surfactants to EHEC leads to an increase in CP. In the presence of NaCl, there is first a decrease and then an increase at higher surfactant concentrations. An increase followed by a decrease in viscosity was also observed. Figure 16a shows the effect of sodium dodecylsulfate (SDS) concentration on the cloud point of a 1 wt % EHEC ($\bar{M}_v$ 150 000) solution. The line separates the two-phase region from the clear solution region. The marked area is the optimal one-phase region with increased viscosity. The rise in viscosity of a system supplemented with 3.0 mM SDS is evident on Figure 16b.

The system EHEC-alkyl betainate-water was used to deliver the anti-glaucome drug timolol maleate [99]. The phase diagram (Fig. 17a) differed from the one with SDS and without the drug. Here, even at room temperature no one-phase region was observed up to about 0.45 wt % surfactant, but a clear gel at body temperature was obtained at 0.475 wt % (dotted line). This system containing 0.34 wt % timolol maleate was assessed in vitro for drug release (Fig. 17b). Sustained release was achieved with the thermogelling EHEC-ionic surfactant system, in comparison with both the simple aqueous EHEC solution and the non-gelling EHEC solution.

The effect is not restricted to EHEC (here of a special grade) but careful formulation is necessary to optimize the gelling capability of the system.

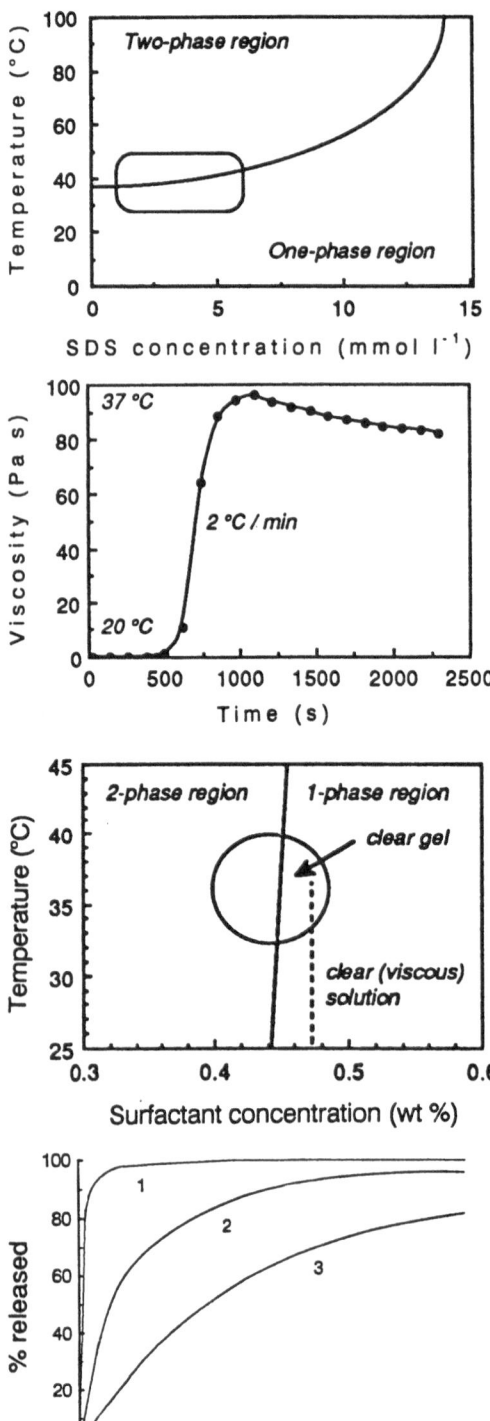

Fig. 16. Phase diagram for 1 wt % EHEC and SDS in water (*top*), and viscosity increase of a system containing 30 mM SDS when heating to 37 °C (*bottom*) [98]

Fig. 17. Phase diagram of the timolol maleate (0.34 wt %)-EHEC (1 wt %)-alkyl betainate-water system (*top*), and *in vitro* release of timolol maleate from aqueous solution (*1*), non-gelling 1 wt % EHEC solution (*2*) 1 wt % EHEC gel (*3*) (*bottom*) (adapted from [99])

3.5 Liquid Crystal Formation

Published observations indicate that at room temperature water-soluble cellulosics form mesophases at a critical volume fraction of polymer generally
ranging from 0.3 to 0.5 for high molecular mass samples. For a given polymer
and solvent, the critical volume fraction decreases with increasing molar mass,
but increases with temperature. Highly polar and acidic solvents favor liquid
crystal formation.

The first report in the open literature concerning cellulose derivatives is that
of Werbowyj and Gray [100]. They observed that 20–50% aqueous solutions of
HPC were birefringent and highly iridescent due to the cholesteric nature of the
liquid crystalline phase. The formation took place at a higher polymer volume
fraction than predicted by the Flory theory, because of some chain flexibility
and of side group interaction.

Since then, liquid crystals have been found with many other water-soluble
(and organosoluble) derivatives [101, 102]. As an example, we have observed
ordered liquid crystal structures, besides HPC (Fig. 18), with MC, HPMC, HEC
and NaCMC, both at room temperature and 37 °C.

The primary factors governing mesophase formation for cellulose derivatives is not only chain stiffness, but also the type and degree of substitution, the
molar mass of the polymer, as well as the solvent and the temperature [103].
Among the water-soluble cellulose biopolymers, HPC is still the most investigated derivative (it forms stable and easy to handle mesophases) and as such will

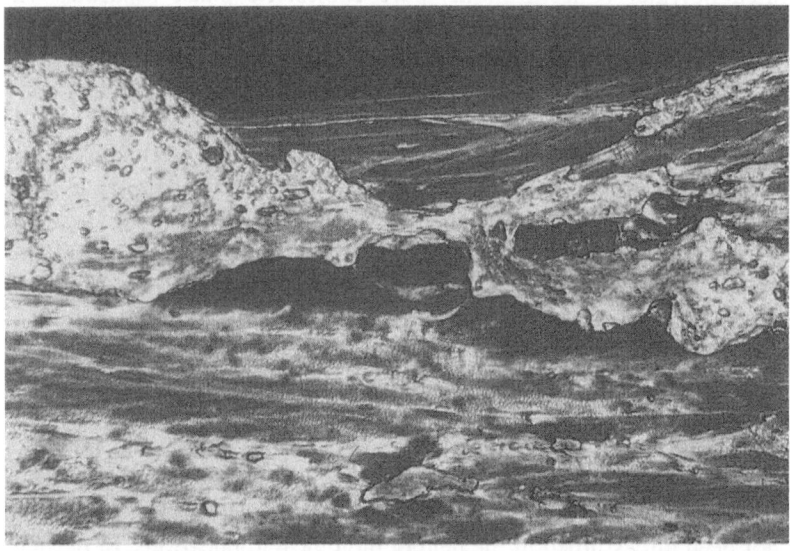

Fig. 18. Cholesteric liquid crystalline structure formed in a 50 wt % HPC gel at room temperature
(Magnification 25x)

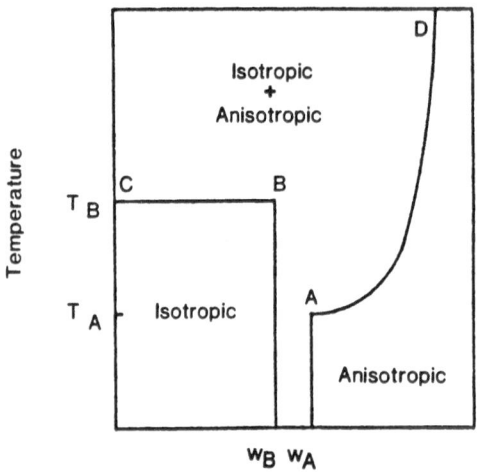

Fig. 19. The limits for the three domains for the HPC-water system [104]

Polymer weight fraction

help to illustrate the incidence of some parameters. Figure 19 is a recent schematic phase diagram of the HPC-water system which delimits three regions: a region with a pure isotropic phase, a two-phase region and a region with a single isotropic phase.

Below temperature T_B (cloud point for low polymer concentrations) and at a polymer weight fraction lower than w_B, isotropic solutions are stable. Above the weight fraction w_A and at a temperature below the line AD (clouds points for concentrated polymer concentrations) pure cholesteric phase separates. All other compositions are biphasic, the liquid crystals being in equilibrium with a dilute isotropic solution. At low temperature this domain exists in the small concentration range delimited by w_A and w_B. Note that T_A may be below body or even room temperature.

The critical concentration w_B necessary for the formation of an anisotropic phase at room temperature has been investigated for HPCs varying in their molar mass ($\bar{M}_w = 60\,000$ to $1\,000\,000$). The values ranged from 39 to 42 wt % and did not change significantly when observations were repeated at 30 and 35 °C. However, at 40 °C (a temperature close to the cloud point) liquid crystal formation took place at a somewhat lower polymer concentration. The nature of the solvent played a greater role. The minimum concentration of a HPC sample with a $\bar{M}_w$ of 100 000, which was 42% in water, rose to 43 °C in methanol and to 47 °C in the less polar ethanol [103]. For another HPC sample, values of 0.21, 0.30, 0.38, 0.42 and 0.43 were found for dichloroacetic acid, acetic acid, dimethyl acetamide, water and ethanol, respectively [105].

The implications of the existence of liquid crystals in water-soluble cellulosic gels has not been fully studied in the field of biological applications. In particular, the models developed for water transport in hydrogels do not consider the presence of mesophases. The same is true for solute (drug) release models where the influence of ordered structures on the diffusibilities is not taken into account.

3.6 Miscellaneous Interaction Phenomena

Other types of interactions (complexation, adsorption, coacervation) may be of concern for the performance of water-soluble cellulose macromolecules. Only illustrative recent examples will be discussed here. The complexation of various additives (low-molecular solutes, surfactants, other macromolecules) with cellulose derivatives has been investigated using a variety of methods including dialysis [106], osmometry [107], and measurements of electrophoretic mobility [108], viscosity [91, 92, 97], cloud point [91, 93, 96], ion activity [92, 96], as well as NMR techniques [109].

Thus, the anionic NaCMC, using dynamic dialysis, has demonstrated the highest limiting binding capacity for the antibiotic chloramphenicol as compared with PEGs, polyvinylpyrrolidones and gelatin [106]. The preservatives parabens sodium influence the viscosity and the zeta potential of NaCMC gels [109]. The interaction of EHEC with various surfactants has been thoroughly examined and previously discussed. According to the authors, the effects produced by adding a surfactant (in particular the change in cloud point and viscosity) imply the formation of micelle-like surfactant clusters on polymer chains and the coupling with segments on other chains [97]. In this respect, the Fourier transform NMR self-diffusion technique revealed that the binding of the dodecyltrimethylammonium ion to EHEC increased with temperature, in contrast to the usual observation. This was interpreted in terms of higher hydrophobicity at elevated temperature [109].

Incorporation of surfactants in HPMC compressed matrices has been shown to cause retardation of drug release [110–112]. The first mechanism postulated was that anionic surfactants were capable of binding to nonionic polymers to increase the viscosity [110]. More recently, drug-surfactant ionic interaction has been put forward instead, resulting in the formation of a complex with low solubility in water [111, 112]. Note that such an effect can unintentionally occur upon putting some surfactant into the release medium in order to increase wettability.

Ionic polymers (e.g. NaCMC) can also be added to HPMC tablets to prolong the sustained release of oppositely charged drugs [113]. In fact, the viscosity of a given cellulose derivative may be increased by admixing with another polymer which is cellulosic or not, due to hydrogen bond-induced crosslinking [114]. This synergistic effect reflects a partial incompatibility of the two types of macromolecules and has been reported many times. It is particularly strong between ionic and nonionic cellulose derivatives, but it is unlikely that the increase in macroviscosity is responsible for the slower drug release from dosage forms.

Water-soluble cellulose derivatives themselves adsorb onto solid particles and may for instance affect the suspension properties of these insolubles. The mechanisms involved are quite complex and depend on the polymer concentration. At low concentrations macromolecules influence the electrophoretic mobility and the flocculation of the particles. At higher concentrations, surface coverage by the adsorbed polymer is sufficient to prevent particle–particle interaction and thus to stabilize the suspension sterically. As an example, the effect of NaCMC (among other polymers) on the zeta potential, flocculation and sedimentation properties of sulfadimidine has been investigated by Kellaway and Najib [115, 116].

The adsorption of nonionic cellulose derivatives (HPMC 2910, HEC, and HPC) and their stabilizing effect on suspensions of polystyrene latices and of the antirheumatismal ibuprofen have been examined by Law and Kayes [117, 118]. Adsorption isotherms (see Fig. 20 for HPMC and ibuprofen) are of a sigmoidal type: rapid initial adsorption, then a plateau corresponding to monolayer adsorption and finally a further increase in adsorption. At the same time, a gradual reduction in the sedimentation volume (decreased flocculation) is observable, while the redispersibility tends to be optimum (minimum number of revolutions necessary) at values of HPMC corresponding to a monolayer surface coverage.

For all cellulose derivatives tested, a reduction of the zeta potential of the suspensions with increased polymer concentration is observed (Table 20). The effect of the molecular mass differs depending on the derivative concerned. For HEC and HPC, the amount adsorbed and the area per molecule decreased as the molecular mass increased, indicating a flatter adsorption conformation. For HPMC, the adsorption increased as the macromolecular chains became longer. Adsorption was maximum for the more hydrophobic HPC.

Recently, EHEC and other hydrocolloids were used for the stabilization of lipid microspheres. The coating reduced the zeta potential of the liposomes to a neutral value and decreased their surface fluidity, as measured by fluorescence

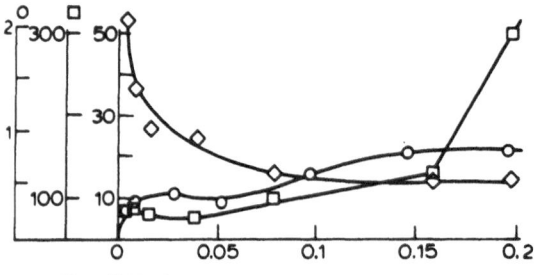

Fig. 20. Effect of HPMC 2910 (3 mPa s) on the characteristics of ibuprofen suspensions at 25 °C. Key: (○), adsorption (mg g^{-1}); (□), redispersibility (number of revolutions); (◇), sedimentation volume (%) [117]

Table 20. Adsorption parameters of cellulose derivatives on ibuprofen crystals and zeta potentials of the suspensions [117, 118]

Derivative	Molecular mass	Plateau adsorption (mg g^{-1})	Area occupied per molecule (nm^2)	Zeta potential (mV)
—	—	—	—	−28.3
HEC	62 000	3.20	18.66	−7.7
	95 000	2.10	43.66	−8.7
HPC	120 000	1.60	72.18	−4.2
	146 000	0.55	255.0	−6.4
HPMC	4000	0.40	9.60	−8.0
	10 000	0.80	12.21	−3.5
	15 000	1.20	12.03	−2.6

polarization. Interestingly, the presence of colloids depressed the aggregation of liposomes induced by calcium ions [119]. The last case of interaction to be discussed will deal with the complex coacervation between NaCMC and gelatin investigated by Koh and Tucker [120, 121]. The system behaves in a similar way to that of the well-known gelatin-acacia complex. With a gelatin of type B having an isoelectric point of 5.0, the low viscosity grade NaCMC showed an optimum coacervation at pH 3.5 and a NaCMC-gelatin weight ratio of 3:7 (Fig. 21). Under these conditions, the negative charges of CMC and the positive charges of the gelatin are maximum, resulting in a strong electrostatic interaction. Turbidity data confirmed viscosimetric results.

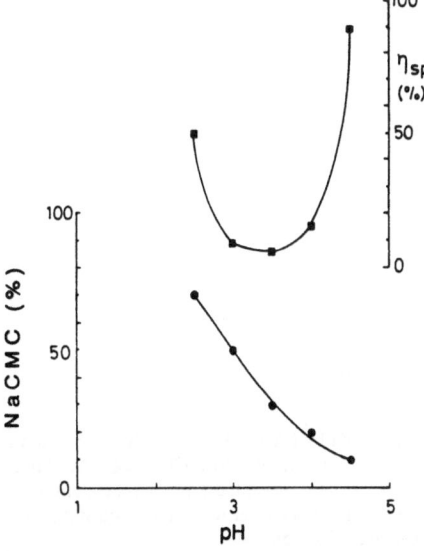

Fig. 21. Colloid mixing ratio expressed as the NaCMC percentage (*left ordinate*) and specific viscosity η_{sp} (*right ordinate*) vs pH of the NaCMC-gelatin coacervation mixtures [120]

This system might prove to be useful for the microencapsulation of various cores (drugs, oils, flavors, etc.), with the further advantage that NaCMC has demonstrated bioadhesive properties.

4 Derivatives Soluble in Organic Solvents

As biopolymers, cellulosics soluble in organic solvents are essentially used as semipermeable membranes (e.g. reverse osmosis separation), and for drug delivery in coating materials for reservoir or osmotic systems and excipients for matrix systems (e.g. films, compressed tablets). Recent advances in the latter fields will be reviewed.

4.1 Aqueous Nanodispersions

Over the last decade, many aqueous dispersions of insoluble polymers have been developed and commercialized for the pharmaceutical industry[1]. These latices have stimulated interest for aqueous coating and granulation because they avoid problems associated with organic solutions such as air pollution, solvent toxicity and fire hazard.

These pseudo-latices, as they are sometimes called, are obtained by direct emulsification of an already existing polymer [122–129], and are therefore free of residual monomers. Basically, these aqueous dispersions are prepared by dissolving the polymer in an appropriate solvent and dispersing the solution in an aqueous phase containing surfactants as emulsion stabilizers. The resulting crude emulsions are then homogenized and the solvent finally evaporated.

In the ternary emulsion system solvent (1)-water (2)-polymer (3), the organic solvent used may be completely immiscible with the aqueous phase or may dissolve some water. For this latter case, we developed a new mathematical model for the equilibrium state of polymer particles containing water [127–129]:

$$\ln K_\infty = \left(\chi_{12} - \chi_{13} + \frac{1}{x_3} - \frac{1}{x_2} \right)(1 - \phi_{2_x}) + \frac{1}{x_2}\chi_{23}\phi_{2_x}(1 - \phi_{2_x})$$

$$- \frac{2V_1\gamma}{r_0 RT}(1 - \phi_{2_x})^{1/3} - \ln(1 - \phi_{2_x}). \tag{24}$$

Here, K_∞ is the infinite dilution partition coefficient of solvent between the

---

[1] The first dispersion marketed under the tradename Aquacoat by FMC (Princeton, NJ) is based on 10 mPa s ethylcellulose. A similar product is under development by Dow Chemical (Midland, MI)/Colorcon (Orpington, U.K.). A cellulose acetate dispersion is presented under experimentation by FMC. Finally, FMC has commercialized a CAP redispersible powder for enteric coating (CAP hydrolyses slowly in an aqueous solution) as Aquateric (see Sect. 5).

polymer and water phases χ_{12}, χ_{13} and χ_{23} are the Flory–Huggins interaction parameters (solvent–water, solvent–polymer, water–polymer, respectively), ϕ_{2_x} is the solubility of water in the polymer particles, V_1 is the molar volume of the solvent, γ is the interfacial tension between the solvent and water, r_0 is the radius of the unswollen (water-free) particles, R is the gas constant and T the absolute temperature. Two normalized volume corrections, x_2 and x_3, were defined as:

$$x_2 = \frac{M_2 \rho_1}{M_1 \rho_2} \quad \text{and} \quad x_3 = \frac{M_3 \rho_1}{M_1 \rho_3} \tag{23}$$

where M and ρ are the molecular mass and the density of each component, respectively. Equation 22 allows us to evaluate the influence of the internal water phase on the distribution of the solvent between the polymeric and aqueous phases.

Film formation from aqueous dispersions necessitates the coalescence of the discrete particles (the coalescence is probably only partial during evaporation of water and is completed on aging). Due to the high glass transition temperature of EC, CA and CAP, plasticizers are required. Whereas ordinary amounts of plasticizer (10–40% of the polymer) are effective for forming films from the EC [130, 131] and CAP [132] dispersions, much higher amounts are necessary with the CA latex [38, 126, 133–135]. This unusual loading is made necessary by the physicochemical properties of CA, in particular by the very high viscosity of its solutions. The use of normal concentrations of hydrophobic plasticizers leads to membranes which crack or crumble in contact with water.

Highly resistant CA membranes of low permeability to water are necessary for producing osmotic drug delivery systems (see Sect. 4.3). Volatile water-miscible plasticizers proved to be satisfactory for circumventing this difficulty with resulting films having useful mechanical and permeability characteristics. Table 21 shows the effect of the nature of the plasticizer and of the initial concentration in coating solution on the maximum tensile strength and elastic modulus of membranes prepared by spraying a cellulose acetate latex onto potassium chloride disks [134]. The weight percentages of leachable material, indicative of the residual plastizers, are also listed. Some comparison is made with organic solutions of CA.

The mechanical strength of the films is highly correlated with the residual plasticizer, as reflected by the amount of leachable materials in water (1.9% sodium dodecylsulfate were also present in the films). The best films were obtained using the highly volatile ethyleneglycol monoacetate (low residual content) which gave similar results with both initial plasticizer concentrations in the coating solution. The permeability of the films appeared to be also highly correlated with the residual plasticizer content. Figure 22 shows the relationship between the apparent mechanical water permeability Lp of the membranes and the weight loss in leachable materials for CA films prepared using various plasticizers and coating conditions [135].

Table 21. Mechanical properties and weight loss of leachable materials in water of membranes obtained by spraying a CA latex onto potassium chloride disks [134]

Plasticizer	Boiling point (°C)	% Plasticizer by weight of CA	Mechanical properties		% Weight loss of leachable materials
			σ_{max}[1] (MPa)	E^2 (MPa)	
Latices					
Ethyleneglycol monoacetate	182	160	34.9	1536	14.25
		320	53.0	1673	14.67
Triethyl phosphate	215	160	23.0	1149	24.00
Diethyl phosphate	280	160	20.3	765	31.83
Organic solutions					
—	—	—	64.7	1948	6.92
Diethyl phosphate	280	42.85	23.0	876	30.36
		66.66	6.9	192	39.08

[1] Maximal tensile strength.
[2] Elastic modulus.

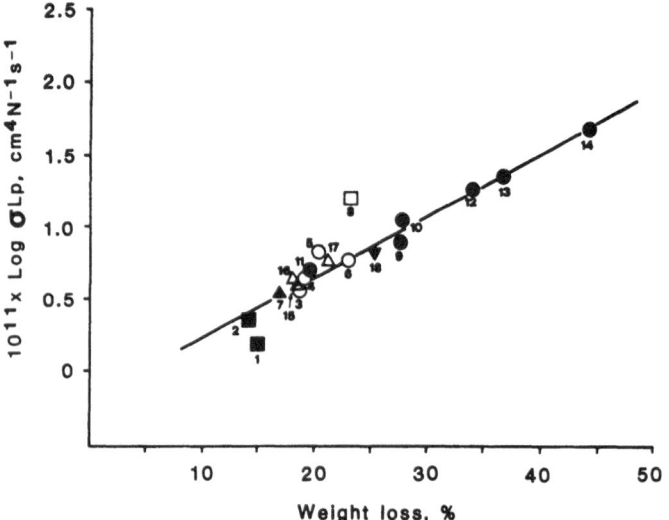

Fig. 22. Apparent mechanical water permeability L_p of sprayed CA latex films vs weight loss of leachable materials. Key: (□) diacetin; (●) diethyl tartrate; (■) ethylene glycol monoacetate; (▲) triethyl phosphate; (○) trimethyl phosphate; (▼) ethylene glycol diacetate + diethyl tartrate; (△) ethylene glycol diacetate + trimethyl phosphate [135]

Interestingly, the permeability varied over a wide range, depending on the plasticizer, its concentration, the temperature, the rate of spraying and the duration of the drying step in the coating process.

A further difficulty of coating with aqueous dispersions comes from the presence of surfactant that can crystallize and form islets, resulting in altered mechanical and permeation properties [136–138]. Under inappropriate conditions, films made with a EC latex exude cetyl alcohol and sodium lauryl sulfate [138]. Sodium dodecyl sulfate sometimes forms small islets in CA films, depending on the initial surfactant concentration, the plasticizer used, and the drying conditions (phase separation is more critical for cast films than for sprayed films) [137]. This problem can be overcome to some extent by the use of a low surfactant concentration, an appropriate plastizer and coating conditions, as has been demonstrated experimentally and through the model proposed [138]. In this respect, the use of a low melting surfactant does not give rise to liquid–solid phase separation but possibly to a liquid–liquid separation into a very large number of tiny droplets. Precipitation of solid surfactants is to be examined carefully, because it considerably impairs film properties.

In order to avoid the presence of true surfactants, a new method for preparing aqueous dispersions has been developed [139, 140]. Here, using a salting-out effect, a solution of the polymer in a water-miscible organic solvent is emulsified in an aqueous phase containing a hydrocolloid as stabilizer and viscosifying agent and saturated with an electrolyte. The emulsion is then diluted so that the organic solvent can diffuse in the outer phase. The latex is washed free of electrolytes and the major part of the hydrocolloid and the polymer particles can be recovered by lyophilization as a redispersible powder. This latter operation is virtually necessary for cellulose esters, because hydrolysis is a real problem (an HPLC method has been recently published for the determination of degradation products of CA, CAB, CAPr and CAP lattices [141]).

4.2 Permeation Properties

The insoluble cellulose derivatives utilized for permeation control of various species (e.g. oxygen and water vapor transport in coated pharmaceuticals, contact lenses, packaging, or water and solute transport through semi-permeable membranes in reverse osmosis, as well as drug release from reservoir systems) differ considerably in their permeability characteristics according to the type and extent of substitution, as well as their molar mass. However, very few comparative data are available from the literature on the polymers actually used in biological applications. Recently, new results have been published. Thus, Sprockel et al. [142] determined the water vapor transmission through various CA, CAT, CAB and CAPr films at different relative humidities (Table 22).

Although they cannot be considered as absolute values because of the experimental setting used, the rate of water vapor transmission generally increased in the order of substituents CAB < CAPr < CA, as could have been expected from the carbon chain lengths ranging from C_2 to C_4. The rates also decreased with increasing substitution, except for CAT. As the solvents used to cast the films varied in some cases, an effect of this factor can be expected. A different technique was used by Bond et al. [143] to assess the permeability of

Table 22. Effect of the type and extent of substitution on the transmission of water vapor WVTR through polymer films at 90% relative humidity [142]

Derivative	% Acetyl	% Butyryl	% Propionyl	% Hydroxyl	WVTR $(mg\,cm^{-2}\,h^{-1})$
CA	32.0	—	—	9.0	5.28
	39.8	—	—	3.4	4.36
CAT	43.7	—	—	0.9	3.84
	43.5	—	—	0.9	4.17
CAB	29.5	17.0	—	1.5	3.90
	13.0	37.0	—	1.5	3.16
	5.0	50.0	—	0.5	3.12
CAPr	—	—	2.5	5.2	3.74

Table 23. Effect of the acetyl content on the coefficients of permeability PC, diffusion D and solubility S of CA membranes [143]

% Acetyl content	$PC \cdot 10^3$ $(cm\,h^{-1})$	$D \cdot 10^4$ $(cm^2\,h^{-1})$	$S \cdot 10^5$ (Pa^{-1})
30.8	2.67	1.18	9.68
39.3	2.62	1.73	6.48
39.6	2.04	1.47	5.95
43.8	1.58	1.55	4.36

CA films with various acetyl contents. The weight change of initially dry films was measured after the passage of moist air (91% relative humidity). The permeability coefficient PC, the diffusion coefficient D and the solubility coefficient S could be calculated (Table 23).

The logical order was again observed, the commonly used secondary (acetone-soluble) cellulose acetate being intermediate between the product with a low acetyl-content and the cellulose triacetate. Note that the WVTR of CAP is about the same as that of other esters, EC being the less permeable insoluble cellulose product.

One problem encountered with polymeric membranes is their physical aging which can alter the permeation and the mechanical properties. Sinko et al. [144] noticed a decrease in the water permeability and in the mechanical rate of relaxation with time, which was proportional to the aging temperature. A relaxation coupling model could be used to predict equilibrium in the structural change profile.

4.3 Solute Osmotic Delivery

Solute (drug) release from a reservoir through insoluble films is known to be diffusion-controlled. Depending on the physicochemical properties and the

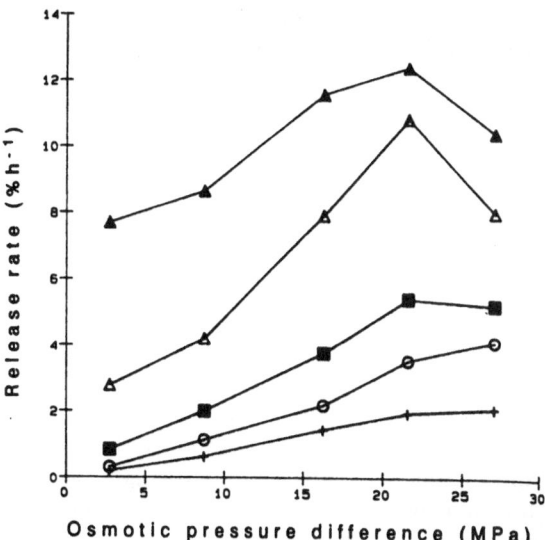

Fig. 23. Release rates vs osmotic pressure differences for potassium chloride tablets coated with EC and HPMC. HPMC concentration: (+) 0%; (○) 18%; (■) 24%; (△) 27%; (▲) 30% [145]

structure of the membrane as well as the solubility and the molecular mass of the solute, release can occur through a continuous plastized phase, plasticizer channels or aqueous pores created by the departure of water-soluble additives. However, recent results have demonstrated that, at least partially, drug release is driven osmotically [145, 146].

Free films of EC containing various amounts of HPMC 2910 as permeability increasing agent, included water but excluded to a large extent the solute potassium chloride. Additionally, cores of potassium chloride were shown to release their content mainly through osmotic pumping, up to a HPMC content of 24%. Only above this limit, diffusion became substantial. Figure 23 depicts the effect of osmotic pressure difference on the release of potassium chloride tablets coated using a EC latex. The release was halted at zero osmotic pressure difference for the first three HPMC contents [145].

The patterns were similar with pellets containing the antihistamine phenyl-propanolamine hydrochloride coated using an EC-based latex plasticized with varying amounts of dibutyl sebacate [146]. The release rate was shown to be highly dependent on the osmotic pressure difference, with some contribution of the diffusion. Pores of 2 μm in diameter were visible, which is consistent with the osmotic pumping mechanism.

The principle of a purely osmotically controlled drug delivery was disclosed some 15 years ago [147]. The so-called elementary osmotic pump consists of a core containing the drug (and if necessary an osmotic agent) and a CA semipermeable membrane with a micro-orifice drilled usually by a laser beam. Drug release is activated by the transport of water through the semipermeable membrane, thus creating an internal osmotic pressure. The zero-order release rate during the steady-state portion of the release profile may calculated from:

$$\frac{dM_t}{dt} = \frac{A \cdot c_s}{L} \cdot L_p \sigma (\Delta\pi - \Delta P) \tag{26}$$

where dM_t/dt is the release rate, A is the surface area of the device, L is the wall thickness, c_s is the solubility of the core mass (drug), L_p is the mechanical water permeability, σ is the reflection coefficient which describes the degree of perm-selectivity (usually assumed to be equal to unity), $\Delta\pi$ is the osmotic pressure difference and ΔP is the hydrostatic pressure difference (usually neglected).

Many advances have been made since that time. Zentner et al. [148] demonstrated that drug release by osmotic control could be achieved through creating controlled porosity by incorporating for example sorbitol in the film instead of drilling an orifice. This could make it easier to produce this type of sustained release dosage form and especially of small multiparticulate systems. Furthermore, this prevents high drug concentrations at the orifice. It was also shown, that CA could be replaced by others polymers such as ethyl cellulose or polymethacrylates [149].

In order to avoid organic solvents for the coating of the osmotic cores, an aqueous CA dispersion has been developed (see Sect. 4.1) and its performance has been investigated as water permeation controlling membrane for potassium chloride tablets [133]. The release rate could be modulated by varying the type and amount of platicizer, the thickness of the membrane or the coating temperature (Fig. 24).

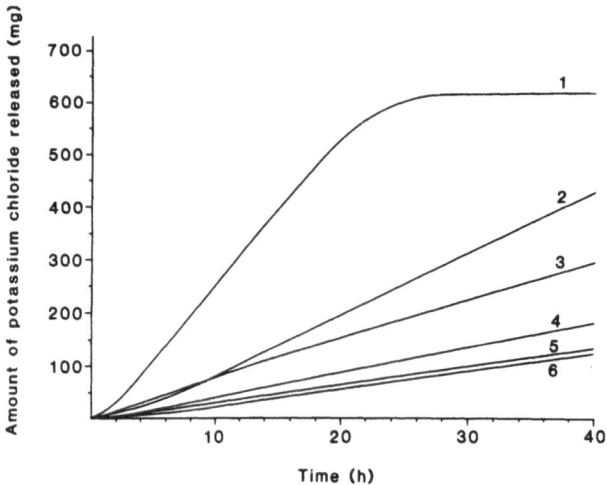

Fig. 24. Release profiles of potassium chloride from osmotic tablets prepared by spraying a CA latex as a function of the plasticizer type and amount, and the coating temperature. Key: (*1*) diethyl tartrate 120%, 50 °C; (*2*) diethyl tartrate 160%, 60 °C; (*3*) trimethyl phosphate 160%, 60 °C; (*4*) ethylene glycol monoacetate 360%, 60 °C; (*5*) ethylene glycol monoacetate 60%, 60 °C; (*6*) organic solution, 35 °C (no plasticizer) [133]

4.4 Solute Release from Matrix Systems

As in the case of hydrophilic (swelling) matrix systems and reservoir (membrane) systems, drug release profiles from insoluble (porous or non porous) systems are most of the time described on a basis of the diffusion theory. This is not true for every situation and we have for example shown that the release from porous matrix systems is dissolution-controlled above the solubility limit of the drug [150]. A simplified equation for the model proposed and for values of $k_d \cdot t > 4$, is:

$$M_t = c_s D_{eff} k_d \left(t + \frac{1}{2k_d} \right) \tag{27}$$

where M_t is the total amount of drug released per unit area at time t, c_s is the drug solubility limit in the release medium, D_{eff} is the effective diffusion coefficient (= the diffusivity in the release medium multiplied by void fraction and divided by a tortuosity factor), and k_d represents the dissolution rate constant of the drug in the release medium.

The drug release rate per unit area is expressed as:

$$\frac{dM_t}{dt} = c_s D_{eff} k_d \tag{28}$$

with the consequence that a zero order release is observed. Experimental results for potassium release from porous EC tablets in water at 37 °C are presented in Fig. 25. Up a critical time t_s, the drug release is dissolution-controlled, and after that time, it is diffusion-controlled.

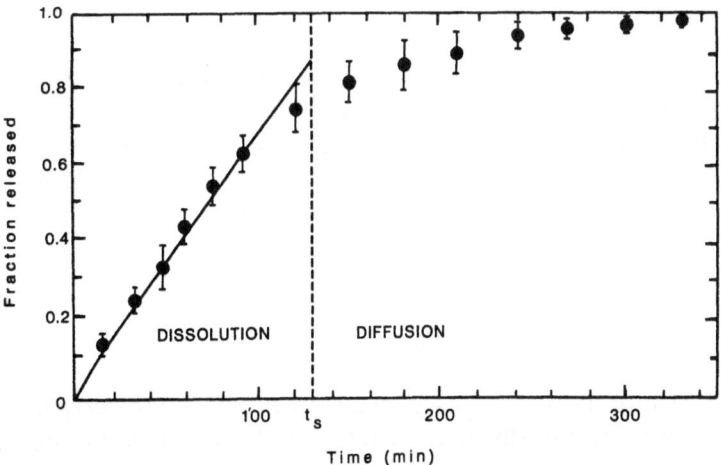

Fig. 25. Release profile of potassium chloride from EC tablets (initial drug loading 0.565 g/cm³, initial porosity = 0.296) [150].

Table 24. Wetting properties (advancing contact angle, hysteresis) and permeability coefficients to SIBA of EC films cast from various solvents [153]

Sample	Wetting properties				Permeability coefficient $\cdot 10^{10}$ $(cm^2 s^{-1})$		
	Glass side		Air side		Glass side	Air side	PC_g/PC_a
	$\theta_a(^\circ)$	H	$\theta_a(^\circ)$	H^1	PC_g	PC_a	
1	47.0	1.18	59.6	4.5	5.9	4.99	1.2
2	62.6	15.73	71.6	13.3	14.5	6.65	2.2
3	54.5	7.06	59.7	12.3	4.46	2.3	1.94

[1] The hysteresis of the polymer-water free energy of adhesion was calculated from $H = \gamma_{wa}$ $(\cos \theta_r - \cos \theta_a)$, where γ_{wa} is the water-air interfacial tension and θ_r the receding contact angle.

In monolithic systems, it is assumed that the structure is homogeneous not only with respect to drug distribution, but also to the polymeric network. For cast films, it has been reported that the solvent used may induce structural differences. Using a proton RMN technique along with scanning electron-microscopy, Azoury et al. [151] were able to show that during subsequent hydration, EC films containing polyethylene glycol and tetracycline have more water binding sites and a thicker water multilayer when cast from ethanol than from chloroform. Additionally, the release of tetracycline was enhanced [152].

A second example of the influence of the casting solvent on the morphology of EC film is given by the work of Rosilio et al. [153]. Monolayer surface pressure-area isotherms, wettability and permeability data clearly indicated that the EC films cast from various solvent systems differ not only in terms of their overall properties but also from one side to the other.

Table 24 summarized the wettability and permeability to 5'-deoxy-5'-S-isobutyladenoside (SIBA) data for EC films cast from chloroform (sample 1), ethanol (sample 2) and chloroform + ethanol 1:1 (sample 3), as measured from both the air side and the glass side (the coating support).

Ethanol used as a casting solvent increased the heterogeneity of both the polymer surface (film 2 is more hydrophilic and displays a greater syneresis) and in its bulk (due to a modification of the crystallinity). Concurrently, release from SIBA-loaded films was also related to the state of the polymeric matrix. Similar observations have been reported several times but they have not received much attention.

4.5 Phase Separation, Coacervation

During the manufacture of a polymeric system (for drug delivery or other) phase separation may occur intentionally or undesirably and it may concern a single polymer and another species or a polymer blend.

The widely used EC is for instance known to be thermodynamically in-compatible not only with the water-soluble cellulose ethers but also with CAP and HPMCP [154]. Recently, the effect of phase separation of EC–CAP blends on the morphology of films cast from methanol-methylene chloride 1:1 has been investigated [155]. The top layer consisted of a EC-rich matrix with CAP-rich dispersions and the reverse held true for the bottom layer. Both the calculated and the experimental data suggested that the solvent system is a cosolvent for CAP and a solvent-nonsolvent mixture for EC. This was demonstrated in particular by the different adsorption characteristics of methanol and methylene chloride on the two polymers.

Phase separation, or coacervation, by adding a nonsolvent is one of the methods for preparing microcapsules or microspheres. The process has been carried out mostly empirically but careful examination of the thermodynamics of the polymer-solvent-nonsolvent system would be helpful for understanding and optimization. Robinson [54] has shown that EC coacervates in a limited number of solvents because coacervation (and gelation or flocculation) occurs closed to the critical interaction parameter ($\chi_c = 0.38$) and in solvents having a solubility parameter close to that of EC. Of course this approach, although of good predictive value, suffers from severe limitations for polymer solutions in strongly hydrogen bonding media.

5 Derivatives Soluble in Nonacidic Media and Organic Solvents

Mixed esters or ether esters of cellulose that are insoluble in water and acidic media, but soluble above neutrality and in organic solvents, have been used for a long time as a sealer subcoat to prevent moisture penetration into the tablet cores and, using larger film thicknesses, for enteric dosage forms (tablets, microcapsules). More recently, the so-called intestine, colon-specific, pH-dependent or pH-activated drug delivery systems have been described but are basically the same. These pH-dependent polymers are also frequently associated with an insoluble polymer in matrix systems. Up to now, CAP (1940) and HPMCPs (since the 1970s) have been the two cellulose derivatives commonly used for enteric coating. Recently, new products have been launched on the market, namely cellulose acetate trimellilate, CAT, and hydroxypropylmethyl-cellulose acetate succinate, HPMCAS. Cellulose hydrogen phthalate has been synthetized and investigated [156], but has not been commercialized. The most relevant characteristics of these poly(carboxylic acids) is their dissolution pH, which is of course related to their pK_a. Values reported in the literature vary depending on the buffers and the ionic strengths used (Table 25).

The advantage of a lower dissolution pH and an increased resistance to hydrolysis has been claimed for the homologues of CAP and the esters of hydroxypropylmethylcellulose over the original product and it may become relevant in some cases. The stability has been reported to be in the order HPMCAS > HPMCP > CAP.

Table 25. Characteristics of enteric polymers

Derivative	% Methoxyl	% Hydroxypropyl	% Acetyl	% Phthalyl	% Trimellityl	% Succinyl	Dissolution pH
CAP	—	—	21.5–26.0[1]	30.0–36.0[1]	—	—	6.2
HPMCP 200731	18.0–22.0[1]	5.0–9.0[1]	—	27.0–35.0[1]	—	—	5.5
HPMCP 220824	20.0–24.0[1]	5.0–9.0[1]	—	21.0–27.0[1]	—	—	5.0
CAT	—	—	18.0–26.0	—	25.0–33.0	—	5.0
HPMCAS-L	20.0–24.0	5.0–9.0	5.0–9.0	—	—	14.0–18.0	5.8
HPMCAS-M	21.0–25.0	5.0–9.0	7.0–11.0	—	—	10.0–14.00	6.2
HPMCAS-H	22.0–26.0	6.0–10.0	10.0–14.0	—	—	4.0–8.0	7.0

[1] USP/NF limits.

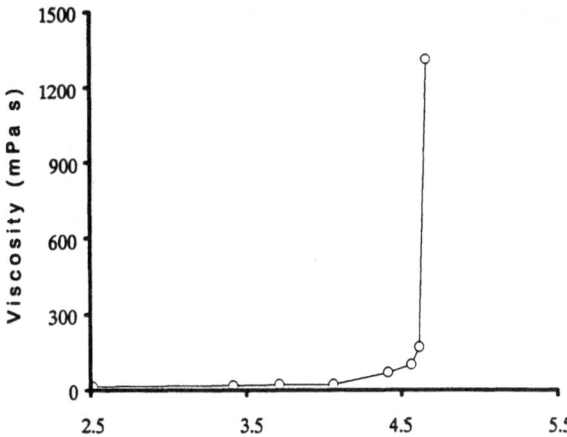

Fig. 26. Viscosity behavior of a 30% CAP latex at a shear rate of 35.5 s^{-1} [159]

In this respect, the development of redispersible powders for aqueous coating is of great interest. Microencapsulation through coacervation (both in aqueous and organic media) has been reported many times for CAP. Recently, phase diagrams were established for the CAT-light mineral oil-solvent (acetone + ethanol 9:3) system indicating the region of microcapsule formation.

The so-called enteric polymers are not only used as coating material, but also as excipients for monolithic systems (often associated with insoluble polymers). Recently, an ophthalmic pseudo-latex triggered by pH has been described [157, 158]. The liquid preparation is easily instilled and a gel forms *in situ* at the lacrymal pH by neutralization of the carboxyl acid groups of the polymer. The contact time of the preparation with the cornea and thus the duration of action are prolonged significantly. Eight polyacids were investigated for their apparent pK$_a$, acid value and maximum buffer capacity. CAP and subsidiary HPMCP were selected. Finally, CAP latices were prepared using various solvent systems selected from a ternary plot of CAP solubility versus fractional solubility parameters. Figure 26 shows the sharp increase in viscosity and the narrow pH range that induces gelification of a CAP latex with polymer particles of 150 to 200 nm in diameter.

In this section, the polymers described were neither soluble in acid media nor in water. Polymers which are soluble in acidic media but insoluble in water also offer good prospects. Such synthetic polymers are available (e.g. polymethacrylates with amine groups), but surprisingly no cellulose derivative with food or pharmacy acceptance is on the market. The purpose of these polymers as coating agents is the protection of sensitive tablets against moisture and more generally the immediate release from gastrosoluble drug delivery systems.

6 Derivatives Insoluble in Water and Organic Solvents

Cellulose derivatives which are insoluble in water (and subsidiarily in organic solvents) but swell in aqueous media have found wide application. In pharmacy, some of these compounds are used as disintegrants for tablets. Others, which display high sensitivity to temperature, pH or ionicity have potentialities as responsive gels for controlled drug delivery. Being general water absorbents, these hydrogels are found in everyday life (disposable diapers, towels), medicine (soft contact lenses), biochemistry or food processing.

Cellulose-based disintegrants are widely used in tablet formulations. Although the exact disintegration mechanism is still subject to controversy, capillary water penetration and swelling seem to be involved most of the time. Table 26 lists some characteristics of tablet disintegrating agents that are either low-substituted or crosslinked derivatives, or CMC divalent salts.

Not many studies have been published on the chemistry and the thermodynamics of disintegrants. The effect of the degree of polymerization and substitution of NaCMC has been investigated [161]. The preferred disintegrant had a high molecular mass, together with a low degree of substitution (0.4). A thermodynamic study of both types of croscarmellose sodium has shown some irreversible swelling [162].

Among the investigations published on the sorption behavior of crosslinked cellulose derivatives, the papers by Harsh and Gehrke [163–165] on temperature-sensitive hydrogels are of utmost interest. Different ether gels were synthetized using divinylsulfone as the crosslinking agent. Their swelling behavior as a function of temperature is depicted in Fig. 27 and it shows a sharper decrease in the degree of swelling for the HPC gel.

The characteristics of these gels are summarized in Table 27, together with the lower critical solution temperature of the original linear polymers. The

Table 26. Properties of cellulosic tablet disintegrating agents

Derivative	DS	θ^5 ($^\circ$)	Swelling in water[6] (%)
CaCMC	0.6 ± 0.1^2	—	—
L-NaCMC	$0.34-0.40^3$	66	36
L-HPC	7–14% HP[4]	50	313
Croscarmellose sodium			
Type A	0.70^2	0	210
Type B[1]	$0.63-0.95^2$	—	—

[1] Not available any more.
[2] No limits are given in the USP/NF.
[3] Fall below the USP/FN limits.
[4] At this low hydroxypropyl content, HPC is not soluble in water but swells.
[5] Contact angle with water reported from [160].
[6] Values reported from [160].

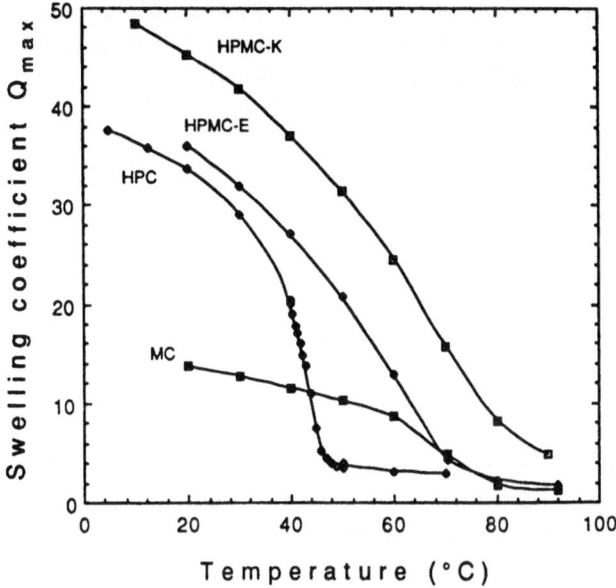

Fig. 27. Swelling behavior of crosslinked HPMC 2208, HPMC 2910, HPC and MC gels vs temperature [165]

Table 27. Effect of polymer type on transition temperature and sharpness of transition of gels crosslinked with divinylsulfone [163]

Derivative	$\bar{M}_n$	LCST[1] (°C)	Crosslinking density × 10^5 (mol cm^{-3})	Sharpness of transition (%/°C)
HPMC 2208	70 000	66	66	4
HPMC 2910	70 000	55	60	2
HPC	370 000	41	43	8
MC	70 000	55	64	3

[1] The lower critical solution temperature corresponds to the cloud point.

effective crosslinking density was determined by uniaxial compression testing [163]. The transition temperature was taken as the inflection point on a normalized swelling curve and the sharpness of the transition was the slope of that curve.

The transition temperature was generally equal or slightly higher (with the exception of MC) than the LCST of the linear polymer. Further, the crosslinking density (5–15% of the theoretical values) affected only the degree of swelling of the swollen gels, but not that of the collapsed gels, nor the transition temperature. Increasing the ionic content of the gels by preparing crosslinked "copolymers" of NaCMC ($\bar{M}_n = 7\,000\,000$) and HPMC 2910 or HPC produced different effects which depended upon the nature of the nonionic derivative. The ionic content of NaCMC–HPMC gels increased the degree of swelling at all temperatures as well as the transition temperature. Only little change in the degree of swelling was observed for NaCMC–HPC gels below the transition temperature, whereas a twelve-fold increase at 80 °C was noted when the NaCMC content passed from 0 to 15%. The transition temperature increased from 43 to 54 °C at a 15% NaCMC content [164, 165].

In a similar study, we investigated the effect of the chemical nature of the crosslinker as well as the crosslinking density on the swelling behavior of gels of MC. Two viscosity-grades of MC were selected (25 mPa s and 4000 mPa s) corresponding to $\bar{M}_n$ values of 18 000 and 86 000, respectively. The crosslinking agents used were divinylsulfone, glyoxal and glutaraldehyde. Crosslinking by divinylsulfone was carried out in a NaOH aqueous solution and in acidified methanol for the two dialdehydes. Three nominal crosslinking ratios (X = 0.02, 0.05 and 0.10) were used. The dried gels were cut into 30×20 mm rectangles (thickness 0.2 mm) and placed in water at 37 °C. The weight, surface and thickness of the samples were measured at various time intervals.

Table 28 reports the maximum swelling coefficient Q_{max}, the time needed to reach this equilibrium state t_{max}, and the constants k and n that characterize the water sorption profiles according to the equation [166]:

$$M_t = k \cdot t^n \tag{29}$$

where M_t is the amount of water absorbed per unit surface area at time t, k is a kinetic constant and n is an exponent characterizing the type of transport (for films: n = 0.5 for Fickian diffusion, 0.5 < n < 1.0 for anomalous transport, and n = 1.0 for case II transport). No results are given for the gels loosely crosslinked by glyoxal because they broke during the swelling experiments.

For both viscosity grades of MC, the swelling coefficients were higher and the times needed to reach equilibrium longer with the five-element crosslinker divinylsulfone. With few exceptions, Q_{max} values were similar for the two dialdehydes but equilibrium was achieved in reversed order: the shortest t_{max} values were observed with glyoxal for the high molecular mass MC but with glutaraldehyde for the low molecular mass linear polymer. Increased crosslinking density led to a reduced degree of swelling but had no significant effect on the kinetics. Finally, swelling was more important for the longer primary molecules, but no definite explanation could be given without experimental determination of the effective crosslinking density.

Quite interestingly, an anomalous water transport was observed for the gels which were the most highly crosslinked with the dialdehydes. Case II transport seemed even to be operating for the 25 mPa s MC gel crosslinked with glyoxal and for the 4000 mPa s MC gel crosslinked with glutaraldehyde.

As a matter of fact, both studies demonstrate that the gel properties can be modulated through the chemical nature and the amount of crosslinking agent, and above all by the linear polymer used.

Table 28. Values of maximum swelling coefficient Q_{max}, time to reach this state t_{max}, and kinetic parameters[1] for two MC hydrogels crosslinked by various agents[2]

Crosslinking agent (crosslinking ratio, X)	MC 25 mPa s				MC 4000 mPa s			
	Q_{max}	t_{max} (min)	k (mg cm^{-2})	n (min^{-n})	Q_{max}	t_{max} (min)	k (mg cm^{-2})	n (min^{-n})
Divinylsulfone								
X = 0.02	21.6	120	49.0	0.54	30.0	120	21.9	0.54
X = 0.05	18.5	120	40.7	0.60	26.0	150	27.5	0.55
X = 0.10	15.0	105	22.4	0.67	21.0	120	17.4	0.64
Glyoxal								
X = 0.05	15.4	75	4.3	0.63	23.6	40	4.8	0.80
X = 0.10	13.4	75	1.7	0.99	16.0	30	7.9	0.90
Glutaraldehyde								
X = 0.02	13.2	35	10.7	0.71	24.2	75	17.8	0.73
X = 0.01	8.1	30	4.9	0.84	18.0	40	13.5	0.78
X = 0.10	3.5	20	3.3	0.90	11.6	35	5.6	0.98

[1] According to Eq. 29 [166].
[2] Unpublished data.

7 Conclusions

The numerous cellulose derivatives commercially available, displaying varying second order transition, molecular mass, viscosity, lyophilicity, thermotropic properties, have considerable potential. Significant advances have been made in the collection of fundamental data, but the interpretation of this information is somewhat more difficult than for synthetic polymers, mostly because of the heterogeneity of the materials, which exhibit a broad molecular mass distribution and an uneven substitution.

Being polymers of natural origin, cellulose derivatives have gained acceptance for food, cosmetic and drug uses (some of them are described in compendia and enjoy the U.S. GRAS ("generally recognized as safe") status and as such are more readily approved by regulatory agencies. This is also true for the crosslinked or grafted cellulosics, which have not revealed all their possibilities, although one can expect of course that the nature of the reactant will be of great importance. On the whole, we are still a long way from a complete understanding of the interesting properties of the cellulose derivatives and their possible applications.

Acknowledgements. The author is grateful to Mrs Josiane Blanc for typing the manuscript.

8 References

1. Mark HF, Gaylor NG, Bikales NM (eds) (1965) Encyclopedia of polymer science and technology, vol 3. J. Wiley, New York, p 325, p 459
2. Bogan RT, Kuo CM, Brewer RJ (1979) In: Kirk-Othmer encyclopedia of chemical technology, vol 5, 3rd edn. J. Wiley, New York, p 118
3. Greminger GK (1979) In: Kirk-Othmer Encyclopedia of chemical technology, vol 5, 3rd edn. J. Wiley, New York, p 143
4. Balser K, Hoppe L, Eicher T, Wandel M, Astheimer HJ (1986) In: Ullmann's encyclopedia of industrial chemistry, vol A5, 5th edn. Verlag Chemie, Weinheim, p 419
5. Brandt L (1986) In: Ullmann's encyclopedia of industrial chemistry, vol A5, 5th edn. Verlag Chemie, Weinheim, p 461
6. Whistler RL, BeMiller JN (eds) (1973) Industrial gums, 2nd edn. Academic, New York
7. Davidson RL (ed) (1980) Handbook of watersoluble gums and resins. McGraw-Hill, New York
8. Kennedy JF, Philips GO, Wedlock DJ, Williams PA (eds) (1985) Cellulose and its derivatives. Ellis Horwood, Chichester
9. Klug ED (1970) Food Technol 24: 51
10. Klug ED (1971) J Polym Sci: (Part C) 36: 491
11. Klug ED, T'Sas HE (1980) in: Neukom H and Pilnik W (eds) Gelier-und Verdickungsmittel in Lebensmitteln (Gelling and thickening agents in foods), Forster Verlag, Zürich, p 163
12. Grosse L (1971) In: Hagers Handbuch der pharmazeutischen Praxis. Springer, Berlin Heidelberg New York, p 111
13. Blaschek W (1990) Pharm unserer Zeit 73: 73
14. Doelker E (1987) In: Peppas NA (ed) Hydrogels in medicine and pharmacy, vol. II: Polymers. CRC Press, Boca Raton, p 115

15. Doelker E (1990) In: Brannon-Peppas L, Harland RS (eds) Absorbent polymer technology (Studies in polymer sciences 8). Elsevier, Amsterdam, p 125
16. Hebeisch A, Guthrie JT (1981) The chemistry and technology of cellulosic copolymers. Springer, Berlin Heidelberg New York
17. Wirick MG (1968) J Polym Sci 6: 1705
18. Wirick MG (1968) J Polym Sci (Part A-1) 6: 1965
19. Dahl TC, Calderwood T, Bormeth A, Trimble K, Piepmeier E (1990) J Control Rel 14: 1
20. Rowe RC (1980) J Pharm Pharmacol 32: 116
21. Sakellariou P, Rowe RC, White EFT (1988) J Control Rel 7: 147
22. Dubernet C, Rouland JC, Benoit JP (1990) Int. J. Pharm. 64: 99
23. Finch CA (1983) Chemistry and technology of water-soluble polymers. Plenum, New York, p 93
24. Tabata Y, Ikada Y (1988) Biomaterials 9: 356
25. Entwistle CA, Rowe RC (1979) J Pharm Pharmacol 31: 269
26. Masilungan FC, Lordi NG (1984) Int J Pharm 20: 295
27. Okhamafe AO, York P (1985) Pharm Res 2: 19
28. Sakellariou P, Rowe R (1985) Int J Pharm 27: 267
29. Okhamafe AO, York P (1988) J Pharm Sci 77: 438
30. Kararli TT, Hurlbut JB, Needham TE (1990) J Pharm Sci 79: 845
31. Johnson K, Hathaway R, Leung P, Franz R (1991) Int J Pharm 73: 197
32. Conte U, Colombo P, Gazzaniga ME, Sangalli ME, La Manna A (1988) Biomaterials 9: 489
33. Rials TG, Glasser WG (1988) J Appl Polym Sci 36: 749
34. Suto S, Kudo M, Karasawa M (1986) J Appl Polym Sci 31: 1327
35. Lippold C, Lippold BH, Sutter BK, Gunder W (1990) Drug Dev Ind Pharm 16: 1725
36. Mandelkern L, Flory PJ (1951) J Am Chem Soc 73: 3206
37. Russel J, van Kerpel RG (1957) J Polym Sci 25: 77
38. King VL, Wheatley TA (1991) Proceed Int Symp Control Rel Bioact Mater 18: 131
39. Sinko CM, Yee AF, Amidon GL (1991) Pharm Res 8: 698
40. Porter SC, Ridgway K (1983) J Pharm Pharmacol 35: 341
41. Dechesne JP, Vanderschueren J, Jaminet F (1984) J Pharm Belg 39: 341
42. Sinko CM, Yee AF, Amidon GL (1990) Pharm Res 7: 648
43. Rogovin ZA (1972) J Polym Sci (Part C) 37: 221
44. Van Krevelen DW, Hoftyzer PJ (1976) Properties of polymers: their estimation and correlation with chemical structure, 2nd edn. Elsevier, Amsterdam
45. Rowe RC (1988) Acta Pharm Technol 34: 144
46. Aulton ME, Houghton RJ, Wells JI (1985) J Pharm Pharmacol 37: Suppl 113P
47. Sakellariou P, Rowe RC, White EFT (1986) Int J Pharm 31: 175
48. Roberts GAF, Thomas IM (1978) Polymer 19: 479
49. Kent DJ, Rowe RC (1978) J Pharm Pharmacol 30: 808
50. Matsuura T, Blais P, Sourirajon S (1976) J Appl Polym Sci 20: 1515
51. Barton A (1983) Handbook of solubility parameters and other cohesion parameters. CRC Press, Boca Raton
52. Rowe RC (1986) J Pharm Pharmacol 38: 214
53. Klein E, Eichelberger J, Eyer C, Smith J (1975) Water Res. 9: 807
54. Rowe RC (1982) Int J Pharm Tech & Prod Mfg 3(1): 3
55. Robinson DH (1989) Drug Ind Pharm 15: 2597
56. Crowley JD, Teagne GS Jr, Lowe JW Jr (1966) J Paint Technol 38: 269
57. Crowley JD, Teagne GS Jr, Lowe JW Jr (1967) J Paint Technol 39: 19
58. Hoernschemeyer D (1974) J Appl Polym Sci 18: 61
59. Rowe RC (1989) Int J Pharm 52: 149
60. Rowe RC (1989) Int J Pharm 53: 75
61. Johnson BA, Zografi G (1986) J Pharm Sci 75: 529
62. Kaelble DH, Moacanin J (1977) Polymer 18: 475
63. Davies MC, Newton JM (1986) Proceed 4th Int Conf Pharm Technol, Paris, 5: 111
64. Okhamafe AO, York P (1987) Int J Pharm 39: 1
65. Sakellariou P, Rowe RC, White EFT (1987) J Appl Polym Sci 34: 2507
66. Okhamafe AO, York P (1989) J Pharm Pharmacol 41: 1
67. Girifalco LA, Good RJ (1957) J Phys Chem 61: 904
68. Wu S (1973) J Adhes 5: 39
69. Rowe RC (1989) Int J Pharm 56: 117

70. Rowe RC (1988) Int J Pharm 41: 219
71. Rowe RC (1990) Int J Pharm 58: 209
72. Parker MD, York P, Rowe RC (1990) Int J Pharm 64: 207
73. Okhamafe AO, York P (1985) J Pharm Pharmacol 37: 385
74. Wan LSC, Prasad KPP (1990) Drug Dev Ind Pharm 16: 945
75. Buri P, Doelker E (1980) Pharm Acta Helv 55: 189
76. Peppas NA, Gurny R, Doelker E, Buri P (1980) J Membr Sci 7: 241
77. Lee PI, Peppas NA (1987) J Contr Rel 6: 287
78. Harland RS, Gazzaniga A, Sangalli ME, Colombo P, Peppas NA (1988) Pharm Res 5: 488
79. Conte U, Colombo P, Gazzaniga A, Sangali ME, La Manna A (1988) Biomaterials 9: 489
80. Vrentas JS, Jarzebski CM, Duda JL (1975) AIChE 21: 894
81. Vrentas JS (1977) J Polym Sci, Polym Phys Ed 15: 441
82. Peppas NA, Lustig SR (1985) Ann N.Y. Acad Sci 446: 26
83. Timmermans J, Moes AJ (1990) Acta Pharm Technol 36: 171
84. Peppas NA, Buri P (1985) J Control Rel 2: 257
85. Smart JD, Kellaway IW, Worthington HEC (1984) J Pharm Pharmacol 36: 295
86. Anders R, Merkle HP (1989) Int J Pharm 49: 231
87. Mikos AG, Peppas NA (1985) Int J Pharm 53: 1
88. Sarkar N (1979) J Appl Polym Sci 24: 1073
89. Touitou E, Donbrow M (1982) Int J Pharm 11: 131
90. Mitchell K, Ford JL, Armstrong DJ, Elliot PNC, Rostron C, Hogan JE (1990) Int J Pharm 66: 233
91. Carlsson A, Karlström G, Lindman B (1986) Langmuir 2: 536
92. Carlsson A, Karlström G, Lindman B, Stenberg O (1988) Colloid Polym Sci 266: 1031
93. Karlström G, Carlsson A, Lindman B (1990) J Phys Chem 94: 5005
94. Rinaudo M, Daune M (1967) J Chim Phys 64: 1761
95. Fagan PG, Harrison PJ, Shankland N (1989) J Pharm Pharmacol 41: 25P
96. Carlsson A, Lindman B, Watanabe T, Shirahama K (1989) Langmuir 5: 1250
97. Carlsson A, Karlström G, Lindman B (1990) Colloids Surfaces 47: 147
98. Carlsson A, Bogentoft C, Lindman B, Andersson L (1991) Procced Intern Symp Control Rel Bioact Mater 18: 455
99. Lindell K, Engström S, Carlsson A (1991) Proceed Intern Symp Control Rel Bioact Mater 18: 265
100. Werbowyj RS, Gray DG (1976) Mol Cryst Liq Cryst 34: 97
101. Gilbert RD, Patton PA (1983) Prog Polym Sci 9: 115
102. Gray DG (1983) J Appl Polym Sci, Appl Polym Symp 37: 179
103. Werbowyj RS, Gray DG (1980) Macromolecules 13: 69
104. Fortin S, Charlet G (1989) Macromolecules 22: 2286
105. Bheda J, Fellers JF, White JL (1980) Colloid Polym Sci 258: 1335
106. Habib FS, Ismail S, El-Shanawany S, Fouad EA (1991) Eur J Pharm Biopharm 37: 38
107. Etter JC, Rey-Bellet C (1980) Pharm Acta Helv 55: 13
108. Rakić R, Jovanović M, Djurić Z (1987) Int J Pharm 39: 47
109. Carlsson A, Karlström G, Lindman B (1989) J Phys Chem 93: 3673
110. Daly PB, Davis SS, Kennerley JW (1984) Int J Pharm 18: 201
111. Feely LC, Davis SS (1988) Int J Pharm 41: 83
112. Ford JL, Mitchell K, Sawh D, Ramdour S, Armstrong DJ, Elliott PNC, Rostron C, Hogan JE (1991) Int J Pharm 71: 213
113. Feely LC, Davis SS (1988) Int J Pharm 44: 131
114. Walker CV, Wells JI (1982) Int J Pharm 11: 309
115. Kellaway IW, Najib NM (1981) Int J Pharm 7: 285
116. Kellaway IW, Najib NM (1981) Int J Pharm 9: 59
117. Law SL, Kay JB (1983) Int J Pharm 15: 251
118. Law SL (1984) Pharm Acta Helv 59: 298
119. Carlsson A, Sato T, Sunamoto J (1989) Bull Chem Soc Jpn 60: 791
120. Koh GL, Tucker IG (1988) J Pharm Pharmacol 40: 233
121. Koh GL, Tucker IG (1988) J Pharm Pharmacol 40: 309
122. Vanderhoff JW, El-Aasser MS, Ugelstadt J (1979) U.S. Patent 4.177.177
123. Banker GS (1980) PCT Int Appl 80.00659
124. Banker GS (1982) U.S. Patent 4.330.338
125. Heinz K (1982) Patentschrift DE 2.926.663 C2

126. Bindschaedler C (1985) Ph.D. thesis, University of Geneva
127. Bindschaedler C, Gurny R, Doelker E, Peppas NA (1985) Makromol Chem, Rapid Commun 6: 267
128. Bindschaedler C, Gurny R, Doelker E, Peppas NA (1985) J Colloid Interface Sci 108: 75
129. Bindschaedler C, Gurny R, Doelker E, Peppas NA (1985) J Colloid Interface Sci 108: 83
130. Banker GS, Peck GE (1981) Pharm Technol 8(4): 55
131. Onions A (1986) Mfg Chem 57(4): 66
132. Gumowski F, Doelker E, Gurny R (1987) Pharm Technol 14(2): 26
133. Bindschaedler C, Gurny R, Doelker E (1986) J Contr Rel 4: 203
134. Bindschaedler C, Gurny R, Doelker E (1987) J Pharm Pharmacol 39: 335
135. Bindschaedler C, Gurny R, Doelker E (1987) J Pharm Sci 76: 455
136. Bindschaedler C, Gurny R, Doelker E (1987) J Appl Polym Sci 34: 2631
137. Bindschaedler C, Gurny R, Doelker E (1989) J Appl Polym Sci 37: 173
138. Lippold BC, Lippold BH, Sutter BK, Grunder W (1990) Drug Dev Ind Pharm 16: 1725
139. Bindschaedler C, Gurny R, Doelker E (1990) U.S. Patent 4.968.350
140. Ibrahim H, Bindschaedler C, Doelker E, Buri P, Gurny R (to be published)
141. Bodmeier R, Chen H (1991) Drug Dev Ind Pharm 17: 1811
142. Sprockel OL, Prapaitrakul W, Shivanand P (1990) J Pharm Pharmacol 42: 152
143. Bond JR, York P, Woodhead PJ (1989) J Pharm Pharmacol 41: 16P
144. Sinko CM, Yee AF, Amidon GL (1991) Pharm Res 8: 698
145. Lindsted B, Ragnarsson G, Hjärlstam J (1989) Int J Pharm 56: 261
146. Ozturk AG, Ozturk SS, Palsson BO, Wheatley TA, Dressman JB (1990) J Control Rel 14: 203
147. Theeuwes F (1975) J Pharm Sci 64: 1987
148. Zentner GM, Rork GS, Himmelstein KJ (1985) J Contr Rel 2: 217
149. Zentner GM, Rork GS, Himmelstein KJ (1985) J Control Rel 1: 269
150. Gurny R, Doelker E, Peppas NA (1982) Biomaterials 3: 27
151. Azoury R, Elkayam R, Friedman M (1988) J Pharm Sci 77: 425
152. Azoury R, Elkayam R, Friedman M (1988) J Pharm Sci 77: 428
153. Rosilio V, Roblot-Treupel, de Lourdes Costa M, Baszkin A (1988) J Control Rel 7: 171
154. Sakellariou P, Rowe RC, White EFT (1986) Int J Pharm 34: 93
155. Sakellariou P, Rowe RC (1991) J Appl Polym Sci 43: 845
156. Benita S, Dor P, Aronhime M, Marom G (1986) Int J Pharm 33: 71
157. Ibrahim H (1989) Ph.D. thesis, University of Geneva
158. Ibrahim H, Gurny R, Bindschaedler C, Doelker E, Buri P (1990) Proceed Intern Symp Control Rel Bioact Mater 17: 303
159. Ibrahim H, Bindschaedler C, Doelker E, Buri P, Gurny R (1991) Int J Pharm 77: 211
160. Gissinger D, Stamm A, Mathis C (1982) Labo-Pharma – Probl Tech 30: 69
161. Shah NH, Lazarus JH, Sheth PR, Jarowski CI (1981) J Pharm Sci 70: 611
162. Gordon RE, Peck GE, Kildsig DO (1984) Drug Dev Ind Pharm 10: 833
163. Harsh DC, Gehrke SH (1990) In: Brannon-Peppas L, Harland RS (eds) Absorbent polymer technology. Elsevier, Amsterdam, p. 103
164. Harsh DC, Gehrke SH (1990) Proceed Intern Symp Control Rel Bioact Mater 17: 383
165. Harsh DC, Gehrke SH (1991) J Control Rel 17: 175
166. Peppas NA (1985) Pharm Acta Helv 60: 110
167. Archer WL (1992) Drug Dev Ind Pharm 18: 599

Received May 15, 1992

Author Index Volume 101–107

Subject Index